Seventh Edition

CONSTRUCTION INSPECTION MANUAL

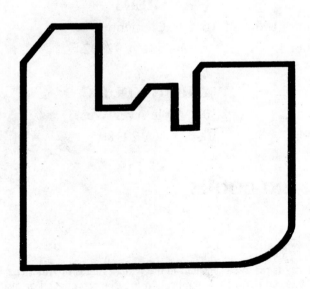

BNi® Building News

TECHNICAL COLLEGE OF THE LOWCOUNTRY
LEARNING RESOURCES CENTER
POST OFFICE BOX 1288
BEAUFORT, SOUTH CAROLINA 29901-1288

Construction Inspection Manual

BNi Building News

Editor-In-Chief
William D. Mahoney, P.E.

Technical Services
Rod Yabut

Design
Robert O. Wright

BNI PUBLICATIONS, INC.

LOS ANGELES	**ANAHEIM**
10801 National Blvd., Suite 100	1612 S. Clementine St.
Los Angeles, CA 90064	Anaheim CA 92802
BOSTON	**WASHINGTON, D.C.**
629 Highland Ave.	502 Maple Ave. West
Needham, MA 02494	Vienna, VA 22180

1-888-BNI-BOOKS

Copyright © 1998 by BNI Publications, Inc. All rights reserved. Printed in the United States of America. Except as permitted under the United States Copyright Act of 1976, no part of this publication may be reproduced or distributed in any form or by any means, or stored in a data base or retrieval system, without the prior written permission of the publisher.

While diligent effort is made to provide reliable, accurate and up-to-date information, neither BNI Publications Inc., nor its authors or editors, can place a guarantee on the correctness of the data or information contained in this book. BNI Publications Inc., and its authors and editors, do hereby disclaim any responsibility or liability in connection with the use of this book or of any data or other information contained therein.

ISBN 1-55701-2571

CONSTRUCTION INSPECTION MANUAL

TABLE OF CONTENTS

PART	TITLE	PAGE
	FOREWORD	V
	ACKNOWLEDGMENTS	VII
	INTRODUCTION	IX
1.	RECOMMENDED DUTIES AND RESPONSIBILITIES	1
1.1	Contract: Definition of Responsibilities	2
1.2	Construction Inspector	3
1.2.1	Qualifications and Requirements	4
1.2.2	Employment, Compensation and Agreements	7
1.2.3	Construction Inspector: Recommended Duties, Responsibilities	10
1.3	Design Professional (Architect/Engineer)	13
1.3.1	Project Design Professional (A/E)	15
1.3.2	Project Designer	16
1.3.3	Project Job Captain	17
1.3.4	Project Specifier	18
1.3.5	Field Administrator	19
1.3.6	Observations and Inspections	20
1.4	Contractor/Subcontractor	21
1.5	Owner/Recommended Duties and Responsibilities	23
1.6	Lawyers Description	26
1.7	Manufacturer's Representative Inspections	27
1.7.1	Required by Contract Documents	28
1.7.2	Required for Warranties and Guarantees	29
1.7.3	Coordination	30
1.8	Others: Special Consultants	31
2.	STANDARDS AND CODES	35
2.1	Contract/Contract Documents/Specifications	37
2.2	Codes and Regulations	38
2.3	Standards	43

3.		**CHECKLIST FOR FIELD INSPECTION**	53
	3.1	Checklist Introduction (Explanation of the Matrix, Primary and Secondary Inspection)	54
	3.2	Field Inspection Checklist	56
4.		**COORDINATION**	207
	4.1	Communication	208
	4.2	Scheduling	210
	4.3	Correspondence and Coordination	211
	4.4	Forms (Overview of Appendix D)	212
5.		**PROJECT LOCATION**	215
	5.1	Office Practice Concerns	216
	5.2	Field Office Concerns	217
	5.3	On-Site Concerns	218
	5.4	Regional Concerns	220
	5.5	Geographic Concerns	221
	5.6	Climate Concerns	222

APPENDICES

- Appendix A - Bibliography
- Appendix B - Terms and Definitions
- Appendix C - Construction Industry Organizations
- Appendix D - Forms
- Appendix E – Masterformat/Inspector's Basic Bookshelf
- Appendix F - Inspector's Basic Tools
- Appendix G - Inspector's Basic Project File
- Appendix H - Miscellaneous Construction Aids
- Appendix I - Mathematics
- Appendix J - Hazardous Materials
- Appendix K - Handy Construction Data

Foreword

The goal of the *Construction Inspection Manual* is to assist the construction industry in improving the inspection procedures on all types of construction work and to achieve a consensus among owners, architects, engineers, contractors and construction inspectors as to the best methods and practices.

Because good plans and specifications are the cornerstone of good construction, this manual also provides valuable guidelines for improvement of the design and specification processes.

The need for consistent field inspection becomes imperative with the welter of overlapping codes, specification standards, and individual interpretations among architects, engineers and agencies, and finally with the ever-more demanding designs and the requirements for precise execution of the work.

It is hoped that this manual will encourage an on-going effort to which all segments of the industry can contribute their expertise to accomplish a more uniform application of inspection methods and thus benefit all parties.

Blair Tulloch

Foreword To The Seventh Edition

Since the original publication of the First Edition of the *Construction Inspection Manual* in 1973, the contents have been periodically reviewed and updated by an Editorial Committee consisting of volunteer members, each a technical expert in their particular field, representing recognized national and local construction industry organizations. Collectively, these organizations embrace all of the disciplines and segments of the construction industry which are included in the text of the *Construction Inspection Manual*. With each succeeding edition, new material has also been added.

The current Editorial Committee has again reviewed and updated this Seventh Edition. Those familiar with previous editions will note a major reorganization along with the addition of a matrix to each checklist indicating the Editorial Committee's opinion regarding which member of the construction team should have primary and secondary responsibility for each checklist item. The intent of the matrix is to assist readers in finding specific items of importance to their particular area of interest, hopefully increasing the value of the Manual to a wider spectrum of the construction industry.

John A. Raeber, FAIA, FCSI, CCS

Acknowledgments

This manual could not have been written without the encouragement and the help of all persons and organizations throughout the construction industry.

The following acknowledgments are made: Blair Tulloch, Tulloch Construction, Inc., Oakland, Calif., and Peter Muller, formerly manager, East Bay District, Associated General Contractors of California, who assisted in the idea and inspiration.

The Construction Industry Advancement Fund, a non-profit Northern California foundation, which provided the original funding grant without which this manual would not have been possible.

The East Bay District, Associated General Contractors of California (C. R. "Tom" Wintch, 1972 chairman), who sponsored the manual.

The Coordinating Committee, whose members volunteered their time and efforts for this edition, their organizations.

Henry W. Berg, CCS	Construction Specifications Institute, East Bay/Oakland Chapter
Norman Broome, SE	Public Agency Representative Office of Statewide Health Planning and Development State of California
Michael Gaddis, CSI	East Bay/Oakland Chapter
Edward J. Gray	American Construction Inspectors Association
Mike Hilliard, AIA	American Institute of Architects, San Francisco Chapter
Richard Hoisington, SE	Public Agency Representative Office of Statewide Health Planning and Development, State of California
Joseph A. McQuillan, PE	Mechanical Engineer
Peter T. Muller	Construction Industry Consultant
George Narancic	American Construction Inspectors Association
Melina Renée, CCS	Construction Specifications Institute, San Francisco Chapter
Murray Slama, FAIA	Former Project Director
Blair Tulloch	Associated General Contractors of California
Stephen Wheeler	Landscape Architect
Craig Wilcox, SE	Structural Engineers Association, Northern California
Larry Wolff, PE	Electrical Engineer, Consulting Engineers and Land Surveyors of California (Cel Soc)
Raúl A. Yáñez, AIA	American Institute of Architects, San Francisco Chapter
William Mahoney, PE	Building News, A Division of BNI Publications, Inc.

The many other people, industry organizations and agencies who supplied invaluable data and reviewed the manual are further acknowledged and thanked.

With this Seventh Edition I would like to acknowledge a few people specifically for their efforts. First, to Sam Jaffe, who was the publisher at BNI Books during the early editions of the Construction Inspection Manual, and who helped nurture it especially relating to the appendix information. I would also like to thank Melina Renee, who took on the responsibility of arranging the committee meetings, keeping everybody informed, and collecting and organizing the information from the various sources. And, to Bill Mahoney a special thanks for his patience with the committee process. Finally, thanks to Murray Slama for his efforts as Project Director through the first six editions.

John A. Raeber, FAIA, FCSI, CCS

Project Director

INTRODUCTION

The purpose of this manual is to supply useful information about construction inspection.

Owners, architects, engineers, contractors and others in the construction industry have discussed the need for a construction inspection manual for years. The original intent of the early editions was to discuss, on a general basis, the duties, qualifications, and abilities of the construction inspector, and the working relationships among the parties during construction.

The Founding Committee, delegated as representatives by the various organizations directly involved in construction, collected data concerning inspection from as many sources as possible. The data was then reviewed, edited, and the most pertinent material was incorporated into a first draft. The draft was distributed to approximately one hundred other individuals or organizations for review and comment before publication.

With the complexity of the topic, it is understandable that some material, information, organizations, or other data may have been overlooked or not included. This omission was not intentional in any respect.

The Coordinating Committee recognizes that the manual may not be comprehensive enough for all types of projects. It was originally written to apply to both public and private "building" construction projects of average complexity, where a full-time construction inspector is engaged. Construction inspection for smaller projects might be the responsibility of a part-time construction inspector or even members of the design team. For highly complex projects, the construction inspection may require a staff of inspectors, including specialized inspectors and assistants.

It did not seem feasible to attempt to discuss all arrangements and methods of construction and inspection within the limited scope of the manual. Traditional methods of construction and inspection can be impacted by many factors. The role of the developer, who can be a contractor, architect, or other party changes the concerns relating to construction inspection. New methods of construction management and design/build also impact the role of construction inspection. Laws governing certain types of public projects may include specific duties for the construction inspector.

With the Twenty-First Century approaching, the Coordinating Committee realized a need to make modifications to the Seventh Edition to reflect some major changes in the construction industry. Recognition of a need for involvement in construction inspection by more members of the construction team resulted in the addition of matrix tables which attempt to provide the Coordinating Committee's recommendation regarding the primary and secondary responsibilities for inspection. These recommendations should be considered a guide only, and intended for the average building construction project. The matrixes will hopefully help readers more easily find information which should be of interest to specific members of the construction team.

Further, it is hoped that as the Seventh and later editions are developed, recognition will be given to a broader understanding of the full extent of the construction process. More and more people are beginning to recognize that the construction process is not limited to just the design and build process. Construction actually begins with conception of a potential building, which includes design. It then includes construction, commissioning, maintenance, remodeling, alterations, potential changes in occupancy type, and only ends when the building is finally demolished. Construction inspection likewise, should be recognized as a part of the complete construction process, from conception through demolition.

Terminology used in the manual may differ in some instances from what some organizations may have established. The reader is referred to Appendix C, "Terms and Definitions."

The reader is cautioned that this publication is not intended as a guide book to legal relationships or as a compendium of forms, legal or otherwise. It is published solely as a working manual, prepared not by lawyers but by a collaborative effort of the construction industry to provide general information regarding construction inspection. Those using the manual should not rely on it to eliminate or solve legal problems. They should consult their legal counsel on all matters involving contractual or other legal relationships.

The masculine gender was utilized in the original 1973 edition of this manual. Where traces still escaped editing, please accept that the feminine gender is intended to be just as applicable in all cases, i.e., he/she, his/her, etc.

Part 1: Recommended Duties and Responsibilities

Today, more than ever, construction requires a team effort. The ever-increasing escalation of costs, the need for faster scheduling, and changes occurring in the industry require full cooperation and understanding among the parties.

Successful production of the work under the traditional process requires the utmost order and efficiency to obtain the highest potential benefits. This goal can be reached only through the understanding that all parties have a mutual function and obligation to perform. Successful construction requires continuous checking, coordination, foresight, good judgment, and an overlapping of efforts by informed and qualified parties.

The following parts reflect the Editorial Committee's recommendations concerning the duties and responsibilities of the construction inspector, design professional (architect/engineer), contractor/subcontractor, owner, lawyer, manufacturer's representative, and special consultants.

Part 1.1
Contract: Definition of Responsibilities

By definition a contract is a legally enforceable agreement that sets forth the obligations of each party to the other. Any violation of these obligations (breach) can expose the party committing it to sanctions of law. The significant benefit of having an agreement is that a mutual understanding by communicating expectations through clear definition of scope and general/terms and conditions is developed.

Contracts come in many shapes and sizes. The most recognized standard document is the Owner-Architect Agreement /AIA B141-1997 Edition but many public Owners, major private enterprises and design professional firms use their own "custom contracts". In order to provide contracts that equitably serve the interest of both parties and meet the specific concerns of the project, amendments to both standard AIA and custom agreements are negotiated.

It is of utmost importance that the project inspector thoroughly review the agreements and the family of documents that define duties and responsibilities among the construction team composed of the Owner, Contractor and Architect. The AIA A201/General Conditions of the Contract for Construction -1997 Edition is the part of this family of Contract Documents that establish the ground rules for this construction team, where contractual relationships exist between the Owner and the Contractor and the Owner and the Architect but not between the Contractor and the Architect. Therefore, the rights, responsibilities and relationships of the parties are defined in this document. Additionally, the Supplementary Conditions are prepared to modify the provisions of the General Conditions to meet the particular requirements of the Project by changing, adding to or deleting this document. The General Conditions can also be drafted as "custom general conditions" by many public owners and major private enterprises.

Roles of the Construction Team

Owner - Entity that furnishes relevant information relative to project requirements, pays for costs associated with the development of project and makes decisions based on recommendations made by the design professional.

Owner's Representative - Entity occasionally designated as the official representative of the Owner. Its extent of project involvement will vary according to the Owner's requirements. It is critical that the extent of its responsibilities and authority is clearly described in its agreement with the Owner, the General Conditions of the Contract for Construction, and coordinated with the inter-related agreements of the other construction team members with the Owner. This will prevent inconsistencies and overlapping of such responsibilities and authority.

Contractor - Entity that prepares bid proposals for the cost of the Work and upon award of contract, directs construction and builds project in conformance with the contract documents.

Design Professional - Entity that designs the Work, prepares construction and bidding documents and administers the construction contract by serving as the Owner's professional adviser and an impartial interpreter of the contract documents.

Part 1.2:
The Construction Inspector

The importance of selecting a competent construction inspector is sometimes misunderstood. Good inspection can be worth many times its cost in preventing errors and omissions of construction that might impair the quality, soundness and durability of the project and interfere with obtaining value for the money invested.

The construction inspector's basic function is to make sure compliance with the construction documents is achieved. In addition, the inspector serves as an extra pair of eyes and should not be satisfied with merely reporting mistakes in the work after they are made.

The inspector can prevent problems and avoid misunderstanding by continually reviewing the construction documents and working in conjunction with the superintendents and the subcontractors.

The inspector should look ahead and be fully acquainted with the construction documents and all phases of the work. Thus avoiding costly and time-wasting mistakes and foreseeing bottlenecks due to delayed delivery of material and improper scheduling of the work.

Prompt inspection of the delivered materials and the preparation and installation can prevent costly tearout, replacement, or redoing of the work.

In these and other ways, the inspector can perform a real service to the owner, design professional, and contractor, and thus becomes an important member of the team needed to procure a smooth running construction process, a sound and properly constructed project, and an on-schedule delivery of work.

Note, however, that the inspector is not responsible for and should not undertake responsibilities that are not a part of the inspection services; for example:

—Trying to tell the contractor how to construct the work. This is the responsibility of the contractor.

—Guaranteeing that the work is constructed in strict compliance with the contract documents. This is the responsibility of the contractor.

—Interpreting or ruling on the intent of the contract documents. This is the responsibility of the design professional.

—Accepting the work or portions of it. The design professional is responsible for recommending this to the owner.

Part 1.2.1:
Qualifications and Requirements

A single, full-time construction inspector may often be employed on building projects of simple or average complexity. In this case, the inspector is expected to be sufficiently qualified to oversee the complete project. This manual refers mainly to these circumstances. On more complex and larger projects, the budget should include an inspection team or staff consisting of a chief inspector, specialized inspectors, secretary, accountant, etc.

The general construction inspector usually is expected to observe mechanical and electrical installations; however, inspectors fully expert in all materials, systems, and methods used in construction today are rare. Specialized inspectors are now available, especially in the fields of mechanical and electrical installation. They are engaged as part of the inspection team on larger or complex projects.

A testing laboratory is normally engaged to provide tests and inspections of materials, installations, or procedures.

Sources

Construction inspectors today are drawn from three general sources:

1. Those having attended or graduated from a college of construction or engineering and obtained practical experience in construction by working with a construction firm or an agency.

2. Those who have risen through the ranks, primarily as a carpenter, up to the capacity of job superintendent and chosen construction inspection as a career. Field experience is often supplemented by attending special courses of instruction.

3. Those who have worked for an architect or engineer, and have been exposed to construction through periodic jobsite observation and have become knowledgeable in construction procedures and techniques.

Construction Inspection Manual

Qualifications

The qualifications of an inspector are made up of three parts — technical, personal, and experience — all equally important.

1. *Technical* — The technical qualifications recommended for construction inspectors of the future could be:

(a) A minimum of two years in a community college or junior college, or an adequate curriculum in instructional courses, with an AA degree and at least two years of practical experience in the construction industry, preferably in a management and on site capacity.

(b) A bachelor's degree from a university or college with a construction curriculum and at least one year of practical experience as in (a). The curriculum should contain sufficient courses in mathematics, physics, and engineering to give the student a reasonable technical, scientific, and engineering background.

2. *Personal* — From a personal standpoint, the "ideal" inspector should:

   (a) Be a diplomat, tactful but candid.
   (b) Have natural ability to work harmoniously without being a "boss."
   (c) Be able to speak and write well.
   (d) Have no connection with and accept no loan, gift, or gratuity directly or indirectly from any contractor or subcontractor company that is in any way connected with the construction contract.
   (e) Never use intoxicants or harmful drugs while on duty or appear for duty under their influence.
   (f) Not suggest or recommend any person for employment to any individual or firm connected with the contract.
   (g) Not assume the responsibility for giving instructions for changes in the work covered by contract without approval of the architect and owner.
   (h) Not assume any duties of or give any orders reserved for the superintendent.
   (i) Have enough personal confidence in his ability that "showing who's boss" is not necessary.
   (j) Always keep in mind that they are the owner's or architect's project representative and acts in their best interests.
   (k) Have thorough knowledge of construction.
   (l) Have ability to enforce rigorously the requirements of the drawings and specifications.
   (m) Use good judgment based on years of experience and education as well as the personal characteristics of honesty, restraints, firmness, fairness, alertness, and patience.
   (n) Have the personal qualifications that will enable them to maintain satisfactory relations with the personnel associated with the project.
   (o) Have an ability to handle the necessary paperwork without creating a "paper war."

3. *Experience* — The experience record of an inspector shows the nature of prior projects and tells whether they have been able to stick with projects rather than jumping around every few months.

More important than a personal interview may be discussions with the owners, architects, and contractors who have dealt with the inspector on past projects. Talking with all three will give one an insight into the inspector's abilities.

One should never hesitate to test or quiz a potential inspector as to what they would do under specific instances or problems relating to construction work.

Other Requirements

To be qualified to serve on certain types of construction — for example, public schools and medical facilities — may be necessary for the construction inspector to be approved by an agency and demonstrate that the inspector can adequately inspect structural materials or systems installations such as concrete, masonry, steel, and wood.

Some states have licensing laws for construction inspectors which may entail specific prerequisites, responsibilities and liabilities.

Model codes contain requirements for special inspection.

Summary

In conclusion, the evidence indicates that inspection is not an easy job. It requires people who are well trained, adequately paid, and properly supported by their employers. It requires properly prepared realistic construction documents, properly and justly administered. It requires capable supervision on the part of the contractor. Proper inspection advocates that the owner gets the value and quality is paying for, within the design intent required by the architect, and acceptable construction industry standards complied with by the contractor.

Part 1.2.2:
Employment, Compensation and Agreement Employment

Here are a few of the many ways to employ the construction inspector:

(a) The owner engages the construction inspector for the project, and the construction inspector works under the direction of the architect. In some instances it is a legal requirement for many public governing boards. The design professional often is required to select the construction inspector.

(b) The design professional engages the construction inspector under a mutually agreed upon extension of the design professional's agreement with the owner. The architect may select and employ the construction inspector for the project. Sometimes the selected person may be a fully qualified member of the design professional's staff.

(c) The owner maintains a staff of inspectors and assigns one or more to the project to provide inspection. Depending upon the owner's organization and methods, the construction inspector may or may not work fully under the direction of the design professional.

(d) On public projects, an agency may provide the owner with a construction inspector who may or may not work fully under the direction of the design professional.

Compensation

The cost of inspection is based on the degree of inspection required, the type, complexity, and size of the project, and the quality of construction inspectors required.

As a rule of thumb, the American Construction Inspectors Association recommends that a full-time construction inspector be compensated at a rate not less than the average of the foreman of the four highest paid trades working on the proposed project. This assumes that compensation is reasonably equivalent to that of the personnel being observed.

Some public agencies classify construction inspectors into categories to suit the position of employment required and establish salary ranges. Factors considered are the type of work, its cost, the qualifications and experience required, the number of projects involved, and supervising capability if a construction inspection staff is involved.

The type and extent of benefits such as mandatory medical insurance, vacation, sick leave, etc., included or not included in the terms of employment are of importance in the consideration of salary or compensation.

Agreement

The types or forms of agreements used to employ a construction inspector are numerous and the terms and conditions of agreements vary considerably. A general outline of items to be considered, if applicable, is as follows:

Date of agreement.

Legal name of the employer and the construction inspector — Definitions of parties.

Definitions of the project — Scope, location, etc.

Intent — Agreement to employ or contract for the project and stipulation that the construction inspector is qualified.

Form of engagement — As an "independent contractor" or an employee (temporary, permanent, etc.)

1. Reference to other documents that may apply, i.e., AIA, Doc. B-352.

2. Owner — Relationship.

3. Design Professional — Relationship, communications, correspondence, notification, etc.

4. Record keeping — Daily log, reports, etc.

5. Legal requirements of agencies — Recordation, submission of documents, information required, performance of required duties, etc.

6. Supervision — Other inspectors or staff.

7. Specialized inspection — By others.

8. Other requirements — Unique to the project, more than one project, etc.

Duration of agreement — Time period (number of months), definition of commencement and completion of services, times of work —

1. Additional services required in disputes, claims, extensions, etc.

2. Absences or sickness — Authority to substitute, deductive compensation.

Terms of payment — Amount, times of payment and commencement —

1. Overtime — Due to unusual circumstances.

2. Additional services required in disputes, claims, extensions, etc.

3. Benefits included if applicable.

Owner's responsibility — Suppliers, forms, materials, postage, equipment, clerical assistance, etc.

Reimbursement — Mileage, photographs, other project expenses authorized.

Termination — Time of notice, terminal expenses, cause, non-penalty, records transfer.

Execution of agreement — Authorized signatures, attestation, approval as-to-form, etc.

The above outline is given for information only. It is recommended that both parties to an agreement retain legal counsel as to type of form and its terms and conditions.

Part 1.2.3:
Construction Inspector: Recommended Duties and Responsibilities

Be completely familiar with the contract documents before commencement of the work. Notify the design professional of any discrepancies observed, and request clarification for all items not fully understood.

Organize a complete system of construction records:

- Daily log book, daily report system.
- Progress report system on a periodic basis (weekly, monthly).
- Correspondence file.
- Payment file.
- Clarifications/ Requests for information (RFI's) file.
- Construction Change Directive file.
- Change order file.
- Shop drawing, product data and sample submittal file.
- Substitutions file.
- Test and inspections file.
- Site conference file.
- Job memo file.
- Visual recordings (photo, video) file.
- City agency correspondence file (or Agency having jurisdiction).
- Electronic transmissions file (E-mail).
- Owner file.
- Owner's representative file.
- Owner's consultants file.
- Owner furnished items file.
- Contractor file.
- Subcontractor file.
- Contractor's design-builders file.
- Design Professional file.
- Close-out file.

Obtain a complete set of contract documents.

Obtain or have access to all codes and standards governing the work.

Comply with requirements to agencies having jurisdiction over inspection, and submit reports required. Remind the contractor of inspections required by agencies.

Determine that a method of procedures is developed concerning communications, correspondence, shop drawings, samples, substitutions, payments, changes, tests, and specialized inspection. Recommend a preconstruction meeting if not specified.

Obtain a schedule of values for progress payment evaluation; also obtain a progress schedule.

Communicate with the various parties of the work as described in "Coordination," Part 1.7.3 of this manual.

Keep in touch with the architect. Notify it about all phases of the work and meetings that may require the inspector's presence at the site.

Keep ahead of the work being performed so as to anticipate items that might tend to interfere with the progress of the construction.

Conduct on-site inspections of the work in progress as a basis to determine compliance with the contract documents.

Report deficiencies observed to the superintendent and the design professional.

Get from the architect further details or information if required for the proper execution of the work.

Help the superintendent to understand the contract documents. Request the design professional's interpretation or decision on all matters needing clarification.

Be familiar with codes applicable to the work. Request interpretation if in doubt.

Generally be acquainted with and have access to referenced standards.

Request manufacturer's literature or printed instructions if referenced and in doubt.

Observe that the testing laboratory performs all tests and inspections required. Keep a record of type and location. Review test results and notify the design professional of observed deficiencies.

Consider suggestions or recommendations made by the contractor, and refer them to the architect.

Review contractor's progress payment applications and report agreement or non-agreement to the architect.

Accompany the design professional's consultants when observing or inspecting the work.

Notify the design professional of material deliveries that are out-of sequence.

Observe actual progress in comparison with estimated progress. Record and report conditions that may cause a delay in completion of the work.

Keep an accurate record of time and materials and force account work. Obtain written concurrence with the contractors representatives, on a daily basis, of labor, materials and equipment being used on and charged to the force account work.

Observe owner occupancy or delivery of owner-furnished equipment before completion. Record and report any damages occurring so that claims can be fully documented.

Do not authorize deviations from the contract documents.

Neither interfere with the work to be performed by the contractor nor assume any responsibility for the performance of the contractor's work.

Do not take instructions from the owner or the owner's employees. Refer all such matters to the design professional.

Do not advise on or issue directions relative to any aspect of construction means, methods, techniques, sequences, or procedures.

Do not assume responsibility for any safety procedures or advise on remedy. Should hazards be observed, report conditions to the superintendent and record them, with a copy to the owner, design professional and contractor. If an emergency situation arises, contact the owner and governing authorities.

Do not stop the work except with written authority of the owner.

Notify contracting parties if the owner occupies the work before completion.

The use of a camera or camcorder serves as an invaluable visual record of job conditions. The use of photographs or videotape to indicate preconstruction conditions, foundation and utilities placement and conditions prior to "closing-in" provides a good record and assists owner in future maintenance.

Be able to affirm that in your best judgment the project conforms to the construction documents at the completion of the work.

Realize that approved shop drawings are not contract documents and are not change orders. They serve to clarify or show more detail, but the drawings and specifications prevail if there is a conflict.

Part 1.3
Design Professional (Architect/Engineer): Recommended Duties and Responsibilities

Today, the construction of a project requires the integrated efforts of a design team composed of competent individuals. The complexity of the numerous disciplines that interact in the construction process, necessitates the division of responsibility of different team members based on area of expertise. Although the project designer conceives the design concept, the actual project realization and delivery requires the technical and management capabilities offered by other team members. Therefore, for the sake of simplicity, the following major positions are represented: Project A/E, designer, job captain, specifier and field administrator.

Thoroughly review the contract documents with the owner to define the scope of the work involved and recommended the necessary procedures to be performed by the owner.

If the Owner agrees, help formulate qualifications for the construction inspector and/or inspection staff. Assist in interviewing and selecting the inspector. Ask the consultants to participate in this selection if necessary.

Establish the nature and extent of services by the testing laboratory and soils engineer and advise the owner of a probable budget. Assist the owner in the selection of a testing laboratory.

Assign a design professional representative who is competent and experienced in the construction process to perform construction administration of the work.

Specify an adequate job office for the construction inspector, with space, equipment, and conditions sufficient for the functions to be performed.

Publish and/or describe the contract administration required for the execution of the work. It is recommended that guidelines tailored for this particular work be developed, and that a pre-construction conference describing them be conducted at the onset of construction. It is further recommended that the owner, contractor, construction inspector, design professional consultants, agency representatives and major subcontractors be present at this conference. Establish pre-installation conferences for specific critical phases of the work and required mock-ups.

Develop an orderly system of reviewing, routing, and distributing submittals, such as shop drawings, product data and samples; process and return to the contractor in a timely fashion. Develop with the owner final color and material selections early in the project so as not to delay the work. Develop a method of correspondence to "put it in writing" observations of the project, intent, interpretations, decisions, memorandums, etc., of meetings and progress of the work. Keep the owner, the construction inspector, the contractor and consultants fully informed and distribute copies of all correspondence, forms, reports, and approvals to all parties.

Review the contractor's schedule of values and generally evaluate whether the breakdown is an adequate representation for payments and cash flow.

Make timely observations of the work with the construction inspector present and promptly notify the contractor of deficiencies observed. Make periodic reports to the owner concerning the progress of the work.

Establish the standards of acceptability.

Receive contractor's applications for payment, and generally evaluate the progress of the work as is claimed. If in order, promptly issue a certificate for payment to the owner.

Make timely decisions concerning interpretation of documents and details of design to the contractor. Do not get involved in the craft jurisdiction. Do not communicate directly with the subcontractors, vendors, or suppliers unless authorized by the contractor. All such communication shall be recorded in writing and distributed.

Require that the design professional's consultants make timely visits to the work to observe the general installation of systems and equipment designed by them.

Return field phone calls promptly. Allow a reasonable number of at least 20 rings before giving up.

Do not issue a "stop work" notice.

Do not expect the contractor to correct deficiencies of the document for which it is not responsible.

Do not instruct the contractor as to methods of job safety. However, if the design architect sees a hazardous condition, tell the contractor or its superintendent at once and instruct the construction inspector to note this in log or daily report.

Inspect the work when the contractor notifies that the date of substantial completion has been attained and attaches the inspection (punch) list. Request the consultants to inspect the work at this time and to modify the contractor's inspection list. Have prepared an inspection list (punch list) of deficient or nonacceptable items and insist the owner, contractor and construction inspector understand or approve the items.

Distribute the inspection list to all parties. If the work at that time (or thereafter) is completed to the extent that the owner may take occupancy, the architect should recommend to the owner to accept the work as substantially completed. At that time and if the remaining items involve a cessation of labor, recommend to the owner, with concurrence of the owner's legal counsel, to have notice of completion processed or take other necessary actions.

Accompanied by the owner, contractor, and construction inspector, make a final inspection of the work after the contractor notifies the owner that the work is totally completed. Promptly process all outstanding change orders and other contract document requirements, and execute a final certificate for payment.

Part 1.3.1:
Project Design Professional (Architect/Engineer) Description

This experienced individual is assigned by the principal-in-charge to manage the project and serve as the owner's contact relative to the day-to-day activities of the project. The design professional's ability to efficiently lead the design team and direct the administrative and technical aspects of the design, is critical to the success of the project. This of course, includes meeting the owner's requirements within the real constraints of cost and time.

Part 1.3.2: Project Designer Description

This individual is assigned by the principal-in-charge to come up with a design concept and solution that meets the project requirements established by the owner. Although the aesthetic aspect of the design is very important, a true solution is tested by attaining the functional requirements of the project. The major refinements of visually exposed elements are comprehensively articulated by the project designer.

Part 1.3.3:
Project Job Captain Description

This experienced individual is selected for a specific project to further develop the design concept from design develop through construction documents. The project job captain's technical ability to be able to integrate the interdisciplinary expertise of the design team, including the project specifier and other technical resources, is critical to the success of the project.

Part 1.3.4:
Project Specifier Description

An experienced individual assigned to prepare the project manual for the project. The most effective use of this person's abilities is to participate in the early formulation of materials, equipment, construction systems and define the quality requirements for the project. The ability for its specifier to be integrated as technical resource throughout the development and delivery of the project is critical to the success of the project.

Part 1.3.5:
Field Administrator (Construction Contract Administrator) Description

This experienced individual, that could be already part of the project team i.e. job captain, is assigned with the responsibility to administer and enforce the requirements of the construction documents during construction. The field administrator's services include certifying pay requests, reviewing contractor's submittals, responding to clarifications, requests for information (RFI's), preparing construction change authorizations/change orders and conducting project site walkthroughs to observe substantial conformance with the contract documents.

Part 1.3.6:
Observations and Inspections
Description

Inspections are conducted by specialized individuals that may be members of an independent testing laboratory hired by the owner, full time representative hired by the owner or architect or officials representing local jurisdictions or authorities.

According to customary practice, architects conduct periodic site visits to observe substantial conformance with the contract documents of visually exposed and accessible conditions noticed but they do not generally conduct inspections since this denotes a level of detailed review and scrutiny not afforded by their scope of work or expertise relative to quantitative assessment and testing.

For the sake of simplicity, where the matrix indicates design professional's (architect/engineer) inspection (AI), it denotes observation work for which the design professional is generally responsible according to prevailing professional standards.

Part 1.4:
Contractor: Recommended Duties and Responsibilities

Carefully study contract documents. Report any error, inconsistency, or omission he, his subcontractors, vendors, and suppliers may discover.

Assign a superintendent and necessary personnel to be on the site during the construction of the work. Delegate sufficient authority to the superintendent to represent and act for the contractor.

Provide subcontractors with construction documents indicating the work of all crafts. Require the superintendent and subcontractors to familiarize themselves fully with the specifications and general and special conditions.

Provide and become familiar with the standards specified. Provide codes at the site, if stipulated.

Notify the testing laboratory in sufficient time to permit proper inspection. Become fully acquainted with all tests, inspections required of the testing laboratory, and special inspections required by agencies. Work requiring continuous inspection should be expedited or scheduled so as to minimize costs. Advise the persons scheduled to inspect of any delay in the work schedule so that unproductive inspection time will be avoided.

Require that subcontractors, vendors, and suppliers communicate through the contractor or its superintendent to the design professional.

Assume fully that the work to be completed and performed is in full compliance with the contract documents. Be responsible for the progress and control of the work and for fully coordinating and overseeing the work of the subcontractors, vendors, and suppliers to meet this obligation.

Submit a schedule of values for the various portions of the work, with breakdowns in sufficient details to meet with the design professional's approval and allow the architect and the construction inspector to rely on the schedule as a basis for progress payments. It is recommended that this schedule follow the CSI MASTERFORMAT.

Prepare a progress schedule for the work, including a schedule of submittals, showing its various phases, and update it periodically.

Maintain on the site a complete set of construction documents (with agency approvals as required) including drawings, specifications, addenda (cut-in to drawings and specifications), approved shop drawings, product data and samples, change orders, and all correspondence concerning the work.

Update the construction documents periodically to record changes.

Prepare a schedule of shop drawings, product data and samples. Coordinate and process submission to allow a reasonable time for review and/or approval by all parties, and maintain proper sequence for timely progress of the work.

Process and note substitutions as are allowed. Be responsible for equivalency and compatibility. Submit in a timely sequence so that the design professional may coordinate color selections and other decisions with the owner.

Require, check, and verify that subcontractors, suppliers, or vendors perform and/or supply their work, materials, and equipment so as to be available in a proper sequence for efficient progress of the work.

Give the construction inspector advance notice of delivery of materials or equipment.

Periodically observe, inspect, and review the work of contractor's employees and the material, equipment, and installation of contractors, subcontractors, vendors, and suppliers so that the construction is in compliance at all times. Promptly correct or remove all defective work.

Comply with the requirements of agencies relating to permits and agency-required inspections during the various phases of the work.

Coordinate mechanical, electrical, and other installation sequences and assign priorities where drawings are diagrammatic.

Promptly make approved payments to subcontractors, vendors, and suppliers in accordance with the contract documents.

Provide changes to the work only through processes allowed in the construction documents.

Notify the owner, through the architect, in writing concerning coordination of owner-furnished equipment, items not-in-contract, etc. Allow sufficient time for orderly sequencing and incorporation of these items into the work.

Notify the design professional in writing when substantial completion is reached, prepare a. punchlist and request inspection by the design professional and its consultants.

Promptly submit all record drawings, reports, instructions, guarantees, and other requirements of the construction documents to the architect.

Accurately and promptly submit in writing any delays due to "Acts of God" (i.e., weather), strikes or any other disruption in the project's completion beyond the Control of the Contractor.

Totally complete the work within the time stipulated after substantial completion and notify the design professional. Request a final inspection in writing.

Part 1.5:
Owner: Recommended Duties and Responsibilities

Establish proof of legal title. Obtain valid site survey and geotechnical investigations necessary for the project design.

Delegate or assign an owner's representative to represent during the construction of the work. Give the owner's representative enough authority to make timely decisions on the part of the owner. Thoroughly review the contract documents to grasp the full concept and scope of the project.

Determine with design professional input the need for construction inspection based upon the type, size, and complexity of the project. Establish a sufficient allowance in the project budget to compensate the construction inspector and/or the construction inspection staff adequately. Consult with the design professional as to how much inspection is required.

Consult with the design professional to establish the qualifications required for inspection of the work. Design professional's consultants should participate in the selection of personnel. Allow sufficient time for recruitment. If interviews are required, allow applicants sufficient time to present their qualifications.

Allow engagement of the construction inspector before construction is begun so that full familiarization of the work is allowed and administrative procedures can be established. Engagement before putting the project out for bid is recommended.

Communicate only through the design professional. The owner's representative, accompanied by the architect, should periodically visit the project to review progress. Do not allow unauthorized persons to interfere with the work by communicating with the design professional, contractor, or construction inspector. All important communications should be in written form.

The owner's representative should make timely decisions based on the recommendations of the design professional, with fair judgment of the conditions involved.

Organize the financing of the project to allow orderly cash flow. Process certificates of payment promptly to avoid undue delay of monies due the contractor.

Realize that the normal "basic" services of the design professional include only "periodic observation" of the work. Do not expect the design professional to observe all the processes, installations, workmanship, etc.

Realize that the function of the design professional during construction is an "interpreter of the documents." The design professional may have to impartially take the side of the owner or the contractor as the documents require.

Allow an adequate budget contingency for changes required by unforeseen factors occurring during construction. It is customary to allow approximately 3 to 10 percent, depending on the size, type and complexity of the project. Where additions and alterations are included, and the project is small, budget a greater contingency.

Process and approve change orders in a timely manner. Where the process for approving change orders is cumbersome or complex due to organizational requirements, it is suggested that the owner's representative should be delegated to approve "field change orders" (within a maximum dollar limitation), since this process appears to have legal precedence for public and private bodies when a contingency is budgeted.

Allow reasonable time extensions for causes beyond the control of the contractor. Realize in some cases that the construction sequence may be so interrupted as to warrant an even longer period than the work stoppage. If necessary, allow extra compensation for making up lost time by shift work or a limited amount of overtime.

Do not request quotations from the contractor for changes not fully expected to be incorporated in the work.

For owner-furnished equipment to be incorporated into work and/or items not-in-contract, it is the owner's responsibility to supply all needed information in a timely fashion so as not to delay construction. This information includes size, weight, points and methods of anchorage, rough-in locations, utility connections and characteristics. etc.

During the preparation of contract documents, consider and allow payment for items suitably stored on-site or off-site and not incorporated into the work when the contractor submits bills of sale or adequacy of storage, insurances, etc. In addition, consider the concept of reducing the withholding or retained monies on the work to 5 percent when the work is 50 percent, completed. This practice is used frequently by public and private agencies.

Do not request the contractor or design professional to perform services that are not included as part of the work without additional compensation.

Do not request the construction inspector to provide services other than those for which he is employed without additional compensation.

Retain legal and insurance counsel knowledgeable in construction practices. Realize that the design professional can only make recommendations on matters within the extent of his professional licensing.

Avoid occupancy of the work until the date of substantial completion. If "beneficial occupancy" is necessary, realize that the owner is responsible for damages and delays occurring during move-in and thereafter which are caused by him or his employees.

Accept the work without undue delay. Accompanied by the architect, the contractor, and the construction inspector, arrange for the inspection and acceptance of the work in an organized manner. It is normally the duty of the architect to rule on acceptability of the work or portions of it and recommend acceptance to the owner.

With the advice of legal counsel do not delay taking recommended actions concerning occupancy and acceptance and executing required documents.

Realize that the work generally has a one-year "correction period" (longer in certain cases and/or where stipulated) from the time of occupancy and acceptance (or use of equipment in some cases). Payment for 100 percent of the contract amount, less the retainage, should be made when the work is substantially completed. In many cases, the work may not be "totally complete" to the last detail. If withholding is necessary, it should be commensurate with the situation.

Investigate as to whether building site or existing facility may contain hazardous materials and have environmental, chemical and pollution testing performed as may be necessary.

Make the final payment promptly upon expiration of lien periods and the issuance of design professional's final certificate of payment.

Lastly, realize that construction of the work is usually a "one-of-a-kind" process and that the parties involved cannot guarantee perfection. The quality of materials and workmanship and their acceptance must be commensurate with the constraints established by the contract documents.

Part 1.6:
Lawyers Description

Lawyers are professionals licensed to practice law and therefore participate in providing legal counsel from contract negotiations relative to drafting custom contracts or amending AIA agreements, general conditions and providing risk management advice primarily when disputes or claims arise.

Due to an increasingly litigious environment their presence and effect has made its indelible mark in the construction industry. The lawyers that are being referred to specialize in construction law and therefore are experts in construction related matters. Their participation in contract formulation and negotiations, and not just conflict resolution after construction, increases the ability to deliver a project that meets the originally intended goals, reduces disputes and encourages early resolution of conflict. In order to avoid escalation of disputes into litigation, mechanisms for alternative dispute resolution like arbitration and mediation are now being used to expedite the settlement process and diminish losses and aggravation. Additionally, new attitudes of collaboration posed by the advent of partnering as a team approach to resolve construction disputes, establishes a collective commitment to work things out to preclude any impending legal action.

Part 1.7:
Manufacturer's Representative Inspections

In order to confirm that the work, as installed meets the manufacturers' recommendations, certain sections of the technical specifications of the project manual may require inspection by a manufacturer's representative. A written confirmation by the manufacturer's representative of the application and installation of its product in an assembly or part thereof, provides additional validation of its final performance. This becomes necessary when certain warranties require manufacturer's representative site inspection and approval endorsing such application.

Part 1.7.1:
Required by Contract Documents

Determine, early on while becoming familiarized with the contract documents, which sections of the project manual will require such inspections by a manufacturer's representative. This serves as a no-cost way to validate the application and address specific site conditions that may have not been detailed in the construction drawings or may have been created during construction.

Part 1.7.2:
Required for Warranties and Guarantees

Generally, specifications for specialty items like roofing and waterproofing that protect the integrity of the building envelope, require an extended warranty by specifying progress inspections by the manufacturer's representative. This could even become more critical when this inspections are not conducted in conjunction with an independent testing laboratory, new products are installed, or field installation susceptable to problems due to specific site concerns.

Part 1.7.3:
Coordination

To get the most of the manufacturer's representative field service, provided at no cost by most manufacturers', advanced notification is required to ensure scheduling of inspections by the manufacturers' representatives and attendance at pre-installation conferences. Be certain the total assembly is considered, and a comprehensive interdisciplinary coordination is made by a team of competent individuals and that the appropriate action is not the determination of a single judgement of an applicator only focused on the task and not the system assembly, or adjacent product, or material compatibility.

Part 1.8:
Others: Special Consultants

With the growing complexity of technology and the resulting specialized areas of expertise, consultants for various fields continue to make their way in the construction industry. Some of these disciplines that may be hired by the owner, design professional or contractor as design-builders are also listed as follows:

DESIGN-BUILD	
1.	Acoustical
2.	Audio/Visual
3.	Data/Communications
4.	Electrical
5.	Energy
6.	Equipment
7.	Exhibit
8.	Fire Protection
9.	Food
10.	Fountain
11.	Furniture and Furnishings
12.	Graphics
13.	GFRC
14.	Hardware
15.	Landscaping
16.	Lighting
17.	Mechanical
18.	Parking
19.	Precast
20.	Plumbing
21.	Roofing
22.	Security
23.	Signage
24.	Specialty Design
25.	Waterproofing
26.	Window Wall

GENERAL CONSULTANTS
1. Acoustical
2. Arborist
3. Art
4. Audio/Visual
5. Civil
6. Cost Estimating
7. Data/Communications
8. Electrical
9. Elevator
10. Energy
11. Equipment
12. Exhibit
13. Fire Protection
14. Fitness
15. Food Service
16. Fountain
17. Furniture and Furnishings
18. Geotechnical
19. Graphics
20. Hardware
21. Historic Preservationist
22. Interior Design
23. Laboratory
24. Landscaping
25. Land Surveyors
26. Library Science
27. Lighting

	GENERAL CONSULTANTS (Cont.)
28.	Mechanical
29.	Medical Equipment
30.	Parking
31.	Physicist
32.	Programming
33.	Plumbing
34.	Public Sector Management
35.	Roofing
36.	Security
37.	Shielding
38.	Signage
39.	Specialty Design
40.	Specifications
41.	Structural
42.	Surgical
43.	Transportation
44.	Water Feature
45.	Waterproofing

Part 2:
Standards and Codes

Many organizations and agencies have developed standards and codes applying to the construction industry. A large portion of a specification is an unwritten part which by reference establishes levels of acceptability in the manufacture, installation, workmanship, or performance of systems, methods, and materials incorporated into the work. These standards and codes often apply on a local, regional, or national basis.

A general listing of the most common agencies that develop these standards appears in Part 2.3.

Many professional and trade organizations and agencies publish material or are organized to assist the construction industry, and they have published documents, forms, manuals, and other useful information that is widely accepted and used by owners and the industry in both a legal and technical sense. Since the scope of the manual is limited, these have not been included in the standards part; however, they and others are listed in Appendix A, "Construction Industry Organizations."

Part 2.3 contains a numerical listing, to the left, of the name of a standard and a CSI MASTERFORMAT section number. In addition, assigned initials are listed before the name. The numerical listing cross-refers to the initials of the standard listed at the top of each page in various sections of Part 3 Checklist for Field Inspection." The CSI MASTERFORMAT section number indicates typical sections to which the standard might be referred.

A general description and listing of major codes appears in Part 2.2.

The *Construction Inspection Manual* was first copyrighted in 1973 which subsequently completed three printings. The first edition, the second edition in 1974 and third edition in 1977, each utilized CSI section numbers as well as standards listed in PSAE (*Production Systems for Architects and Engineers, Masterspec*) and PIC (*Pacific International Computing Corporation, Comspec*). The fourth edition in 1983 also identified specifying aids such as Masterspec, Comspect and Spectext and included references to CSRF (*Construction Science Research Foundation*). Each edition was updated to accommodate CSI section numbers as stated in MP-2-1 editions in 1972 (MP-2A), 1975 (MP-2A), 1978 (MP-2-1) and 1983 (MP-2-1). The current *Construction Inspection Manual*, likewise, utilizes the latest edition for MASTERFORMAT Section titles and numbers (MP-1-2-95).

While the *Construction Inspection Manual* has attributed the section numbers to CSI (*Construction Specifications Institute*) it is emphasized that the MASTERFORMAT, the official title since 1978, is a joint publication with CSC (*Construction Specifications Canada, Document 004E*). Prior to MASTERFORMAT, the official titles were known as the CSI Format for

Construction Specifications, the Uniform System for Construction Specifications, Data Filing and Cost Accounting and the Uniform Construction Index. The formulation of the MASTERFORMAT predecessors involved American Institute of Architects, American Society of Landscape Architects, Associated General Contractors of America Inc., Associated Specialty Contractors Construction Products Manufacturing Council, National Society of Professional Engineers as well as CSI and CSC. Among others, the U.S. Department of Defense has adopted MASTERFORMAT for use on Naval, Air Force and Army Corps of Engineers projects. McGraw-Hill Information System Co. adopted MASTERFORMAT for Sweets Catalog Files. MASTERFORMAT is the framework for SPECTEXT, the CSRF master guide specification system marketed by CSI and for other commercially available master guide specification systems.

Part 2.1:
Contract/Contract Documents/Specifications

As previously indicated in Chapter 1.1, a contract is a legally enforceable agreement that sets forth the obligations of each party to the other. Any violation of these obligations (breach) can expose the party committing it to sanctions of law. The significant benefit of having an agreement is that a mutual understanding by communicating expectations through clear definition of scope and general/terms and conditions is developed.

Contract documents are composed of the agreement between the owner and contractor, conditions of the contract (general, supplementary and others when applicable), drawings, specifications, Addenda and any other modifications issued after the signing of the contract.

Technical specifications are the technical portions of the project manual that do not include the "front end": bidding requirements, contract forms, contract conditions, and Division 1 sections. It is within a specification section that reference standards are included to define the level of quality and industry accepted practices relative to installation or construction of a component or system assembly. For example, the installation of metal flashings require conformance with the Architectural Sheet Metal Manual by SMACNA (Sheet Metal and Air Conditioning Contractors National Association).

All the above documents provide the basis for the enforcement of the agreed upon contract requirements for project delivery during construction. It is highly encouraged that the inspector develops an intimate familiarity with the contract documents to better serve the owner's interest and achieve the successful execution of the design intent.

Part 2.2:
Codes and Regulations

The proliferation of codes and regulations applying to construction continues year by year. They are adopted by federal, state, and local governments and can be supplemented by countless agencies.

Model codes are developed by various forms of consensus processes, usually chaired/controlled by organizations for code enforcement officials such as the Building Officials and Code Administrators (BOCA), the Southern Building Code Congress, International (SBCCI), the International Conference of Building Officials (ICBO), or the National Fire Protection Association (NFPA). Presently these organizations each publish their own particular model codes. However, in the year 2000 we will see the beginning of a new era in model codes, a single set of model codes published jointly by BOCA, SBCCI, and ICBO, the "International Codes," some of which have already been published, such as the "International Plumbing Code" and the "International Mechanical Code."

Other codes and regulations may be developed based on actual or perceived problems, and may be incorporated with no more input than the power of some special interest group's political savvy. Among the national regulations impacting construction are the Occupational Safety and Health Act (OSHA), the federal Safety Glazing Standards, the Fair Housing Accessibility Guidelines, and the Americans with Disabilities Act Accessibility Guidelines.

State and regional requirements vary as widely as do the interests of people within those states and regions. Some states adopt voluminous sets of codes and regulations based loosely on one of the model codes. Other states adopt one of the model codes only to ensure that various jurisdictions throughout the state maintain a consistent quality of construction. The same can be said for local government agencies.

It is important that the design professional (architect, engineer), contractor and building inspector be aware of codes applicable to a particular project. Additionally, it is important to use the appropriate edition, because changes are constantly occurring, and the most recent edition is not always the applicable edition.

The three major national model codes listed below, with their related codes, are often referenced or apply to the contract documents.

BOCA – Building Officials and Code Administrators, International, Inc., 4051 W. Flossmoor Road, Country Club Hills, IL 60477-5795, (708) 799-2300, publishes the following:

    National Building Code
    National Fire Prevention Code
    National Property Maintenance Code
    International Mechanical Code
    International Plumbing Code
    BOCA Energy Conservation Code
    With certain exceptions, this code is generally adopted in the Northeast and Midwest.

SBCCI – Southern Building Code Congress, International, Inc., 900 Montclair Road, Birmingham, AL 35213-1206, (205) 591-1853, publishes the following:

    Standard Building Code
    Standard Fire Prevention Code
    Standard Gas Code
    International Mechanical Code
    International Plumbing Code
    Standard Housing Code

With certain exceptions, SBCCI codes are generally adopted in the Southeast.

ICBO – International Conference of Building Officials, 5360 South Workman Mill Road, Whittier, CA 90601-2298, (562) 699-0541, publishes the following:

    Uniform Building Code
    International Mechanical Code
    Uniform Housing Code
    Uniform Code for the Abatement of Dangerous Buildings
    Uniform Sign Code
    Uniform Fire Code
    Uniform Code for Building Conservation
    With certain exceptions, ICBO codes are generally adopted west of the Mississippi River.

IAPMO – International Association Of Plumbing and Mechanical Officials, 2001 South Walnut Drive, Walnut, CA 91789-2825

    Uniform Plumbing Code

NFPA – National Fire Protection Association, 1 Batterymarch Park, P.O. Box 9101, Quincy, MA 02269-9101, (617) 770-3000, (800) 344-3555, publishes the following:

National Electric Code, NFPA 70

Life Safety Code, NFPA 101

The National Electric Code is universally adopted throughout the United States. The Life Safety Code is often adopted by state and local fire departments, and can present problems regarding differences of interpretation with similar requirements in the model building codes relating to life safety.

In many states it is usually necessary to comply with more than a single code or enforcing agency to design and construct the work. The work (especially public or institutional) may also be governed by one or more of the following codes:

Federal

FHA – Federal Housing Administration, U. S. Department of Housing and Urban Development (HUD). 451 Seventh Street, S. W. Washington, D. C. 20410-0500, (202) 275-7543, publishes the following:

Minimum Property Standards for Multi-family Housing
Minimum Property Standards for One and Two Living Units
Manual of Acceptable Practices
The publications apply usually to federally assisted residential construction.

FHA - Federal Housing Administration, U.S. Department of Housing and Urban Development (HUD), 451 Seventh Street, S.W., Washington D.C. 20410-0500, (202) 708-2618, publishes the following:

FHAG - Fair Housing Accessibility Guidelines

The publications result from the Fair Housing Amendments Act of 1988, concerning accessibility for persons with disabilities to both public and private housing including requirements relating to design and construction.

OSHA – Occupational Safety and Health Administration, United States Department of Labor, 200 Constitution Avenue, N.W., Room N3101, Washington D.C. 20210, publishes the following:

Construction Industry Standards. 29 CFR 1926, 29 (CFRP 1900-1910)

The publications result from the Occupational Safety and Health Act of 1970, concern health and safety requirements for employees and apply to design and construction.

Americans with Disabilities Act, Civil Rights Division, U.S. Department of Justice, Washington D.C..20530, (202) 514-0381, publishes the following:

ADAAG, Americans with Disabilities Act Accessibility Guidelines
Americans with Disabilities Act Handbook

These publications result from the Americans with Disabilities Act of 1990, concerning accessibility for persons with disabilities including requirements to state and local government services (Title II) and to public accommodations in commercial facilities (Title III) both relating to design and construction.

ANSI A.117.1 – Accessible and Usable Buildings and Facilities. American National Standards Institute, Inc., 10 East 40th Street, New York, N.Y. 10018.

In the past various editions of this publication have been adopted by some state and local agencies, although the common practice has changed to accept the ADAAG and FHAG as the standards for accessibility for persons with disabilities.

States: In addition, many states adopt their own versions of the model codes, the following is an example of the California Codes of Regulations

CCR – California Code of Regulations General Services. Publication Section, P.O. Box 1015, Sacramento, CA 95660, publishes the following, applicable to building construction:

Title 8 – Industrial Relations

Part 1, Chapters 3.2, 3.3, 3.5 Occupational Safety and Health Regulations, Appeals and Standards; Chapter 4, Division of Industrial Safety (includes all Safety Orders).

Title 17 – Public Health
Title 19 – Public Safety
Title 21 – Public Works
Title 22 – Health Planning and Facility Construction; Division 7 and 8, T22.
Title 24 – California State Building Standards Codes
   Part 1, Administration, Building Standards Commission
   Part 2, California Building Code
   Part 3, California Electrical Code
   Part 4, California Mechanical Code
   Part 5, California Plumbing Code
   Part 6, Special Building Regulations
   Part 7, Elevator Safety Regulations
   Part 8, State Historical Building Code
   Part 9, California Fire Code
   Part 10, State Uniform Code Building Conservation
   Part 12, Reference Standards Code

Title 25, Housing and Community Development Chapter 1, State Housing Law and Earthquake Protection Regulations

Chapter 3, Factory Built Housing, Mobile Homes, Recreational Vehicles and Commercial Coaches.

These state codes may or may not apply to all design and construction. Most are applicable to institutional and/or public buildings; however, private buildings often are subject to the codes. Local building departments and fire marshals can indicate applicable titles, chapters or parts.

Public Safety: The construction inspector is not directly responsible for public safety. However, unsafe conditions cannot be ignored. And, codes and regulations are typically developed in order to ensure the public health, safety and welfare. If the construction inspector observes anything which might be unsafe, it should be immediately reported to the contractor's superintendent. If member of the general public might be in danger it would be appropriate for the construction inspector to stay and do whatever is possible to prevent a potential accident.

One public safety requirement which can be especially important is the safety standards for architectural glazing materials. These federal regulations, which have been incorporated into the model building codes, requires safety glazing materials be used in doors and in areas where there is potential impact by people. Safety glazing is easily identified by a permanent mark required in the lower corner of each safety glazing lite, usually noting the glass is safety glass or conforms with ANSI A97.1. Only wired glass used in fire rated doors and sidelights is exempt from the safety glazing requirements.

Life Safety: The design professionals should include life safety requirements related to codes and regulations in the contract documents. As such, the construction inspector can be responsible for verification of compliance. Some knowledge of the codes and regulations in addition to the contract documents would be especially useful.

Egress becomes the first concern regarding life safety. A movie called "Towering Infernal" made years ago included an exit stair landing with a wheelbarrow of concrete tipped over, blocking the exit door into the stair. With all the potential inspectors, including fire department inspections, such a problem should be almost impossible. However, blocked exits, operable hardware which doesn't operate properly, and construction debris blocking exits in remodeling projects and after beneficial occupancy can be prevented by construction inspection.

Access for Persons with Disabilities: Design of proper accessibility should be part of the construction documents. Verification of compliance should, however, be part of the construction inspector's observations. Access regulations such as ANSI A117.1 and the Americans with Disabilities Act Accessibility Guidelines can provide some understanding of how to ensure accessibility in a construction project. Refer also to Appendix J.

Hazardous Materials: Many public agencies now require material safety data sheets (MSDS) be kept on-site. The construction inspector may or may not have some responsibility for the MSDS, but even where there is no direct responsibility, it would be appropriate for the construction inspector to know where these sheets are kept. Refer also to Appendix K.

Part 2.3:
Standards

NO.	STANDARD		CSI SECTION	
1.	*AIMA* — Acoustical and Insulating Materials Association, 205 W. Touhy Ave., Park Ridge, Il. 60068, (630) 553-0129.			09800
2.	*AA* — Aluminum Association, 900 19th St., N.W. Washington, DC 20006, (202) 862-5100. Fax (202) 862-5164	05500 07600 08400 08900 10750	05100 05700 08100 08500 10400 12490	07400 08300 08600 10550 14200
3.	*AAN* — American Association of Nurserymen, 1250 1 Street, N.W., Suite 500 Washington, DC 20005, (202) 789-2900.			02480
4.	*AASHTO* — American Association of State Highway and Transportation Officials, 444 N. Capitol St., N.W. #249, Washington. DC 20001, (202) 624-5800.			02300 02700
5.	*ACI* — American Concrete Institute, P.O. Box 19150, Redford Station, Detroit, MI 48219, (313) 532-2600.	02450 02930 03200 03370	02500 03100 03300 03400	02400 02900 03150 03360 03500
6.	*AGA* — American Gas Association, 1515 Wilson Boulevard. Arlington, VTA 22209, (703) 841-8400.	15010 15600	10960 15050 15650	15400 15800
7.	*AHA* — American Hardware Association, 20 N. Wacker Drive Chicago, IL 60606.			08700

NO.	STANDARD	CSI SECTION		
8.	*AHMA* — American Hardware Manufacturers' Association, 801 N Plaza Dr., Schaumburg, IL 60173, (847) 605-1025.			08700
9.	*AHDGA* — American Hot Dipped Galvanizers Association, 1000 Vermont Ave., N.W., Washington, DC 20005, (202) 628-4634.	03200 05400	02530 05120 05500	02900 05300 07400
10.	*AISC* — American Institute of Steel Construction, Inc., 1 E Wacker Dr., Chicago, IL 60601, (312) 670-2400.	03100 05300	05120 05400	02300 05200 05500
11.	*AITC* — American Institute of Timber Construction, 7012 S. Revere Parkway #140, Englewood, CO 80112, (303) 792-9559, FAX: (303) 792-0669.	06130	06100 06170	06070
12.	*AIA* — American Insurance Association, 1130 Connecticut Ave., Suite 1000, Washington, DC 20036, (202) 828-7100.	04200	08100	04090
13.	*AISI* — American Iron & Steel Institute, 1101 17th St. NW., Washington, DC 20036, (202) 452-7100 (800) 545-2433.	03200 04200 05300 07400 10750	03400 05100 05400 09800 10800	04150 05200 05700 10150 14200
14.	*ALA* — American Library Association, 50 E. Huron St., Chicago, Il 60611, (312) 944-6780.		11650	
15.	*API* — American Petroleum Institute, 1220 L Street, N. W., Washington, DC 20005 (202) 682-8000.		02550	15600
16.	ANSI, ANS — American National Standards Institute, Inc., 11 W 42nd St., 13th Fl., New York. NY 10036, (212) 642-4900.			
17.	APA — American Plywood Association, P.O. Box 11700, Tacoma, WA 98411, (253) 565-6600.	06100	06200	03100 06400

NO.	STANDARD	CSI SECTION		
18.	*APWA* — American Public Works Association, 1313 E. 60th St., Chicago, IL 60637, (773) 667-2200, FAX: (312) 667-2304.			02500
19.	*ASTM* — American Society for Testing and Materials, 100 Barr Harbor Dr. West Conshohocken, PA 19428-2959 (610) 832-9585.			
20.	*ASAHC* — American Society of Architectural Hardware Consultants, 1815 N. Ft. Myer Dr., Ste. 412, Arlington, VA 22209	08100	08400	08700
21.	*ASHRAE* — American Society of Heating, Refrigeration and Air Conditioning Engineers, Inc., 1791 Tullie Cir. N.E., Atlanta, GA 30329, (800) 527-4723.		15700	15600 15800
22.	ASME — American Society of Mechanical Engineers, 345 E. 47th St., New York, N.Y. 10017, (212) 705-7722 (800) THE-ASME.			15600
23.	*AWS* — American Welding Society, 550 N.W. 42nd Avenue, Miami, FL 33126 (305) 443-9353, (800) 334-9353	02400 02900 05120 05400 15600	02530 02930 05200 05700	02870 03200 05300 15400 15700
24.	*AWWA* — American Water Works Association, 6666 W. Quincy Avenue, Denver, CO 80235, (303) 794-7711.	05120	15400 05200	15600 05300
25.	AWPA American Wood Preservers Association, P.O. Box 2865, Woodstock, MD 21163 (410) 465-3169.	06200	06300	06100 06400
26.	AAMA — Architectural Aluminum Manufacturers' Assoc., 35 E. Wacker Dr., Chicago, IL 60601, (312) 782-8256.	05500 08300	05700 08500	08100 08900

NO.	STANDARD	CSI SECTION		
27.	*AWI* — Architectural Woodwork Institute, 1952 Isaac Newton Square W. Reston, VA 20191, (703) 733-0660.	06400	08200	06200 12490
28.	*AI* — Asphalt Institute, Research Park Dr. Lexington, KY 40512, (606) 288-4961			02500
29.	*BIA* — Brick Institute of America, 11490 Commerce Park Dr, Reston, VA (703) 620-0010.	04500	04060 04900	04090 04550
29A.	*BHMA* — Builders Hardware Manufacturers Association, 11490 Commerce Park Dr. Reston, VA 20191 (703) 620-0010		08700	
30.	*CRA* — California Redwood Association, 405 Enfrente Dr. Suite 200, Novato, CA 94949 (415) 382-0662, FAX (415) 382-8531.	02530 06170	03100 06200	06100 06400
31.	*CSS* — California Standard Specifications, Division of Highways, Documents Section, Sacramento, CA 94807.		02300	02500
32.	*CRI* — Carpet & Rug Institute, P.O. Box 2048, Dalton, GA 30722, (706) 278-3176.		09680	12480
33.	*CTI* — Ceramic Tile Institute, 700 North Virgil Avenue, Los Angeles, CA 90029, (213) 660-1911.	05120	05200	05300
34.	*CLFMI* — Chain Link Fence Manufacturer's Institute, 1776 Massachusetts Ave, NW No. 500 Washington, DC 20036, (202) 659-3537.			02530
35.	*CS* — Commercial Standard (U.S. Dept. of Commerce), Govt. Printing Office, Washington, DC 20402.	06200	08100	02530 08200
36.	CRSI Concrete Reinforcing Steel Institute, 933 No. Plum Grove Road, Schaumburg, IL 60173-4758, (708) 517-1200 (800) 465-2774.			03200

NO.	STANDARD	CSI SECTION		
37.	*CDA* — Copper Development Association, 260 Madison Ave., New York, N.Y. 10016 (212) 251-7200.	05500 10350	05700 14200 16400	07600 15700 15770
38.	*DFPA* — Douglas Fir Plywood Association (see American Plywood Association), P.O. Box 11700 Tacoma, WA 98411, (202) 272-2283.	03100	06100	06200
39.	*FTI* — Facing Tile Institute, c/o Box 8880, 04150 Canton, OH 44711, (216) 488-1211.			04050
40.	*FM* — Factory Mutual Systems Division, 1151 Boston - Providence Highway, Norwood, MA 02062, (781) 762-4300.	07500	05300 08100	07200 14200
41.	*FS* — Federal Specifications, Supt. of Documents, Government Printing Office, Washington, DC 20234.			
42.	*FGMA* — Flat Glass Marketing Association, 3310 S.W. Harrison St., Topeka, KS 66611, (913) 266-7013, FAX (913) 266-0272.		08800	
43.	*FPL* — Forest Products Lab., U.S. Dept. of Agriculture, Madison, WI 53705.	06100	06130	06070
44.	*GTA* — Glass Association of North America, White Lakes Professional Bldg., 3310 S.W. Harrison, Topeka, KS 66611, (785) 266-7064.			08800
45.	*GA* — Gypsum Association, 810 1st St, NE Suite 510, Washington, DC 20002, (202) 289-5440		09100	09250
46.	*HPMA* — Hardwood Plywood Manufacturers' Association, P.O. Box 2789, Reston, VA 22090, (703) 435-2900.	06200	06400	12500
47.	*HI* — Hydraulic Institute, 9 Sylvan Way, Parsnippany, NJ 07054, (201) 267-9700.	02700		15400

NO.	STANDARD	CSI SECTION		
48.	*ILIA* — Indiana Limestone Institute of America, Inc., Stone City Bank Bldg., Suite 400, Bedford, IN 47421, (812) 275-4426.			04400
49.	*WLPDIA* — Western Lath/Plaster/Drywall Industries Assn., 8635 Navajo Road, San Diego, CA 92119, (619) 466-9070.			
50.	*MFMA* — Maple Flooring Manufacturers 60 Revere Dr. #500, Northbrook, IL 60062, (708) 480-9138.	09620	09640	
51.	*MIA* — Marble Institute of America, 30 Eden Alley St 301, Columbus Ohio 43215 (614) 228-6190.			04400
52.	*MI* — International Masonry Institute, 815 15th Street, N.W., Washington, DC 20005, (202) 783-3908.	04200	04060 04500	04090
53.	*MLA* — Metal Lath Steel Framing Division of National Association of Architectural Metal Manufacturers, 8 S. Michigan Suite 1000, Chicago, IL 60603, (312) 456-5590.			09100
54.	*NAAMM* — National Association of Architectural Metal Manufacturers, 8 South Michigan, Suite 1000, Chicago, IL 60603 (312) 332-0405.	05700 08400 10400	08100 08500 10700	05500 08300 08900 14200
55.	*NBHA* — National Builders Hardware Association, 1815 N. Ft. Myer Dr., Ste. 412, Arlington, VA 22209, (703) 527-2060.	08100	08400	08700
56.	*NBGQA* — National Building Granite Quarries Association, 369 North State St., Concord, NH 03301, (603) 225-8397.			04900
57.	*NBS* — National Bureau of Standards, U.S. Dept. of Commerce, Government Printing Officer, Washington, DC 20234.			

NO.	STANDARD	CSI SECTION		
58.	*NCMA* — National Concrete Masonry Association, 2302 Horse Pen Road, Herndon, VA 20171, (703) 713-1900.	04070	04090	04200
59.	*NEC* — National Electric Code, National Fire Protection Assn., One Batterymarch Park, Quincy, MA 02269, (617) 770-3000.	16100	16400	14200 16500
60.	*NEMA* — National Electrical Manufacturers' Association, 2101 L St., N.W. Suite 300 Washington, DC 20037; (202) 457-8474, FAX (202) 457-8474.	06200 10750 16100	08200 11650 16400	10150 14200 16500
61.	*NEMI* — National Elevators Manufacturing Industry, Inc., 600 Third Ave., New York, NY 10016, (212) 986-1545.			14200
62.	*NFMA* — National Fan Manufacturers' Association, 5-157 General Motors Building, Detroit, MI 48202.			15800
63.	*NFPA* — National Fire Protection Association, One Batterymarch Park, Quincy, MA 02269, (617) 770-3000 (800) 344-3555.	08100 09300 09650 16100	08200 09400 10250 16400	07200 08500 09800 14200 16500
64.	*AFPA* — American Wood Council (Successor NFoPA), 1111 19th St., N.W. #800, Washington, DC 20036, (202) 463-2700.	06100	06130	08100
65.	*NOFMA* — National Oak Flooring Manufacturers' Assoc., P.O. Box 3009, Memphis, TN 38173-0009, (901) 526-5016.		09640	
66.	*NPVLA* — National Paint & Coatings Association, 1500 Rhode Island Avenue, N.W., Washington, DC 20005, (202) 462-6272.		09960	09900
67.	*NPA* — National Particleboard Association, 18928 Premiere Ct., Gaithersburg, DM 20879, (301) 670-0604.	06200	06400	06100 09680

NO.	STANDARD	CSI SECTION		
68.	*NRMCA* — National Ready Mixed Concrete Association, 900 Spring St., Silver Springs, MD 20910, (301) 587-1400.	02500	03300	
69.	*NSF* — N S F International, NSF Building, 3475 Plymouth Rd. Ann Arbor, MI 48105, (313) 769-8010.			
70.	*NTMA* — National Terrazzo and Mosaic Assc. 110E Market Pl. Suite 200A, Leesburg, VA 20176 (703) 779-1022.	05120	09400 05200	05300
71.	*NWMA* — National Woodwork Manufacturers' Association, 205 W. Touhy Ave., Park Ridge, IL 60068, (312) 823-6747.	06200	06400	08200
72.	*PI* — Perlite Institute, 88 New Dorp Plaza., Staten Island, NY 10306-2994, (718) 351-5723	07200	09100	03300 09960
73.	*PPI* — Plastics Pipe Institute, 1275 K St., N.W. #400, Washington, DC 20005, (202) 429-2039.	02700		15400
74.	*PEI* — Porcelain Enamel Institute, 4004 Hillsboro Pike, Suite 224B Nashville TN, 37125 (615) 385-0758.	08900	10100	07400 10150
75.	*PCA* — Portland Cement Association, 5420 Old Orchard Rd. PO Box 726, Skokie, IL 60076, (708) 966-6288.	03300 03200	02500 03360 03300	02900 03370 03400
76.	*PCI* - Prestressed Concrete Institute, 175 W Jackson Blvd., Chicago, IL 60604 (312) 786-0300.	03200	03300	03400
77.	*PS* — Products Standards Section, U.S. Dept. of Commerce, 14th and Constitution Ave., N.W., Washington, DC 20230. (800) 638-2772	06100	06200	03100 10100

NO.	STANDARD	CSI SECTION		
78.	*RCSHSB* — Cedar Shake and Shingle Bureau. 515 116th Ave., N.E. Suite 275, Bellevue, WA 98004, (206) 543-1323, FAX (206) 455-1314.	06130	06400	02530 07300
79.	*RIS* — Redwood Inspection Service, c/o California Redwood Assn, 405 Enfrente Dr Suite 209, Navato CA 94949 (415) 352-0662		06100	06200
80.	*RFCI* — Resilient Floor Covering Institute, 1030 15th St., N.W., Ste. 350, Washington, DC 20005, (202) 833-2635.		09650	09680
81.	*RMA* — Rubber Manufacturers' Association, 1400 K St., N.W. #900, Washington, DC 20005, (202) 682-4800.			09650
82.	*SMACNA* — Sheetmetal & Air Conditioning Contractors' National Association, 4201 Lafayette Center Dr. Chantilly, VA 20151 (703) 803-2980.	07600	05700 08600	07400 15800
83.	*SPR* — Simplified Practice Recommendation, U.S. Dept. of Commerce, 14th & Constitution Ave., N.W., Washington, DC 20230.	06100	06200	03100 10100
84.	*SPI* — Society of the Plastics Industry, Inc., 355 Lexington Ave., New York, NY 10017, (212) 351-5400.	06200	06500	06600
85.	*SPA* — Southern Forest Products Association, P.O. Box 641700, Kenner, LA 70064, (504) 443-4464.	06170	06100 06200	06130 06400
86.	*SDI* — Steel Deck Institute, P.O. Box 9506, Canton, OH 44711 (330) 493-7886.			05300
87.	*SDI* — Steel Door Institute, 30200 Detroit Rd, Cleveland, OH 44145 (216) 899-0010.			08100
88.	*SJI* — Steel Joist Institute, 3217 10th Ave Mertle Beach, SC 29577 (803) 626-1995.			05200

NO.	STANDARD	CSI SECTION		
89.	*SSPC* — Steel Structures Painting Council 4516 Henry St., #301, Pittsburgh, PA 15213.	05200 08100	02400 05500 08900	05100 05700 09860
90.	*SWI* — Steel Window Institute, 1300 Summer Avenue, Cleveland, OH 44115, (216) 241-7333.		08500	08900
91.	*TCA* — Tile Council of America Research Center, 100 Clemson Research Blvd. Anderson, SC 29625 (864) 646-8453			09300
92.	*TPI* — Truss Plate Institute, 583 D'Onofrio Dr Suite 200, Madison, WI 53719 (608) 833-5900			06170
93.	*UL* — Underwriters' Laboratories, 333 Pfingsten Rd., Northbrook, IL 60062, (708) 272-8800.			
94.	*WCLIB* — West Coast Lumber Inspection Bureau, P.O. Box 23145, Portland, OR 97281, (503) 639-0651, FAX (503) 684-8928.			03100
95.	*WRCLA* — Western Red Cedar Lumber Association, 1500 SW Ave, Suite 500, Portland, OR 97204, (503) 224-3930.		06100	06200
96.	*WWPA* — Western Wood Products Assn. 1500 Yeon Blvd, 522 SW Ave, Portland, OR 97204 (503) 224-3930 (800) 825-0100.		06100	06130 06200 06400
97.	*WFI* — Wood Flooring Institute, 1800 Pickwick Ave., Glenview, IL 60025.			09640
98.	*WIC* — Woodwork Institute of California, 3164 Industrial Blvd, PO Box 980247 West Sacramento, CA 95691 (916) 372-9943	06200 08300	06400 08600	08200 12300

Part 3: Checklist for Field Inspection

The Checklist for Field Inspection has been developed to list items that a full-time construction inspector might reasonably be expected to check and observe during the construction phase of a project.

This checklist compiles a list of items pertaining to various divisions and sections of the work from known data sources and from practical comments made by members of the construction industry. The listings of certain activities into particular categories is solely for the purpose of agreeing with the Construction Specifications Institute format and is not intended to imply that certain activities are required to be performed by any particular trade or craft. Further, it is impossible to provide a fully comprehensive checklist covering every facet of construction. Certain materials listed in the "Bibliography" (Appendix A) and possibly other publications provide a much more detailed inspection procedure for particular phases of the work. Many items will seem to be redundant to the contract documents; however, they are listed to serve as reminders.

It should not be construed in any manner that this part of the manual implies any legal or contractual obligation or responsibility on the part of any party or person. It is to be used only as a guide or aid by the inspector. The contract documents and various legal agreements govern the execution of the work. Many of the items listed herein could be construed to be duties or responsibilities of other parties; however, the whole purpose and intent of this checklist is to discover and avoid misunderstanding as early as possible in the best interests of the work and for the benefit of all parties involved.

"As required" may imply that the items is "as specified," "as approved," "as indicated," required by code, or customary in normal good practice of the industry.

Part 3.1:
Checklist Introduction- How To Use The Checklist

A good way to use the checklist is as follows:

Read the specifications and review the drawings with the checklist at hand, to see which items are applicable and to make a note of other items relative to the project.

A list of standards that may be applicable is provided at the top of the first page in each section of the Technical Items Checklists. If the inspector is not familiar with the standards indicated, they should be reviewed since they may modify, or affect the materials, installation, or performance of the work. (See also "Standards and Codes," Part 2.)

Give attention to substitutions, shop drawings, or samples that have been approved, since again they may affect the work.

Review codes referred to in the contract documents and note relevant items of information.

Because it is important for the inspector to be fully familiar with the specifications at all times, it is recommended that the applicable specification sections should be re-read at the start of each new phase of the work. Many inspectors recommend, as a good practice, that at the start of each phase they meet with the superintendent and the subcontractor foremen to completely review this phase.

HOW TO READ THE MATRIX

A matrix indicating which construction entities would make the primary and secondary inspections for each line item activity has been placed on each page of the Field Inspection Checklist. The design professional, contractor, subcontractor, owner, or special consultant can quickly scan the columns to find the activities relating to their type of business.

The first four columns represent four main inspection groups: Special Inspections (SI), Architectural Inspections (AI), Engineering Inspections (EI), and Other Inspections (OI). The last column contains the line number of the activity.

Construction Inspection Manual

Within each main group, there are secondary inspections indicated by the following designations:

Special Inspections - SI

    Testing Laboratory -   [T] [t]

    Hazardous Materials [H] [h]

    Safety   [Y] [y]

Architectural Inspections - AI (no subgroups but includes Landscape Architect, Interior

    Designers, Architects) [X] [x]*

Engineering Inspections - EI

    Civil   [V] [v]
    Structural   [S] [s]
    Mechanical   [M] [m]
    Electrical   [E] [e]

Other Inspections - Other - OI

    Owner   [O] [o]
    Contractor   [C] [c]
    Subcontractor [B] [b]
    Legal   [L] [l]
    Government   [G] [g]
    Fire Protection [F] [f]
    Plumbing   [P] [p]
    Acoustical   [A] [a]
    Roofing   [R] [r]

* If the activity can be applicable to all subgroups of a main group [X] or [x] is used.

An **UPPERCASE BOLD LETTER** indicates the primary responsibility or inspection. A lowercase letter indicates a secondary inspection. Note that there can be several inspections listed for one activity. Use the footer on each page as a convenient guide to the matrix.

Part 3.2:
Field Inspection Checklist

CHECKLIST INDEX

DIV.	CSI	SECTION	PAGE
0		**BIDDING AND CONTRACTING REQUIREMENTS**	58
1		**GENERAL REQUIREMENTS**	59
	01100	Summary	58A
	01300	Administrative requirements	58B
	01400	Quality Requirements	58C
	01500	Temporary Facilities and Controls	58D
	01600	Product Requirements	58E
	01700	Execution Requirements	58F
	01770	Closeout Procedures	59
	01800	Facility Operation	59A
2		**SITEWORK**	
	02300	Earthwork	61
	02450	Foundation and Load-Bearing Elements	65
	02475	Caissons	67
	02530	Sanitary Sewerage and Drainage	68
	02700	Bases, Ballasts, Pavements, and Apportenances	70
	02810	Irrigation Systems	75
	02900	Planting	76
3		**CONCRETE**	
	03100	Concrete Forms and Accessories	79
	03200	Concrete Reinforcement	82
	03300	Cast-in-Place Concrete	85
	03400	Precast Concrete	88
4		**MASONRY**	
	04200	Masonry Units	90
5		**METALS**	
	05120	Structural Steel	94
	05200	Metal Joists	97
	05300	Metal Deck	98
	05500	Metal Fabrications	100

Construction Inspection Manual

DIV.	CSI	SECTION	PAGE
6		**WOOD AND PLASTICS**	
	06100	Rough Carpentry	101
	06170	Prefabricated Structural Wood	103
	06200	Finish Carpentry	104
	06400	Architectural Woodwork	106
7		**THERMAL AND MOISTURE PROTECTION**	
	07100	Waterproofing	107
	07200	Thermal Protection	109
	07300	Shingles, and Roof Tiles, and Roof Covering	113
	07500	Membrane Roofing	116
	07600	Flashing and Sheet Metal	119
	07810	Cementitious Fireproofing	XXX
8		**DOORS AND WINDOWS**	
	08100	Metal Doors and Frames	123
	08200	Wood and Plastic Doors	125
	08400	Entrances and Storefronts	127
	08500	Metal Windows	129
	08700	Hardware	131
	08800	Glazing	135
	08900	Glazed Curtain Walls	137
9		**FINISHES**	
	09200	Lath and Plaster	139
	09250	Gypsum Board	143
	09300	Tile	147
	09400	Terrazzo	151
	09640	Wood Flooring	152
	09650	Resilient Flooring	154
	09680	Carpet	156
	09500	Acoustical Treatment	158
	09900	Paints and Coatings	160

DIV.	CSI	SECTION	PAGE
10		**SPECIALTIES**	
		(No Sections Included)	
11		**EQUIPMENT**	
		(No Sections Included)	
12		**FURNISHING**	
	12300	Manufactured Casework	162
	12500	Window Treatments	163
13		**SPECIAL CONSTRUCTION**	
	13900	Suppression	164
14		**CONVEYING SYSTEMS**	
		(No Sections Included)	
15		**MECHANICAL**	
	15400	Fixtures and Equipment	166
	15550	Heat Generation (Including Liquid Heat Transfer)	172
	15650	Refrigeration	175
	15800	Air Distribution	178
16		**ELECTRICAL**	
	16050	Basic Electrical Materials and Methods	182
	16400	Low Voltage	189
	16500	Lighting	191
	16700	Communications	193

BIDDING AND CONTRACTING REQUIREMENTS

SI	AI	EI	OI	
[ ]	[X]	[ ]	[L] [O]	1. Verify the agreement between Contractor and Owner establishes kind of Contract, i.e. Stipulated Sum, Cost Plus Fee, Construction Management or other.
[ ]	[X]	[ ]	[ ]	2. Verify that the required list of Unit Prices and Allowances as shown in Bid Form has been submitted.
[X]	[X]	[X]	[X]	3. Read the General Conditions as applied to Contract and the Supplementary Conditions or Special Conditions as applied to specific project.
[ ]	[X]	[ ]	[1] [O]	4. Verify that Bonds and Certificates are properly filed for Faithful Performance, Labor and Materials Payment, Maintenance, Insurance.
[ ]	[X]	[ ]	[ ]	5. Verify that Addenda and other modifications to the Contract are included.
[ ]	[X]	[ ]	[1] [O]	6. Verify that the date of start of construction is established by Notice to Proceed or other instrument in writing.
[ ]	[X]	[ ]	[1] [O]	7. Verify the time period for construction in writing.
[ ]	[X]	[ ]	[C]	8. Ensure the Contract Documents are identified and available onsite.
[ ]	[x]	[ ]	[C]	9. Ensure regulatory agencies are notified if necessary.
[ ]	[x]	[ ]	[C]	10. Verify permits are obtained and provided on site as required.

Legend: Upper Case Letter and **BOLD** = Primary Inspection; Lower Case = Secondary Inspection
Main Groups SI: Special Inspections, **AI:** Architectural Inspections, **EI:** Engineering Inspections, **OI:** Other Inspections
Sub-Groups   **SI:** [T] Testing Laboratory, [H] Hazardous Materials, [Y] Safety   **AI:** No Sub-Groups
EI: [V] Civil, [S] Structural, [M] Mechanical, [E] Electrical   **OI:** [O] Owner, [C] Contractor, [B] Subcontractor,
[L] Legal, [G] Government, [F] Fire Protection, [P] Plumbing, [A] Acoustical

Construction Inspection Manual

DIVISION 1 — GENERAL REQUIREMENTS - SUMMARY 01100

It is recommended that a pre-construction meeting be held to assure that, at a minimum, the information required under this checklist is available, coordinated clearly understood by all construction process participants.

SI   AI   EI   OI

[ ] [X] [ ] [C]   1. Ensure that the scope of work performed under each prime contract is properly distinguished.

[ ] [x] [ ] [C]   2. Confirm requirements for access to the construction site.

[ ] [x] [ ] [O]
            [C]   3. Ensure that procedures for accounting for Owner requested allowances are established.

[ ] [X] [ ] [O]   4. Ensure that submission and acceptance procedures for Owner requested alternates are established.

[ ] [X] [ ] [ ]   5. Ensure that procedures for making changes in the Contract Documents are established.

[ ] [X] [ ] [ ]   6. Ensure that procedures for measuring unit price quantities are established.

[ ] [X] [ ] [ ]   7. Ensure that procedures for submitting application for payment are established.

[ ] [X] [ ] [ ]   8. Ensure that requirements for breakdown of line item amounts in the Schedule of Values are established.

Legend: Upper Case Letter and **BOLD** = Primary Inspection; Lower Case = Secondary Inspection
Main Groups SI: Special Inspections, **AI:** Architectural Inspections, **EI:** Engineering Inspections, **OI:** Other Inspections
Sub-Groups   SI: [T] Testing Laboratory, [H] Hazardous Materials, [Y] Safety   **AI:** No Sub-Groups
EI: [V] Civil, [S] Structural, [M] Mechanical, [E] Electrical   **OI:** [O] Owner, [C] Contractor, [B] Subcontractor, [L] Legal, [G] Government, [F] Fire Protection, [P] Plumbing, [A] Acoustical

Construction Inspection Manual

DIVISION 1 — GENERAL REQUIREMENTS - ADMINISTRATIVE REQUIREMENTS 01300

SI AI EI OI

[ ] [x] [ ] [C]   1. Obtain a complete directory of names, addresses, phone and fax numbers; e-mail addresses, etc., of all entities involved in the Work. Obtain emergency telephone numbers for owner, contractor.

[ ] [ ] [ ] [C]   2. Ensure that site is secure. Confirm that required fencing, locks, and security guards are in place.

[ ] **[X]** [ ] [C]   3. Ensure that current construction progress schedule is available, which indicates the anticipated times for starting and completing the various elements of the work. Include submittal approval period within schedule.

[X] [X] [ ] [C]   4. Ensure that all samples which require testing are submitted in sufficient time not to impede the construction schedule.

[ ] **[X]** [ ] [C]   5. Ensure that approved submittals, samples, and color schedules are available on site.

[ ] **[X]** [ ] [C]   6. Ensure that there is a clear delineation of responsibility for coordinating the Work.

[ ] **[X]** [ ] [C]   7. Confirm requirements for Project meetings.

[X] [X] [ ] [C]   8. Confirm requirements for Project site administration. Ensure that a complete set of contract documents is at the job site and is available to the inspector.

[ ] [x] [ ] [C]   9. Ensure that all materials are delivered in time to suit the construction sequence.

[ ] [ ] [ ] [C]   10. Verify that all product accessories and parts are on site before installation is begun on that specific portion of the work.

[ ] [ ] [ ] [C]   11. Ensure that adequate equipment, tools and workers are available for proper and timely execution of the work.

Legend: Upper Case Letter and <u>BOLD</u> = Primary Inspection;   Lower Case = Secondary Inspection
Main Groups SI: Special Inspections, **AI:** Architectural Inspections, **EI:** Engineering Inspections, **OI:** Other Inspections
Sub-Groups   SI: [T] Testing Laboratory, [H] Hazardous Materials, [Y] Safety   **AI:** No Sub-Groups
EI: [V] Civil, [S] Structural, [M] Mechanical, [E] Electrical   **OI:** [O] Owner, [C] Contractor, [B] Subcontractor, [L] Legal, [G] Government, [F] Fire Protection, [P] Plumbing, [A] Acoustical

SI	AI	EI	OI	
[ ]	**[X]**	[ ]	[C]	12. Confirm requirements for construction progress documentation.
[ ]	**[X]**	[ ]	[C]	a. Photographs.
[ ]	**[X]**	[ ]	[C]	b. Progress Reports.
[ ]	**[X]**	[ ]	[C]	c. Record Documents.
[ ]	**[X]**	[ ]	[C]	13. Review distinction between substitutions and submittals. Confirm submittal procedures.
[X]	**[X]**	[ ]	[C]	14. Ensure that all applicable codes and referenced standards are readily available.

Construction Inspection Manual

DIVISION 1 — GENERAL REQUIREMENTS - QUALITY REQUIREMENTS 01400

SI AI EI OI

[ ] [X] [ ][C] 1. Confirm requirements of regulatory agencies. Verify that permits and approvals have been obtained from all jurisdictional agencies. Verify that approvals are readily available at jobsite for inspection.

[X] [X] [ ][C] 2. Confirm that regulatory agency inspections have been performed. Accompany agency inspector.

[ ] [ ] [ ][C] 3. Ensure that applicable hazardous material regulations, requirements and restrictions are being observed.

[ ] [X] [ ][C] 4. Ensure that applicable accessibility regulations and requirements are being observed.

[X] [X] [ ][C] 5. Testing agency

[X] [X] [ ][C]   a. Establish requirements of testing agency.

[X] [X] [ ][C]   b. Review qualifications for testing agency.

[X] [X] [ ][C]   c. Select testing agency.

[X] [X] [ ][C]   d. Review testing laboratory required services.

[X] [x] [ ][C] 6. Observe field tests; record locations and conditions under which tests are performed.

[X] [ ] [ ][C] 7. Verify that proper test equipment is on site.

[X] [ ] [ ][C] 8. Record who has taken a particular sample. Verify how samples are taken and storage requirements for samples.

[X] [x] [ ][C] 9. Collect, review and file copies of all tests, certificates, reports, delivery tags, etc., issued by onsite and offsite entities, agencies, and other inspectors.

[ ] [ ] [ ][C] 10. Review requirements for manufacturer field services.

Legend: Upper Case Letter and **BOLD** = Primary Inspection; Lower Case = Secondary Inspection
Main Groups SI: Special Inspections, **AI:** Architectural Inspections, **EI:** Engineering Inspections, **OI:** Other Inspections
Sub-Groups   **SI:** [T] Testing Laboratory, [H] Hazardous Materials, [Y] Safety   **AI:** No Sub-Groups
EI: [V] Civil, [S] Structural, [M] Mechanical, [E] Electrical   **OI:** [O] Owner, [C] Contractor, [B] Subcontractor, [L] Legal, [G] Government, [F] Fire Protection, [P] Plumbing, [A] Acoustical

SI	AI	EI	OI	
[X]	[X]	[X]	[C]	11. Ensure that work is inspected in a timely manner sot that deficiencies in materials or methods can be discovered at the earliest possible moment.
[X]	[X]	[X]	[C]	a. Review requirements for field tests and reports
[X]	[X]	[X]	[C]	b. Review requirements for plant inspections and reports
[X]	[X]	[X]	[C]	12. When required by either code or contract documents, notify the architect or design group leader before portions of the work requiring their inspection are covered.

Legend: Upper Case Letter and <u>BOLD</u> = Primary Inspection; Lower Case = Secondary Inspection
Main Groups SI: Special Inspections, **AI:** Architectural Inspections, **EI:** Engineering Inspections, **OI:** Other Inspections
Sub-Groups **SI:** [T] Testing Laboratory, [H] Hazardous Materials, [Y] Safety   **AI:** No Sub-Groups
EI: [V] Civil, [S] Structural, [M] Mechanical, [E] Electrical   **OI:** [O] Owner, [C] Contractor, [B] Subcontractor, [L] Legal, [G] Government, [F] Fire Protection, [P] Plumbing, [A] Acoustical

Construction Inspection Manual

DIVISION 1 — GENERAL REQUIREMENTS - TEMPORARY FACILITIES AND CONTROLS 01500

<u>SI</u>   <u>AI</u>   <u>EI</u>   <u>OI</u>

[ ] [**X**] [ ] [C]   1. Confirm source and connection requirements for utilities used in the construction.

[x] [x] [ ] [C]   2. Verify location and suitability of construction facilities, temporary construction, construction aids and temporary controls.

[x] [x] [ ] [C]   a. Ensure that Inspector and Contractor have adequate field offices.

[x] [ ] [ ] [C]   b. Ensure adequacy of sanitation facilities.

[x] [ ] [ ] [C]   c. Verify and coordinate location of temporary ramps, overpasses, barriers, enclosures, etc.

[ ] [ ] [ ] [C]   d. Verify and coordinate location of special construction elevators, hoists, and cranes.

[ ] [x] [ ] [C]   e. Verify and coordinate location of vehicular access and parking.

[ ] [x] [**V**] [C]   f. Ensure existence of a plan for erosion and sediment control.

[ ] [**X**] [ ] [C]   g. Protect identification signage is provided and located as required by the contract documents and jurisdictional authorities.

[**X**] [ ] [ ] [C]   h. Ensure that a job safety program is developed and maintained.

Legend: Upper Case Letter and <u>BOLD</u> = Primary Inspection; Lower Case = Secondary Inspection
Main Groups SI: Special Inspections, **AI:** Architectural Inspections, **EI:** Engineering Inspections, **OI:** Other Inspections
Sub-Groups   <u>**SI**</u>**:** [T] Testing Laboratory, [H] Hazardous Materials, [Y] Safety   <u>**AI**</u>**:** No Sub-Groups
  <u>**EI**</u>**:** [V] Civil, [S] Structural, [M] Mechanical, [E] Electrical   <u>**OI**</u>**:** [O] Owner, [C] Contractor, [B] Subcontractor,
[L] Legal, [G] Government, [F] Fire Protection, [P] Plumbing, [A] Acoustical

DIVISION 1 — GENERAL REQUIREMENTS - PRODUCT REQUIREMENTS 01600

SI AI EI OI

[ ] **[X]** [ ] [C]   1. Determine procedures for submission of proposed substitute products or methods of installation.

[ ] **[X]** [ ] [C]   2. Determine requirements for installation of products furnished by the Owner.

[ ] [x] [ ] [C]   3. Verify that basic procedures are established for delivery of products to the site and acceptance of products.

[X] [x] [x] [C]   a. Ensure that all materials delivered to the site are inspected for damage.

[X] [x] [x] [C]   b. Verify that materials containers are sealed and are identified by tags, markings, stamps, and are of type, size, material, gauge, weight, grade, treatment, finish, pattern, color, as required by the contract documents and approved submittals.

[X] [X] [X] [C]   c. Verify that products are new, unless otherwise permitted by the contract documents.

[x] **[X] [X]** [C]   d. Obtain certificates which are required by the contract documents at time of delivery.

[x] [x] [ ] [C]   4. Verify that basic procedures are established for storage and handling of products on site.

Legend: Upper Case Letter and **BOLD** = Primary Inspection; Lower Case = Secondary Inspection
Main Groups SI: Special Inspections, **AI:** Architectural Inspections, **EI:** Engineering Inspections, **OI:** Other Inspections
Sub-Groups   SI: [T] Testing Laboratory, [H] Hazardous Materials, [Y] Safety   **AI:** No Sub-Groups
  EI: [V] Civil, [S] Structural, [M] Mechanical, [E] Electrical   **OI:** [O] Owner, [C] Contractor, [B] Subcontractor,
  [L] Legal, [G] Government, [F] Fire Protection, [P] Plumbing, [A] Acoustical

Construction Inspection Manual

DIVISION 1 — GENERAL REQUIREMENTS - EXECUTION REQUIREMENTS 01700

SI AI EI OI

[ ] [ ] [ ] [C] 1. Verify that adjacent connection points are in place.

[B] [x] [ ] [B] [C] 2. Ensure that surfaces over which materials are to be installed are suitable for such installation.

[ ] [x] [ ] [B] [C] 3. Ensure that climatic and temperature conditions are as required for installation of the work.

[ ] [ ] [ ] [B] [C] 4. Ensure that protection of adjacent surfaces is in place before commencing, portions of the work. Ensure that protective procedures remain in place during the course of construction.

[ ] [ ] [ ] [B] [C] 5. Ensure that adequate lighting and other working conditions are provided to facilitate good workmanship and worker safety.

[ ] [X] [ ] [C] [O] 6. Determine requirements for salvaging existing improvements.

[ ] [x] [ ] [C] [O] a. Determine requirements for removal of salvaged materials.

[ ] [x] [ ] [C] [O] b. Determine requirements for storage of salvaged materials.

[ ] [x] [ ] [C] [O] c. Determine requirements for re-installation of salvaged materials.

[ ] [x] [v] [C] 7. Verify requirements for layout of the construction and requirements for field engineering.

[X] [X] [V] [C] 8. Ensure that survey information is available and accurate. Verify that there are sufficient bench-marks, monuments and stakes to identify key points.

[X] [X] [V] [O] [C] 9. Obtain a geotechnical investigation, including sub-surface ground moisture, bearing capacity of the soils, geologic features, chemical composition, and other data pertaining to the sub-surface conditions.

[ ] [x] [ ] [C] 10. Verify basic requirements for cutting and patching of the work.

[ ] [X] [ ] [C] 11. Determine requirements for protecting installed construction.

[ ] [X] [ ] [C] 12. Verify basic requirements for progress cleaning and final cleaning.

Legend: Upper Case Letter and **BOLD** = Primary Inspection; Lower Case = Secondary Inspection
Main Groups SI: Special Inspections, **AI:** Architectural Inspections, **EI:** Engineering Inspections, **OI:** Other Inspections
Sub-Groups   SI: [T] Testing Laboratory, [H] Hazardous Materials, [Y] Safety   **AI:** No Sub-Groups
EI: [V] Civil, [S] Structural, [M] Mechanical, [E] Electrical   **OI:** [O] Owner, [C] Contractor, [B] Subcontractor, [L] Legal, [G] Government, [F] Fire Protection, [P] Plumbing, [A] Acoustical

SI   AI   EI   OI

[ ] [X] [ ][C]   13. Verify that Contract close-out procedures are established and general requirements for close-out submittals.

[ ] [x] [ ][C]   a. Verify punchlisting procedures

[ ] [x] [ ][C]   b. Verify that maintenance binders, warranties and guarantees, are bound and presented in accordance with contract requirements.

Construction Inspection Manual

DIVISION 1 — GENERAL REQUIREMENTS - CLOSEOUT PROCEDURES 01770

<u>SI</u> <u>AI</u> <u>EI</u> <u>OI</u>

[X] [X] [X] [C] 1. After written notification of readiness is provided by the contractor and the construction inspector is satisfied that work is complete, the written notification of readiness is provided by the contractor and the date of substantial completion is established. Architect inspection is made with contractor's and owner's representative present, and forms are processed as required.

[x] [X] [X] [C] 2. Inspection List (punch list) of items, deficient or still required is made first by the contractor and then amended by the architect at substantial completion. This includes lists furnished to the contractor by its architect's consultants and promptly distributed to all parties.

[ ] [X] [ ] [C] 3. Again review drawings, specifications, addenda, change orders, etc., for work to be done and note.

[ ] [X] [ ] [C] [O] 4. If owner has occupied portions of work before substantial completion (beneficial occupancy), note all items not the responsibility of contractor.

[x] [X] [ ] [C] [O] 5. If required and through legal advisement, record date and information concerning processing and recording of notice of completion, other notice of lien documents or lien waiver as advised.

[ ] [x] [X] [C] 6. Verify coordination for final utility and service connections meters, etc., has been made.

[X] [ ] [M] [C] 7. Verify sterilization of plumbing systems has been performed if required.

[ ] [X] [ ] [C] [O] 8. Verify final agency inspections have been arranged and permission to occupy work is obtained as may be required.

[ ] [X] [ ] [C] [O] 9. Verify owner-furnished equipment and furnishings is coordinated and placed as required.

[X] [x] [M] [C] 10. Verify operational tests of systems and equipment have been performed as required.

Legend: Upper Case Letter and <u>BOLD</u> = Primary Inspection; Lower Case = Secondary Inspection
Main Groups SI: Special Inspections, **AI:** Architectural Inspections, **EI:** Engineering Inspections, **OI:** Other Inspections
Sub-Groups **SI:** [T] Testing Laboratory, [H] Hazardous Materials, [Y] Safety **AI:** No Sub-Groups
EI: [V] Civil, [S] Structural, [M] Mechanical, [E] Electrical **OI:** [O] Owner, [C] Contractor, [B] Subcontractor,
[L] Legal, [G] Government, [F] Fire Protection, [P] Plumbing, [A] Acoustical

SI	AI	EI	OI	
[X]	[x]	[M]	[C]	11. Verify systems adjustments, such as balancing, equipment operations, etc., have been performed. Ensure reports have been submitted.
[X]	[X]	[X]	[C] [O]	12. Verify owner's personnel are instructed in system and equipment operations as required.
[X]	[X]	[ ]	[C]	13. Verify schedule for corrections, deficiencies, and items to be supplied is established by contractor. Assist contractor and trades as to location of specific defects if necessary.
[x]	[x]	[ ]	[C]	14. Verify removal of contractor's temporary work, verify cleanup and debris removal are performed.
[X]	[X]	[X]	[C]	15. Verify record drawings (as-built) requirements are performed.
[ ]	[X]	[ ]	[C]	16. Verify final change orders are processed.
[ ]	[X]	[X]	[C] [O]	17. Verify guarantee/warrantee one year correction period requirements are met. Determine and provide list of extended warranties.
[ ]	[X]	[ ]	[C] [O]	18. Verify instruction, manuals, guides, and charts are transmitted to owner.
[ ]	[X]	[ ]	[O]	19. Verify insurance coverage is transferred as required.
[X]	[X]	[ ]	[C] [O]	20. Ensure that air quality requirements are met.
[X]	[X]	[ ]	[C] [O]	21. Verify permanent keying, keys, and keying instructions have been performed and owner-coordinated maintenance is scheduled.
[ ]	[X]	[ ]	[C] [O]	22. Verify extra materials, specified overage, and spares are delivered to owner.
[X]	[X]	[ ]	[C] [O]	23. Verify final inspection date is established, final inspection is made, and owner's acceptance is made.
[ ]	[X]	[ ]	[C] [O]	24. Verify final payment forms are being processed.
[X]	[X]	[ ]	[O]	25. Verify all records, reports, files, documents of construction inspector are in order and turned over to owner as arranged.
[ ]	[X]	[X]	[C] [o]	26. Verify post-contract maintenance conditions, such as equipment, landscaping are arranged and the owner is notified of the arrangement.

Legend: Upper Case Letter and **BOLD** = Primary Inspection; Lower Case = Secondary Inspection
Main Groups **SI:** Special Inspections, **AI:** Architectural Inspections, **EI:** Engineering Inspections, **OI:** Other Inspections
Sub-Groups **SI:** [T] Testing Laboratory, [H] Hazardous Materials, [Y] Safety   **AI:** No Sub-Groups
EI: [V] Civil, [S] Structural, [M] Mechanical, [E] Electrical   **OI:** [O] Owner, [C] Contractor, [B] Subcontractor, [L] Legal, [G] Government, [F] Fire Protection, [P] Plumbing, [A] Acoustical

DIVISION 1 — GENERAL REQUIREMENTS - FACILITY OPERATION 01800

FACILITY OPERATION

SI **AI** **EI** **OI**

[X] [X] [X] [C]   1. Verify start-up procedures for equipment.

DIVISION 2 — SITEWORK - EARTHWORK 02300

STANDARDS: AASHTO (4) APWA (18) ASTM (19) CSS (31)

SI	AI	EI	OI	
[x]	[ ]	[V]	[C]	1. Verify soil information report is on job and reviewed. Note elevation of water table.
[x]	[ ]	[V]	[B] [C]	2. Review job survey monuments and stakes. Verify limits of work are established.
[x]	[x]	[V]	[C]	3. Observe removal of existing buildings and foundations.
[x]	[ ]	[V]	[C]	4. Note condition of, or photograph, offsite and onsite improvements to remain, such as paving curbs, gutters, and walks before work begins.
[ ]	[ ]	[V]	[C]	5. Verify existing vegetation to remain is protected.
[x]	[ ]	[V]	[B] [C]	6. Verify existing utility lines to remain are located, staked, and protected. Observe conditions of uncovered lines. Verify utility companies have been notified. Verify lines to be removed or abandoned are properly capped. If unknown lines are encountered, notify appropriate parties.
[ ]	[x]	[V]	[C]	7. Verify adjacent property is protected. Verify whether adjacent property owner is notified as required by work or code.
[x]	[ ]	[V]	[B] [C]	8. Verify shoring and underpinning is provided if required.
[ ]	[ ]	[V]	[B] [C]	9. Verify extent of grubbing and removal of stumps and matted roots is performed. Depressions are properly filled and compacted.
[Y]	[ ]	[V]	[C]	10. Ensure spillage of materials or soil on streets and sidewalks is promptly removed for public safety. Verify hazardous material handling regulations are followed.
[ ]	[ ]	[V]	[B] [C]	11. Ensure spillage of gas, oil, and slurry is prevented in areas to be planted or near existing vegetation to be retained.

Legend: Upper Case Letter and **BOLD** = Primary Inspection; Lower Case = Secondary Inspection
Main Groups **SI**: Special Inspections, **AI**: Architectural Inspections, **EI**: Engineering Inspections, **OI**: Other Inspections
Sub-Groups   **SI**: [T] Testing Laboratory, [H] Hazardous Materials, [Y] Safety   **AI**: No Sub-Groups
EI: [V] Civil, [S] Structural, [M] Mechanical, [E] Electrical   **OI**: [O] Owner, [C] Contractor, [B] Subcontractor, [L] Legal, [G] Government, [F] Fire Protection, [P] Plumbing, [A] Acoustical

Construction Inspection Manual

SI	AI	EI	OI	
[Y]	[x]	[V]	[C]	12. Verify contractor provides public safety methods such as protective covers, fences, barricades, lighting, warning devices, and signs as required.
[x]	[ ]	[V]	[C]	13. Verify dust control is provided as required.
[x]	[ ]	[V]	[B] [C]	14. Verify deleterious material is removed from site and/or otherwise properly disposed.
[ ]	[ ]	[V]	[C]	15. Verify stripping of site, preservation and depth of removal of topsoil, and location of stockpile are performed and established.
[x]	[ ]	[V]	[B]	16. Observe that topsoil is not contaminated with subsoil and is free from roots, stones, and other deleterious materials.
[ ]	[ ]	[V]	[B] [C]	17. Check that satisfactory materials are used and unsuitable materials are disposed of properly and legally in waste areas. Do not allow contamination.
[ ]	[ ]	[V]	[C]	18. Observe removal of material and note unusual conditions. Observe subsoil conditions for irregularities such as soft spots, springs, and previous debris.
[ ]	[ ]	[V]	[C]	19. Verify excavation is performed in scheduled sequence if required.
[ ]	[ ]	[V]	[C]	20. Verify excavating does not cause unusual rutting and appears adequate for work to be performed.
[x]	[ ]	[V]	[B] [C]	21. Observe that over-excavation does not occur.
[ ]	[ ]	[V]	[C]	22. Verify drainage is provided continuously as excavation progresses, other dewatering methods such as well points are provided, drainage ditches are maintained, and ponding does not occur.
[T]	[x]	[V]	[B] [C]	23. Verify testing, inspection, and compaction are performed during excavation and filling.
[ ]	[ ]	[V]	[C]	24. Verify that borrow excavation procedures and materials are adequate.
[ ]	[ ]	[V]	[B] [C]	25. Verify source and type of imported materials are as approved. Verify samples are tested and approved.

Legend: Upper Case Letter and **BOLD** = Primary Inspection; Lower Case = Secondary Inspection
Main Groups **SI:** Special Inspections, **AI:** Architectural Inspections, **EI:** Engineering Inspections, **OI:** Other Inspections
Sub-Groups **SI:** [T] Testing Laboratory, [H] Hazardous Materials, [Y] Safety **AI:** No Sub-Groups
EI: [V] Civil, [S] Structural, [M] Mechanical, [E] Electrical **OI:** [O] Owner, [C] Contractor, [B] Subcontractor, [L] Legal, [G] Government, [F] Fire Protection, [P] Plumbing, [A] Acoustical

Construction Inspection Manual

SI	**AI**	**EI**	**OI**	
[x]	[ ]	[V]	[B] [C]	26. Verify compaction is performed in lifts as required.
[x]	[x]	[V]	[B] [C]	27. Verify building layout is properly established, setbacks are observed, and batterboards and elevations are established. See that compacted material extends beyond foundation line as required.
[Y]	[ ]	[V]	[B] [C]	28. Inspect foundation excavating for adequacy, bracing, form clearances and type of soil.
[x]	[ ]	[V]	[B] [C]	29. See that corrective measures are performed where over-excavation occurs.
[x]	[ ]	[V]	[C]	30. Observe methods of dewatering foundation excavations and see that footing beds are not disturbed or softened. Verify methods for surface drainage are provided.
[ ]	[ ]	[V]	[B] [C]	31. Verify footing drains are installed in manner specified.
[X]	[ ]	[V]	[B] [C]	32. Verify backfill materials are from approved source. Verify they are installed in specified layers and adequate compaction equipment is used. Verify relative density of backfill is checked. Verify walls are properly cured before backfilling.
[ ]	[ ]	[V]	[C]	33. Verify waterproof membranes are protected against damage during backfilling operations.
[ ]	[ ]	[V]	[C]	34. Observe fine grading, deposits of top soil in scheduled areas, proper drainage conditions, and job cleanup.
[ ]	[ ]	[V]	[B] [C]	35. Verify soil poisoning, performed if required, uses approved method and materials.
[X]	[ ]	[V]	[B] [C]	36. Ensure preservation of monuments and markers is observed. Verify construction survey grading stakes are in place prior to grading and are protected or replaced to ensure proper site grading elevations are achieved.
[X]	[x]	[V]	[C]	37. Verify record survey of site is performed if required.
[H] **[Y]** **[T]**	[ ]	[V]	[B] [C]	38. Have soils suspected to be contaminated with hazardous materials tested. Verify that suspect soils have been revised on site until soils tests confirm compliance with legal standards.

Legend: Upper Case Letter and <u>BOLD</u> = Primary Inspection;   Lower Case = Secondary Inspection
  Main Groups **SI**: Special Inspections, **AI**: Architectural Inspections, **EI**: Engineering Inspections, **OI**: Other Inspections
  Sub-Groups   **SI**: [T] Testing Laboratory, [H] Hazardous Materials, [Y] Safety   **AI**: No Sub-Groups
    EI: [V] Civil, [S] Structural, [M] Mechanical, [E] Electrical   **OI**: [O] Owner, [C] Contractor, [B] Subcontractor,
      [L] Legal, [G] Government, [F] Fire Protection, [P] Plumbing, [A] Acoustical

Construction Inspection Manual

SI AI EI OI

[ ] [ ] [V] [B] 39. Have soils engineering representative on site as needed, to certify
        [C]   compaction and method of soil work.

[X] [ ] [V] [B] 40. If archeological remains are encountered, stop that portion of work until
        [C]   qualified people can analyze remains.

[ ] [ ] [V] [B] 41. Review erosion control plan and ensure that appropriate measures are
        [C]   undertaken during the wet season. Make sure erosion control devices are
                replaced at the end of each day's work.

[x] [ ] [V] [C] 42. Check to see if contractor has permit from Underground Service Alert or other appropriate agency or authority governing excavation near existing underground utilities. (Underground Service Alert, "USA," is a central agency which alerts all utilities who have underground services at least 48 hours before any excavation work is commenced at a particular site, so that such utilities. may mark their services and post appropriate surface warning signs. In California, notification of and obtaining a permit number from "USA" is mandatory by law (per Section 4216 and 4217 of Governing Code) at least 48 hours before work commences. The "USA" toll free number for California is 1-800-422-4133. For other areas, call information.)

[ ] [ ] [V] [C] 43. Check to see if a National Pollution Discharge Efficiency System (NPDES) Notice of Intention plan has been filed, if required and is being complied with.

[ ] [ ] [V] [C] 44. Verify import or export haul routes have been approved by local agencies, if required.

Legend: Upper Case Letter and **BOLD** = Primary Inspection; Lower Case = Secondary Inspection
  Main Groups SI: Special Inspections, **AI:** Architectural Inspections, **EI:** Engineering Inspections, **OI:** Other Inspections
  Sub-Groups **SI:** [T] Testing Laboratory, [H] Hazardous Materials, [Y] Safety **AI:** No Sub-Groups
    EI: [V] Civil, [S] Structural, [M] Mechanical, [E] Electrical **OI:** [O] Owner, [C] Contractor, [B] Subcontractor,
      [L] Legal, [G] Government, [F] Fire Protection, [P] Plumbing, [A] Acoustical

DIVISION 2 — FOUNDATION AND LOAD - BEARING ELEMENTS 02450

STANDARDS: ACI (5) AISC (10) ANSI (16) ASTM (19) AWPA (25) AWS (23) SSPC (89)

SI	AI	EI	OI	
[x]	[ ]	[V]	[B] [C]	1. Check whether test piles are require
[x]	[ ]	[V]	[B] [C]	2. Verify known underground utilities are protected.
[ ]	[ ]	[V]	[B] [C]	3. Note conditions of nearby structures. Record and report to architect if cracks and other deficiencies are evident.
[x]	[ ]	[V] [s]	[C]	4. Verify that the parameters for "refusal" of this work is understood by the contractor.
[ ]	[ ]	[V]	[C]	5. If predrilling is proposed, verify that approval has been granted.
[ ]	[ ]	[V]	[B] [C]	6. Check whether jetting is allowed.
[x]	[ ]	[V]	[B] [C]	7. Precast concrete piling — verify length, dimensions, delivered condition, mix utilized test run, type and time of curing. Note inspector's marks if plant inspected.
[x]	[ ]	[V]	[B] [C]	8. Wood piling — verify species, treatment, length, circumference, and diameter at tip and 3 feet from butt; straightness, presence and severity and shalies, banding and tip protection.
[x]	[ ]	[V]	[B] [C]	9. Steel piling — verify dimensions of "H" or pipe piling; test reports as required. Verify thickness of pile and condition of interlock for steel sheet piling.
[x]	[ ]	[V]	[B] [C]	10. Steel shell piling — verify gauge, diameter and length of sections to be driven and filled with concrete. Check fit of driving mandrel.
[ ]	[ ]	[V]	[C]	11. Mark piles as necessary to observe compliance with requirements.
[ ]	[ ]	[V]	[C]	12. Verify waterproofing at joints is provided if required on shells.
[ ]	[ ]	[V]	[C]	13. Verify concrete delivery method is approved — tremie, pump, and limited free fall.
[ ]	[ ]	[V]	[B] [C]	14. Verify proper driving block or shoe is used to avoid end damage.

Legend: Upper Case Letter and **BOLD** = Primary Inspection; Lower Case = Secondary Inspection
Main Groups SI: Special Inspections, AI: Architectural Inspections, EI: Engineering Inspections, OI: Other Inspections
Sub-Groups **SI**: [T] Testing Laboratory, [H] Hazardous Materials, [Y] Safety **AI**: No Sub-Groups
EI: [V] Civil, [S] Structural, [M] Mechanical, [E] Electrical **OI**: [O] Owner, [C] Contractor, [B] Subcontractor, [L] Legal, [G] Government, [F] Fire Protection, [P] Plumbing, [A] Acoustical

Construction Inspection Manual

SI AI EI OI

[x] [ ] [V] [B]   15. Record:
        [C]

                Make and method of hammer energy source

                Weight of striking parts*

                Height of fall of striking parts*

                Weight of pile Blows per unit of penetration

                Abnormal difference in sound of driving

                *For double-acting pile driver, record energy per blow in ft. lb. from manufacturer's data sheet. Check required steam or water pressure for rated energy during driving. Check height of fall for diesel hammers.

[ ] [ ] [V] [B]   16. Verify plumbness before driving.

[x] [ ] [V] [B]   17. Verify as-driven location is within tolerances required.
        [C]

[ ] [ ] [V] [B]   18. Report excessive vibration during driving.

[x] [ ] [V] [B]   19. Visually check steel shells in place for plumbness, straightness, and damage from driving — look for water in pipe and; metal shells to be filled with concrete.
        [C]

[x] [ ] [V] [C]   20. Observe condition of tops of precast and wood piles after Ail driving.

[x] [ ] [V] [B]   21. Observe heaving; check as required.

[x] [ ] [V] [B]   22. Rebar cages — verify size, spacing, steel grade, clearances, and dowel and tendon extensions.
        [C]

[x] [ ] [V] [B]   23. Verify agency inspections were made if required.
        [C]

[ ] [ ] [V] [C]   24. Check driving equipment to ensure it meets or is capable of meeting specifications for driving.

Legend: Upper Case Letter and <u>BOLD</u> = Primary Inspection; Lower Case = Secondary Inspection
Main Groups <u>SI</u>: Special Inspections, **AI**: Architectural Inspections, **EI**: Engineering Inspections, <u>OI</u>: Other Inspections
Sub-Groups   <u>SI</u>: [T] Testing Laboratory, [H] Hazardous Materials, [Y] Safety   <u>AI</u>: No Sub-Groups
  <u>EI</u>: [V] Civil, [S] Structural, [M] Mechanical, [E] Electrical   <u>OI</u>: [O] Owner, [C] Contractor, [B] Subcontractor,
  [L] Legal, [G] Government, [F] Fire Protection, [P] Plumbing, [A] Acoustical

Construction Inspection Manual

DIVISION 2 — SITEWORK — CAISSONS 02475

STANDARDS: ACI (5) AISC ASTM (19)

SI	AI	EI	OI	
[x]	[ ]	[V]	[B] [C]	1. Verify drilling apparatus for diameter and bucket and reamer, if belled bottoms are required.
[ ]	[ ]	[V]	[B] [C]	2. Verify material removed correlates with soil report.
[Y]	[ ]	[V]	[B] [C]	3. Observe stability of material; caving.
[ ]	[ ]	[V]	[B] [C]	4. Verify sufficient quantity of casing available at site.
[ ]	[ ]	[V]	[B] [C]	5. If water is encountered; record depths.
[x]	[ ]	[V]	[B] [C]	6. Rebar cages: verify size, length, and grade of verticals, and size and spacing of ties.
[ ]	[ ]	[V]	[C]	7. Verify concrete delivery method is approved — tremie, pump, and limited free fall.
[x]	[ ]	[V]	[B] [C]	8. Verify holes are dry. If wet, check that an approved method of handling water is used — pump, bail, dry mix, and tremie.
[x]	[ ]	[V]	[B] [C]	9. Check bottom of hole for cleanliness; use light or downhole inspection if necessary.
[x]	[ ]	[V]	[B] [C]	10. Record location, depth, and diameter.
[x]	[ ]	[V]	[B] [C]	11. Observe placement of steel. Check clearances and dowel projections.
[x]	[ ]	[V]	[B] [C]	12. Observe placement of concrete. Check coordination of casing withdrawal, where required, and placement of concrete. Casing should remain below surface of concrete at all times.
[ ]	[ ]	[Y]	[B] [C]	13. Check for straightness and plumbness.
[x]	[ ]	[V]	[B] [C]	14. Verify agency inspections are made if required.

Legend: Upper Case Letter and **BOLD** = Primary Inspection; Lower Case = Secondary Inspection
Main Groups SI: Special Inspections, **AI:** Architectural Inspections, **EI:** Engineering Inspections, **OI:** Other Inspections
Sub-Groups   **SI:** [T] Testing Laboratory, [H] Hazardous Materials, [Y] Safety   **AI:** No Sub-Groups
   EI: [V] Civil, [S] Structural, [M] Mechanical, [E] Electrical   **OI:** [O] Owner, [C] Contractor, [B] Subcontractor,
   [L] Legal, [G] Government, [F] Fire Protection, [P] Plumbing, [A] Acoustical

Construction Inspection Manual

DIVISION 2 — SITEWORK — SANITARY SEWERAGE 02530

STANDARDS: ASSHTO (4) APWA (18) ASTM (19)

<u>**SI**</u> <u>**AI**</u> <u>**EI**</u> <u>**OI**</u>

[ ] [ ] [V] [B] [C]   1. Familiarize self on where all surface site drainage is directed. In general, all site drainage will be directed to underground storm drain systems which will carry it offsite. If there are areas of the site drainage directly onto adjacent sites and/or if catch basins located in low points would back storm water into a structure if the catch basin becomes blocked, bring these facts to the architect's attention.

[ ] [ ] [V] [B] [C]   2. Familiarize self on where all underground site drainage is directed, making sure that there is no reverse flow in pipes. Review plans for possible conflicts between underground storm drain pipes and other utility lines, especially other gravity flow pipes (sanitary sewer lines). If possible conflicts, expose conflict before storm drain excavation is begun and have surveyor verify that no vertical conflicts exist.

[ ] [ ] [V] [B] [C]   3. Review standards for catch basins and manholes (or junction boxes). Require submittal of catalog cut be submitted to architect if structures are precast.

[x] [ ] [V] [B] [C]   4. Verify elevations of tops and bottoms of drainage structures and flow lines of storm drain pipes against layout survey stakes.

[X] [ ] [V] [B] [C]   5. Review backfill standards for drainage structures and storm drain lines. Have soils engineer test and certify backfill methods and compaction.

[Y] [ ] [V] [B] [C]   6. Familiarize self with types of storm drain pipes allowed. If ACP is allowed, special handling, cutting and disposal requirements apply, as asbestos is considered a hazardous material. Obtain manufacturer's recommendations on jointing methods. Review bedding against local codes and manufacturer's recommendations.

[Y] [ ] [V] [B] [C]   7. Review safety shoring requirements for excavations over 5 feet in depth.

[ ] [ ] [V] [B] [C]   8. Familiarize self with concrete mix and strength requirements. Collect delivery tags. Obtain certificate of conformance to specifications. Take concrete slump tests and cylinders, if required.

Legend: Upper Case Letter and <u>**BOLD**</u> = Primary Inspection; Lower Case = Secondary Inspection
  Main Groups SI: Special Inspections, **AI:** Architectural Inspections, **EI:** Engineering Inspections, **OI:** Other Inspections
  Sub-Groups   **SI:** [T] Testing Laboratory, [H] Hazardous Materials, [Y] Safety   **AI:** No Sub-Groups
     EI: [V] Civil, [S] Structural, [M] Mechanical, [E] Electrical   **OI:** [O] Owner, [C] Contractor, [B] Subcontractor,
       [L] Legal, [G] Government, [F] Fire Protection, [P] Plumbing, [A] Acoustical

SI	AI	EI	OI	
[X]	[ ]	[V]	[B] [C]	9. Test site for finish drainage slopes by water testing all site areas and looking for "bird baths."
[x]	[ ]	[V]	[C]	10. Have contractor clean all catch basins and flush all storm drain lines prior to job acceptance.
[ ]	[ ]	[V]	[B] [C]	11. Review inside of all drainage structures for rock pockets and defects in workmanship. All forms to be removed.
[ ]	[ ]	[V]	[C]	12. Ensure that an encroachment permit is obtained by the contractor, if required, for tying-in site drainage storm drain lines into the jurisdiction's main lines.
[ ]	[ ]	[V]	[B] [C]	13. Inspect all storm drain lines for joint integrity, that lines are true for slope and horizontal alignment, and that bedding is proper prior to allowing contractor to backfill storm drain lines and structures.

Legend: Upper Case Letter and **BOLD** = Primary Inspection; Lower Case = Secondary Inspection
Main Groups **SI:** Special Inspections, **AI:** Architectural Inspections, **EI:** Engineering Inspections, **OI:** Other Inspections
Sub-Groups **SI:** [T] Testing Laboratory, [H] Hazardous Materials, [Y] Safety   **AI:** No Sub-Groups
EI: [V] Civil, [S] Structural, [M] Mechanical, [E] Electrical   **OI:** [O] Owner, [C] Contractor, [B] Subcontractor, [L] Legal, [G] Government, [F] Fire Protection, [P] Plumbing, [A] Acoustical

Construction Inspection Manual

DIVISION 2 — SITEWORK — BASES, BALLASTS, PAVEMENTS AND APPORTENANCES 02700

STANDARDS: ASSHTO (4) ACI (5) AI (28) APWA (18) ASTM (19) CSS (31) NRMCA (68) PCA (75)

Subgrades and Bases

SI AI EI OI

[x] [ ] [V] [B]   1. Verify subgrade is to proper elevation and cross-section.
            [C]

[x] [ ] [V] [B]   2. Verify subgrade is dense and properly compacted.
            [C]

[ ] [ ] [V] [C]   3. Verify drains, utilities, and other underground construction are in place.

[ ] [ ] [V] [B]   4. Verify trench backfilling is performed as required.
            [C]

[x] [ ] [V] [B]   5. Verify control testing of subgrade and subgrade materials is being
            [C]   performed and recorded if required.

[x] [ ] [V] [B]   6. Verify subbase and base courses are of source, type, thickness and
            [C]   material specified.

[X] [ ] [V] [B]   7. Ensure source of material is sampled and approved by testing laboratory.
            [C]

[ ] [ ] [V] [B]   8. Verify materials delivered are of uniform quality.
            [C]

[ ] [ ] [V] [B]   9. Verify equipment is suitable.
            [C]

[ ] [ ] [V] [B]   10. Verify hauling equipment does not produce ruts in subgrade.
            [C]

[ ] [ ] [V] [B]   11. Verify location of all manholes, outlets, and surface features is known.
            [C]

Priming and Tack Coats

[ ] [ ] [V] [B]   12. Verify soil sterilization is provided if required.
            [C]

[h] [ ] [V] [B]   13. Verify prime coat is provided and applied as specified. (Verify whether
            [C]   prime coat is required by local governing authorities, e.g. Air Quality
                       Management Board.)

[ ] [ ] [V] [B]   14. Verify prime coat properly seals the surface voids and provides proper
            [C]   binding.

Legend: Upper Case Letter and **BOLD** = Primary Inspection; Lower Case = Secondary Inspection
Main Groups SI: Special Inspections, AI: Architectural Inspections, EI: Engineering Inspections, OI: Other Inspections
Sub-Groups   **SI:** [T] Testing Laboratory, [H] Hazardous Materials, [Y] Safety   **AI:** No Sub-Groups
    EI: [V] Civil, [S] Structural, [M] Mechanical, [E] Electrical   **OI:** [O] Owner, [C] Contractor, [B] Subcontractor,
    [L] Legal, [G] Government, [F] Fire Protection, [P] Plumbing, [A] Acoustical

SI	AI	EI	OI	
[ ]	[ ]	[V]	[B] [C]	15. Verify prime coat is applied to waterfree surface free of objectionable substances.
[ ]	[ ]	[V]	[B] [C]	16. Verify prime coat is applied uniformly and receives proper curing time.
[ ]	[ ]	[V]	[B] [C]	17. Verify pack coat is applied to all conforming concrete and asphalt surfaces.

Asphalt Paving

[ ]	[ ]	[V]	[B] [C]	18. Verify concrete against which paving is to be placed is to be at least 7 days old.
[ ]	[ ]	[V]	[B]	19. Verify truck beds are tight and smooth.
[ ]	[ ]	[V]	[B] [C]	20. Verify suitable covers are provided to protect mix, if required or because of climatic conditions.
[ ]	[ ]	[V]	[B] [C]	21. Verify spreading equipment is suitable.
[ ]	[ ]	[V]	[B] [C]	22. Verify size and type of roller and paving equipment is as specified.
[ ]	[ ]	[V]	[B] [C]	23. Verify asphalt is of proper mix and approved for this work.
[x]	[ ]	[V]	[B] [C]	24. Verify plant inspection has been made if required.
[x]	[ ]	[V]	[B] [C]	25. Verify temperature of mix when delivered is within limits required.
[ ]	[ ]	[V]	[B] [C]	26. Verify weather limitations are observed.
[x]	[ ]	[V]	[B] [C]	27. Verify records of placement and suspension of operations are kept.
[x]	[ ]	[V]	[B] [C]	28. Verify headers and screeds are properly installed for thickness control and as required.
[ ]	[ ]	[V]	[B] [C]	29. Verify materials are properly spread, raked, and placed for thickness and uniformity. Raking to be kept at a minimum.
[ ]	[ ]	[V]	[B] [C]	30. Verify rollers are operated within speed range and have proper backup features, drum scrapes, and wetting devices. Ensure roller is backed only over compacted areas.

Legend: Upper Case Letter and **BOLD** = Primary Inspection; Lower Case = Secondary Inspection
Main Groups **SI**: Special Inspections, **AI**: Architectural Inspections, **EI**: Engineering Inspections, **OI**: Other Inspections
Sub-Groups **SI**: [T] Testing Laboratory, [H] Hazardous Materials, [Y] Safety **AI**: No Sub-Groups
EI: [V] Civil, [S] Structural, [M] Mechanical, [E] Electrical **OI**: [O] Owner, [C] Contractor, [B] Subcontractor, [L] Legal, [G] Government, [F] Fire Protection, [P] Plumbing, [A] Acoustical

Construction Inspection Manual

SI AI EI OI

[ ] [ ] [V] [B]   31. Verify correction of humps and depressions to obtain smoothness and
         [C]   uniformity is to be performed before the start of compaction.

[ ] [ ] [V] [B]   32. Verify temperature after final rolling is. within limits.
         [C]

[ ] [ ] [V] [B]   33. Verify tie-ins to adjacent surfaces are as required.
         [C]

[ ] [ ] [V] [B]   34. Verify damage of adjacent. surfaces is corrected.
         [C]

[x] [ ] [V] [B]   35. Verify drainage tests are made.
         [C]

[ ] [ ] [V] [B]   36. If required, verify seal coat is properly applied and of material called for,
         [C]   and with proper curing period before and after installation.

[ ] [ ] [V] [B]   37. Verify special topping surfaces, color coatings, and striping, are applied
         [C]   of materials required; proper curing times are observed.

[ ] [ ] [V] [B]   38. Verify a copy of all truck delivery tags is collected as trucks enter site
         [C]   and kept for reference. Verify the total weight of asphalt delivery is
                 equal to or over the calculated weight required for the project. (If
                 underweight, possible thin areas in the paving may exist and core
                 thickness tests may be warranted).

Concrete Paving

Refer to Sections 03100, 03200 and 03300 where applicable.

[x] [ ] [V] [B]   39. Verify requirements for mixes, batching plant, and mixing plant are met
         [C]   and approved if specified. Verify admixtures are as approved.

[ ] [ ] [V] [B]   40. Verify all paving equipment is available on job, is in good condition,
         [C]   and meets requirements.

[ ] [ ] [V] [B]   41. Verify base course is maintained in a firm, moist condition and is as
         [C]   required.

[ ] [ ] [V] [B]   42. Verify all forms, headers, outlets, boxes; and equipment are in place
         [C]   before pouring.

[ ] [ ] [V] [B]   43. Verify all embedded items., sleeves, dowels, and reinforcement are as
         [C]   required. Verify that reinforcement is not shifted or forced to the
                 bottom of the pour.

Legend: Upper Case Letter and **BOLD** = Primary Inspection; Lower Case = Secondary Inspection
Main Groups SI: Special Inspections, **AI:** Architectural Inspections, **EI:** Engineering Inspections, **OI:** Other Inspections
Sub-Groups   **SI:** [T] Testing Laboratory, [H] Hazardous Materials, [Y] Safety   **AI:** No Sub-Groups
     EI: [V] Civil, [S] Structural, [M] Mechanical, [E] Electrical   **OI:** [O] Owner, [C] Contractor, [B] Subcontractor,
     [L] Legal, [G] Government, [F] Fire Protection, [P] Plumbing, [A] Acoustical

Construction Inspection Manual

SI	AI	EI	OI	
[ ]	[ ]	[V]	[B] [C]	44. Verify joint methods and materials are provided and observed.
[x]	[ ]	[V]	[B] [C]	45. Verify grade, slope, pitch, and thickness control is provided as required.
[ ]	[ ]	[V]	[B] [C]	46. Verify concrete is deposited, rodded, and vibrated to suit conditions. See that reinforcement is maintained at elevation required.
[ ]	[ ]	[V]	[B] [C]	47. Verify time interval between pours allows for continuous working.
[ ]	[ ]	[V]	[B] [C]	48. Verify controls joints, construction joints, and expansion joints are provided as required.
[ ]	[ ]	[V]	[B] [C]	49. Verify color, if required, is of proper type, tone, and amount.
[ ]	[ ]	[V]	[B] [C]	50. Verify finishing treatment and texture is as required.
[ ]	[ ]	[V]	[B] [C]	51. Verify curing provisions are as required and work is properly protected.
[ ]	[ ]	[V]	[B] [C]	52. Ensure over-troweling is avoided.
[ ]	[ ]	[V]	[B] [C]	53. Verify jointing of old concrete to new work is properly performed.
[ ]	[ ]	[V]	[B] [C]	54. Ensure forms are not removed until minimum required time after placement has elapsed.
[ ]	[ ]	[V]	[B] [C]	55. Verify sawed joints are made at proper time and are properly aligned.
[ ]	[ ]	[V]	[B] [C]	56. Verify sawed joints are of proper width and depth.
[ ]	[ ]	[V]	[B] [C]	57. Verify joints are cleaned and cured as required.
[ ]	[ ]	[V]	[B] [C]	58. Verify joints are sealed properly as required.
[ ]	[ ]	[V]	[B] [C]	59. See that concrete is protected from damage during backfilling.
[x]	[ ]	[V]	[B] [C]	60. Verify tests for drainage and surface variation are made.
[ ]	[ ]	[V]	[B] [C]	61. Verify agency requirements are met for design regarding sidewalks, curbs, gutters, and aprons.
[ ]	[ ]	[V]	[C]	62. Ensure that contractor protects against vandalism while concrete is still green and able to be marked. Inspect first thing next morning for marks which can then be stoned out.

Legend: Upper Case Letter and **BOLD** = Primary Inspection; Lower Case = Secondary Inspection
Main Groups **SI**: Special Inspections, **AI**: Architectural Inspections, **EI**: Engineering Inspections, **OI**: Other Inspections
Sub-Groups **SI**: [T] Testing Laboratory, [H] Hazardous Materials, [Y] Safety **AI**: No Sub-Groups
EI: [V] Civil, [S] Structural, [M] Mechanical, [E] Electrical **OI**: [O] Owner, [C] Contractor, [B] Subcontractor, [L] Legal, [G] Government, [F] Fire Protection, [P] Plumbing, [A] Acoustical

SI	AI	EI	OI	
[ ]	[X]	[V]	[B] [C]	63. Verify location and layout of parking for disabled persons.
[ ]	[X]	[V]	[B] [C]	64. Verify wheelchair curb-cuts, slopes of sidewalks, curb-cuts, ramps, marking of stairs for visually impaired.

DIVISION 2 — SITEWORK — IRRIGATION SYSTEM 02810

STANDARDS: AAN (3)

Irrigation

<u>**SI**</u> <u>**AI**</u> <u>**EI**</u> <u>**OI**</u>

[ ] [X] [ ] [B] 1. Verify layout of main service heads, valves, controllers, vacuum breakers, is as required. Record field conditions as required.
        [C]

[ ] [x] [V] [B] 2. Verify trenching depth and backfill are as required. Refer to Section 15400, "Plumbing," for applicable items concerning installation of piping, connectors and equipment.
        [C]

[ ] [X] [ ] [B] 3. Verify pressure tests are performed.
        [C]

[ ] [X] [ ] [B] 4. Verify electrical work is provided as required. Observe work related to lighting of landscape features.
        [C]

[ ] [X] [V] [B] 5. Verify that the controller is installed as required. Observe installation.
        [C]

[ ] [X] [V] [B] 6. Observe operation, adjustment and coverage of completed irrigation system, including controller timing.
        [C]

Legend: Upper Case Letter and <u>BOLD</u> = Primary Inspection; Lower Case = Secondary Inspection
Main Groups SI: Special Inspections, **AI:** Architectural Inspections, **EI:** Engineering Inspections, **OI:** Other Inspections
Sub-Groups   **SI:** [T] Testing Laboratory, [H] Hazardous Materials, [Y] Safety   **AI:** No Sub-Groups
    EI: [V] Civil, [S] Structural, [M] Mechanical, [E] Electrical   **OI:** [O] Owner, [C] Contractor, [B] Subcontractor,
    [L] Legal, [G] Government, [F] Fire Protection, [P] Plumbing, [A] Acoustical

Construction Inspection Manual

DIVISION 2 — SITEWORK — PLANTING 02900

STANDARDS: AAN (3)

Sitework and Existing Vegetation

SI AI EI OI

[ ] [X] [ ] [B] 1. Verify existing trees and other vegetation to remain are protected as required. Verify barricades and fencing are provided if required.
          [C]

[ ] [X] [ ] [C] 2. Verify traffic, parking, storage of materials, or debris is not allowed within drip line of trees.

[ ] [X] [ ] [B] 3. Verify existing trees and other vegetation are maintained by regular feeding and watering during construction, if required.
          [C]

[ ] [X] [ ] [B] 4. Verify pruning of branches is performed only by qualified persons.
          [C]

[ ] [X] [ ] [B] 5. Observe excavation adjacent to existing trees. Unless otherwise required, do not allow exposing of root systems; keep equipment beyond drip line.
          [C]

[ ] [X] [ ] [B] 6. Verify ponding around base of existing trees does not occur.
          [C]

[ ] [X] [ ] [B] 7. Verify existing vegetation not to be saved is removed as required.
          [C]

[x] [X] [ ] [B] 8. Verify depth of cuts and other conditions required are met for retainage and/or reuse of on-site topsoil. Observe required stockpiling and that topsoil is not intermixed with deleterious materials.
          [C]

[ ] [x] [V] [B] 9. Verify subgrade is carried sufficiently below finish grade to provide depth of topsoil required. Do not allow in subgrade debris that is detrimental to landscaping.
          [C]

[ ] [X] [ ] [B] 10. Verify elevations of manholes, catch basins, valves, and boxes are coordinated to finished grades required on landscaping plans. Verify coordination to other drawings.
           [C]

[ ] [X] [ ] [B] 11. Verify backfill against foundation walls that receive planting is clean sand free of rocks, concrete, and debris.
           [C]

Legend: Upper Case Letter and **BOLD** = Primary Inspection; Lower Case = Secondary Inspection
Main Groups **SI:** Special Inspections, **AI:** Architectural Inspections, **EI:** Engineering Inspections, **OI:** Other Inspections
Sub-Groups **SI:** [T] Testing Laboratory, [H] Hazardous Materials, [Y] Safety   **AI:** No Sub-Groups
EI: [V] Civil, [S] Structural, [M] Mechanical, [E] Electrical   **OI:** [O] Owner, [C] Contractor, [B] Subcontractor, [L] Legal, [G] Government, [F] Fire Protection, [P] Plumbing, [A] Acoustical

SI AI EI OI

[ ] [X] [ ] [C]   12. Verify paints, solvents, oils, old plaster, and other debris is not placed in areas to receive planting. Do not allow maintenance of construction equipment to be performed in areas to receive planting.

[ ] [X] [ ] [C]   13. In general, ensure that all deep utility work is to be done before soil preparation, all shallow utilities such as irrigation laterals are done after soil preparation.

Finish Grading

[ ] [X] [ ] [B] [C]   14. Verify subgrade is scarified as required and prepared to receive finish grading.

[ ] [X] [ ] [B] [C]   15. Verify stockpiled topsoil is distributed, prepared, and of depths required.

[T] [x] [ ] [B] [C]   16. Verify import, borrow, or selected off-site soil is used as required. Get certificates, and weight tags if required. Verify source is as required.

[T] [ ] [ ] [B] [C]   17. Verify topsoil mixture including amendments and preparation is as required.

[ ] [x] [V] [B] [C]   18. Verify installation of sleeves, raceways, boxes, and piping required for irrigation, drainage, and electrical is provided as required, and is installed in coordination with site improvements, paving, and walks.

[ ] [X] [ ] [B] [C]   19. Verify areas to receive planting are not excessively compacted by traffic storage or equipment.

[ ] [X] [ ] [C]   20. Verify scheduling of phases of landscaping work is in accordance with overall construction. Verify work is scheduled to avoid out-of-phase sequences that could cause damage or require rework. In general, ensure that all deep utility work is to be done before soil preparation — all shallow' utilities such as irrigation laterals are done after soil preparation.

Landscape Construction

[ ] [X] [ ] [B] [C]   21. Verify layout of landscape construction such as walls, fences, paving, paths, and benches is as required.

[ ] [x] [V] [B] [C]   22. Verify grades and elevations. Record field conditions as required.

Legend: Upper Case Letter and **BOLD** = Primary Inspection; Lower Case = Secondary Inspection
  Main Groups SI: Special Inspections, **AI:** Architectural Inspections, **EI:** Engineering Inspections, **OI:** Other Inspections
  Sub-Groups   **SI:** [T] Testing Laboratory, [H] Hazardous Materials, [Y] Safety   **AI:** No Sub-Groups
    EI: [V] Civil, [S] Structural, [M] Mechanical, [E] Electrical   **OI:** [O] Owner, [C] Contractor, [B] Subcontractor,
    [L] Legal, [G] Government, [F] Fire Protection, [P] Plumbing, [A] Acoustical

Construction Inspection Manual

Planting

<u>SI</u> <u>AI</u> <u>EI</u> <u>OI</u>

[ ] [X] [ ] [B]   23. Verify planting areas are in horticultural condition required.
            [C]

[ ] [X] [ ] [B]   24. Verify that required drainage conditions are provided.
            [C]

[ ] [X] [ ] [B]   25. Verify layout of major plant materials and adjustment to field conditions
            [C]   is provided as required.

[ ] [X] [ ] [B]   26. Verify plant materials are as approved and inspected before installation
            [C]   as required.

[ ] [X] [ ] [B]   27. Verify staking, pruning, and spraying are as required.
            [C]

[ ] [X] [ ] [B]   28. Verify plants are watered in after planting to settle soil around root ball.
            [C]

[ ] [X] [ ] [B]   29. Verify plants are protected from acid backsplash used to clean or etch
            [C]   concrete. Do not allow diluted acid near root zones of trees or shrubs.

[ ] [X] [ ] [C]   30. To avoid scorching of foliage, ensure hot tar boilers are not allowed near
                       trees or vegetation.

[ ] [X] [ ] [C]   31. Verify maintenance is provided as required during construction, and
                       arrangements with permanent maintenance are coordinated as required.

Legend: Upper Case Letter and <u>BOLD</u> = Primary Inspection; Lower Case = Secondary Inspection
  Main Groups SI: Special Inspections, **AI:** Architectural Inspections, **EI:** Engineering Inspections, **OI:** Other Inspections
  Sub-Groups   SI: [T] Testing Laboratory, [H] Hazardous Materials, [Y] Safety   **AI:** No Sub-Groups
    EI: [V] Civil, [S] Structural, [M] Mechanical, [E] Electrical   **OI:** [O] Owner, [C] Contractor, [B] Subcontractor,
      [L] Legal, [G] Government, [F] Fire Protection, [P] Plumbing, [A] Acoustical

DIVISION 3 — CONCRETE — CONCRETE FORMS AND ACCESSORIES 03100

STANDARDS: ACI (.5) AISC (10) APA (17) CRA (30) DFPA (38) PS (77) SPR (83) WCLIB (95)

SI   AI   EI   OI

[x] [ ] [S] [B]
          [C] 1. Verify location, dimensions, and grades are as required.

[ ] [ ] [S] [B]
          [C] 2. Verify formwork materials are as specified. Formwork is properly sealed, oiled, or wetted and treatment is compatible with other materials to be applied.

[ ] [ ] [S] [B]
          [C] 3. Verify reused formwork is properly reconditioned and treated for reuse.

[ ] [ ] [S] [B]
          [C] 4. Verify construction provides mortar-tight condition, free from offsets and defects.

[X] [ ] [S] [B]
          [C] 5. Verify completed formwork provides structural sections required.

[X] [ ] [S] [B]
          [C] 6. Observe type of form spacers, ties, and bracing used. Verify ties indicated to be set to pattern are of proper type and spacing. Verify ties are arranged to be withdrawn or snapped off to leave no metal within specified or required distance to surface of concrete.

[ ] [ ] [S] [B]
          [C] 7. Verify temporary spreaders are arranged for easy removal. Verify removal.

[ ] [ ] [S] [B]
          [C] 8. Verify foundation forms are deepened as required for pipes, conduits, soft spots, etc.

[ ] [ ] [X] [B]
          [C] 9. Verify sleeves for piping and conduits are provided as required.

[ ] [ ] [X] [B]
          [C] 10. Verify provisions are made for anchors, hanger wires, inserts, cans, bucks, etc.

[ ] [ ] [S] [B]
          [C] 11. Verify forms are secured against movements during placing operations. Reference lines are established. Screeds are set high to allow for deflection, and deflection is checked during pour.

[ ] [X] [ ] [B]
          [C] 12. Verify chamfer strips, nailer strips, chases, and rustication strips are accurately placed, aligned adequately, fastened, and protected.

[ ] [ ] [S] [B]
          [C] 13. Verify cleanouts are provided at ends and low points of forms. Ports are provided in high forms. Verify number and location.

Legend: Upper Case Letter and **BOLD** = Primary Inspection; Lower Case = Secondary Inspection
Main Groups SI: Special Inspections, AI: Architectural Inspections, EI: Engineering Inspections, OI: Other Inspections
Sub-Groups   **SI:** [T] Testing Laboratory, [H] Hazardous Materials, [Y] Safety   **AI:** No Sub-Groups
EI: [V] Civil, [S] Structural, [M] Mechanical, [E] Electrical   **OI:** [O] Owner, [C] Contractor, [B] Subcontractor,
[L] Legal, [G] Government, [F] Fire Protection, [P] Plumbing, [A] Acoustical

Construction Inspection Manual

SI   AI   EI   OI

[x] [x] [S] [B] 14. Verify expansion, construction, and contraction joints are provided as required or indicated.
             [C]

[ ] [ ] [S] [B] 15. Verify filler is installed and securely fastened in expansion joints.
             [C]

[ ] [ ] [S] [B] 16. Verify free movement of expansion and contraction joints can occur. Verify there is no reinforcement or fixed metal continuous through the joint.
             [C]

[ ] [ ] [S] [B] 17. Verify form cut-offs and bulkheads are established in locations approved.
             [C]

[ ] [ ] [S] [B] 18. Verify forms are in alignment, especially at top of walls.
             [C]

[ ] [ ] [S] [B] 19. Verify forms provide for depressed slab areas, cut outs, curbs, etc.
             [C]

[ ] [X] [ ] [B] 20. Verify form surface treatment and pattern for architectural concrete is as required, non-corrosive nails are used to hold curtains.
             [C]

[ ] [X] [ ] [B] 21. Verify non-corrosive nails are used to hold curtains.
             [C]

[ ] [X] [ ] [B] 22. Verify forms are taped and prepared as required for architectural concrete. Verify joint patterns are formed as indicated.
             [C]

[ ] [ ] [S] [B] 23. Verify keys are provided in construction joints. Verify provisions for waterproofing are met.
             [C]

[ ] [ ] [S] [B] 24. Verify forms allow sufficient space and openings for depositing of concrete. Contact architect if field conditions appear difficult.
             [C]

[ ] [ ] [S] [B] 25. Verify forms provide for features such as doors, windows, openings, etc., and they are removable. Ensure cross bracing is provided in formed openings to prevent bowing.
             [C]

[ ] [ ] [S] [C] 26. Verify proper provisions are made during pour for sawed joints; location, equipment, personnel, and timing.

[ ] [ ] [S] [B] 27. Verify forms are properly cleaned of all debris and surface-treated as required prior to pour.
             [C]

[ ] [ ] [S] [B] 28. Note appearance of obvious defects in formwork that might impair its ability to withstand concrete pressure without excessive deflection.
             [C]

Legend: Upper Case Letter and **BOLD** = Primary Inspection; Lower Case = Secondary Inspection
  Main Groups SI: Special Inspections, **AI:** Architectural Inspections, **EI:** Engineering Inspections, **OI:** Other Inspections
  Sub-Groups **SI:** [T] Testing Laboratory, [H] Hazardous Materials, [Y] Safety   **AI:** No Sub-Groups
    EI: [V] Civil, [S] Structural, [M] Mechanical, [E] Electrical   **OI:** [O] Owner, [C] Contractor, [B] Subcontractor,
    [L] Legal, [G] Government, [F] Fire Protection, [P] Plumbing, [A] Acoustical

Construction Inspection Manual

SI	AI	EI	OI	
[ ]	[ ]	[S]	[B] [C]	29. Check shoring for location and bearing. Check shores for settlement during pouring.
[ ]	[ ]	[X]	[B] [C]	30. Verify trench ducts, boxes, cleanouts, flanges, etc., are set at proper elevation to allow for flush or required installation of finish floor. Sufficient anchorage is provided to avoid movement.
[ ]	[ ]	[S]	[B] [C]	31. Verify scaffolds or other accessories are adequate and do not affect form work.

Form Removal

[ ]	[ ]	[S]	[B] [C]	32. Verify concrete is sufficiently strong before removal of forms and surfaces are not damaged or spalled by too early removal.
[ ]	[ ]	[S]	[B] [C]	33. Ensure forms remain in place until expiration of curing period required.
[ ]	[ ]	[S]	[B] [C]	34. Ensure forms are tightened and maintained snug against concrete surfaces.
[ ]	[ ]	[S]	[B] [C]	35. Verify reshoring requirements are observed.
[ ]	[ ]	[ ]	[C]	36. Verify wood formwork is completely removed above and below grade.
[ ]	[X]	[s]	[B] [C]	37. Ensure preparations for patching are made as soon as practicable and methods are approved if required.

Legend: Upper Case Letter and **BOLD** = Primary Inspection; Lower Case = Secondary Inspection
Main Groups SI: Special Inspections, **AI:** Architectural Inspections, **EI:** Engineering Inspections, **OI:** Other Inspections
Sub-Groups **SI:** [T] Testing Laboratory, [H] Hazardous Materials, [Y] Safety   **AI:** No Sub-Groups
EI: [V] Civil, [S] Structural, [M] Mechanical, [E] Electrical   **OI:** [O] Owner, [C] Contractor, [B] Subcontractor,
[L] Legal, [G] Government, [F] Fire Protection, [P] Plumbing, [A] Acoustical

Construction Inspection Manual

DIVISION 3 — CONCRETE — CONCRETE REINFORCEMENT 03200

STANDARDS: ACI (5) AGDGA (9) ASTM (19) AWS (23) CRSI (36) PCI (76)

SI   AI   EI   OI

[x]	[ ]	[S]	[B] [C]	1. Verify grade of steel delivered to jobsite is identified as required.
[x]	[ ]	[S]	[B] [C]	2. Verify steel is installed in accordance with approved drawings and of grade and sizes indicated: bar length, bar diameter, bar-to-bar spacing. Verify spacing is coordinated to suit units where masonry walls are scheduled.
[ ]	[ ]	[S]	[B] [C]	3. Verify required clearance of steel from earth and forms is provided. See Appendix I.
[x]	[ ]	[S]	[B] [C]	4. Verify splice lapping, length of splices, and staggering of splices is as required.
[ ]	[ ]	[S]	[B] [C]	5. Verify bends are within radii and tolerance required and are uniformly made.
[ ]	[ ]	[S]	[B] [C]	6. Ensure additional bars, such as wall intersection bars, trim bars, at openings, and corner bars, are provided.
[x]	[ ]	[S]	[B] [C]	7. Verify reinforcement is cleaned of all loose or flaky rust or scale, dried concrete, oil, bond-breaker, and other foreign material that might reduce bond. Verify Tags are removed.
[ ]	[ ]	[S]	[B] [C]	8. Verify dowels are provided for marginal bars at wall ends and openings.
[x]	[ ]	[S]	[B] [C]	9. Verify beam and slab bars are spliced near supports for bottom steel and near midspan for top steel.
[ ]	[ ]	[S]	[B] [C]	10. Ensure no bent bars and tension members are installed except where approved.
[ ]	[ ]	[S]	[B] [C]	11. Verify vertical bars at wall ends are installed inside horizontal bars unless otherwise required.
[ ]	[ ]	[S]	[B] [C]	12. Verify extra turns are provided at top and bottom of spiral columns.
[ ]	[ ]	[S]	[B] [C]	13. Verify reinforcement is tied and supported securely to prevent displacement during placement.

Legend: Upper Case Letter and **BOLD** = Primary Inspection; Lower Case = Secondary Inspection
Main Groups SI: Special Inspections, **AI:** Architectural Inspections, **EI:** Engineering Inspections, **OI:** Other Inspections
Sub-Groups   SI: [T] Testing Laboratory, [H] Hazardous Materials, [Y] Safety   **AI:** No Sub-Groups
EI: [V] Civil, [S] Structural, [M] Mechanical, [E] Electrical   **OI:** [O] Owner, [C] Contractor, [B] Subcontractor, [L] Legal, [G] Government, [F] Fire Protection, [P] Plumbing, [A] Acoustical

Construction Inspection Manual

SI	AI	EI	OI	
[x]	[ ]	[S]	[B] [C]	14. Verify reinforcement spacers, tie wires, chairs, and supports are provided of type, size, and finish required.
[ ]	[ ]	[X]	[B] [C]	15. Verify conduits are separated from other conduit and rebar by three diameters minimum.
[ ]	[ ]	[S]	[B] [C]	16. Ensure no conduit or piping is placed below rebar mat in suspended slabs unless approved by consultant.
[ ]	[ ]	[X]	[B] [C]	17. Verify no secured pipes are embedded in concrete (liquid carrying).
[x]	[ ]	[S]	[B] [C]	18. Verify no contact of bars is made with dissimilar metals.
[ ]	[ ]	[S]	[B] [C]	19. Verify bars are not placed near surfaces that might allow rusting.
[ ]	[ ]	[S]	[B] [C]	20. Ensure rebars are welded only with consultant's approval, in accordance with reinforcing steel welding code, AWS D1.4.
[ ]	[ ]	[S]	[B] [C]	21. Verify adequate space and clearance are provided for proper deposit of concrete and use of vibrators.
[ ]	[ ]	[S]	[B] [C]	22. Verify embedded items are securely anchored in place and in proper location.
[ ]	[ ]	[S]	[B] [C]	23. Unless approved, ensure boxing-out is not allowed for subsequent grouting-in.
[ ]	[ ]	[S]	[B] [C]	24. Verify embedded items are supplied and installed as required. Generally review drawings for anchor bolts, piping, sleeves, conduits, boxes, special items, etc. Verify embedded items are suitably protected from damage during placement operations.
[ ]	[ ]	[S]	[B] [C]	25. If conflict occurs between embedded items and reinforcing bars, allow no cutting, bending, or omission without consultant's approval.
[ ]	[ ]	[S]	[B] [C]	26. Check placing tolerances of reinforcing. (See Appendix.)
[x]	[ ]	[S]	[B] [C]	27. Use the following rules of thumb for bar splices: For 24d lap: multiple bar size by 3 = lap in inches. For 32d lap: multiple bar size by 4 = lap in inches. For 40d lap: multiple bar size by 5 = lap in inches.

Legend: Upper Case Letter and **BOLD** = Primary Inspection; Lower Case = Secondary Inspection
Main Groups SI: Special Inspections, AI: Architectural Inspections, EI: Engineering Inspections, OI: Other Inspections
Sub-Groups **SI:** [T] Testing Laboratory, [H] Hazardous Materials, [Y] Safety   **AI:** No Sub-Groups
EI: [V] Civil, [S] Structural, [M] Mechanical, [E] Electrical   **OI:** [O] Owner, [C] Contractor, [B] Subcontractor, [L] Legal, [G] Government, [F] Fire Protection, [P] Plumbing, [A] Acoustical

SI	AI	EI	OI	
[ ]	[ ]	[S]	[B] [C]	28. Verify the rebar is not bent excessively ("hickeying"). To determine proper bending use Max slope =1:6. Field bending of partially embedded bar is done with consultants approval.
[X]	[ ]	[S]	[B] [C]	29. Ensure agency inspection is performed if required.

Legend: Upper Case Letter and <u>BOLD</u> = Primary Inspection; Lower Case = Secondary Inspection
  Main Groups SI: Special Inspections, **AI:** Architectural Inspections, **EI:** Engineering Inspections, **OI:** Other Inspections
  Sub-Groups   <u>SI</u>: [T] Testing Laboratory, [H] Hazardous Materials, [Y] Safety   **<u>AI</u>:** No Sub-Groups
    <u>EI</u>: [V] Civil, [S] Structural, [M] Mechanical, [E] Electrical   **<u>OI</u>:** [O] Owner, [C] Contractor, [B] Subcontractor,
      [L] Legal, [G] Government, [F] Fire Protection, [P] Plumbing, [A] Acoustical

DIVISION 3 — CONCRETE CAST-IN-PLACE CONCRETE 03300

STANDARDS: ACI (5) ASTM (19) NRMCA (68) PCI (76) PI (72) VI (94)

SI	AI	EI	OI	
[x]	[ ]	[S]	[C]	1. Ensure agency approval of forms and rebar is obtained prior to pour, if required.
[t]	[ ]	[S]	[C]	2. Verify requirements of concrete specifications have been met before delivery or placement of concrete — tests, mix design, ingredients, inspections, etc.
[t]	[ ]	[S]	[C]	3. Ensure testing laboratory is notified prior to pour, and testing is arranged at plant and site.
[ ]	[ ]	[S]	[B][C]	4. Verify areas to receive concrete are cleaned, wetted, or otherwise prepared as required. Verify foundations are free of frost and water. Verify previously placed concrete is properly prepared to receive new.
[ ]	[ ]	[S]	[B][C]	5. Verify vibrators, standby vibrators, and other necessary tools are available and in working condition. Check frequency and amplitude if required.
[ ]	[ ]	[S]	[B][C]	6. Verify that conveying equipment and depositing equipment is capable of reaching all areas of placement without segregation, loss of ingredients, formation of air pockets, or cold joints.
[ ]	[ ]	[S]	[B][C]	7. Confirm adequate manpower is available for timely placement.
[ ]	[ ]	[S]	[B][C]	8. Verify temporary form openings, tremies, chutes, etc., are provided.
[ ]	[ ]	[S]	[B][C]	9. Verify "pockets" are vented to prevent air entrapment.
[x]	[ ]	[S]	[B][C]	10. Verify subbase and capillary fills are compacted and membrane is provided and installed as required.
[ ]	[ ]	[S]	[B][C]	11. Confirm arrangements have been made for specified curing, sawed joints, and protection including cold weather protection, if needed.
[t]	[ ]	[S]	[B][C]	12. Verify cylinders, measuring equipment, and slump cone are at site and samples are properly taken.
[t]	[ ]	[S]	[C]	13. Confirm time interval between adding water to concrete and placement in final position is known.

Legend: Upper Case Letter and **BOLD** = Primary Inspection; Lower Case = Secondary Inspection
Main Groups **SI**: Special Inspections, **AI**: Architectural Inspections, **EI**: Engineering Inspections, **OI**: Other Inspections
Sub-Groups **SI**: [T] Testing Laboratory, [H] Hazardous Materials, [Y] Safety   **AI**: No Sub-Groups
EI: [V] Civil, [S] Structural, [M] Mechanical, [E] Electrical   **OI**: [O] Owner, [C] Contractor, [B] Subcontractor, [L] Legal, [G] Government, [F] Fire Protection, [P] Plumbing, [A] Acoustical

Construction Inspection Manual

SI   AI   EI   OI

[ ] [ ] [S] [C] 14. Confirm delivery of concrete and sequence of delivery is scheduled to allow continuous placement to prevent cold joints.

[t] [ ] [S] [C] 15. Verify age of concrete is within specified or required time limit and delivery tickets contain proper information.

[ ] [ ] [S] [B] [C] 16. Verify modified grout is provided at first lift and where rebar congestion occurs, as required.

[ ] [ ] [S] [B] [C] 17. Ensure layers are kept approximately horizontal and do not exceed required lifts.

[ ] [ ] [S] [B] [C] 18. Verify bolts and loose items to be embedded are properly located and installed.

[ ] [ ] [S] [C] 19. Record date and location of pours.

[x] [ ] [S] [B] [C] 20. Verify grades, elevations, alignment, form adjustment, and supports are being checked during pouring.

[ ] [ ] [S] [B] [C] 21. Verify time delay is made for concrete in columns, piers, walls, and openings to allow concrete to settle before placing concrete above them; however, initial set is to be avoided.

[ ] [ ] [S] [B] [C] 22. Verify vibration is performed properly, using correct equipment, and forms are not damaged.

[ ] [ ] [S] [B] [C] 23. Verify instruction concerning watering and drainage at site.

[t] [ ] [S] [B] [C] 24. Ensure cylinders are cast and stored and slump and other tests are performed as required (shrinkage bars, air entrainment, unit, etc.).

Finishing and Curing

[ ] [X] [ ] [B] [C] 25. Verify type of finishes as on unformed surfaces such as smooth, rubbed, broomed, nonslip, exposed aggregate, and colored is checked and provided as required.

[ ] [ ] [S] [B] [C] 26. Verify screeds are provided as required and wood screeds are removed.

[ ] [ ] [S] [B] [C] 27. Verify over-troweling is avoided and troweling is not performed while bleed water is on surface.

Legend: Upper Case Letter and **BOLD** = Primary Inspection; Lower Case = Secondary Inspection
Main Groups SI: Special Inspections, AI: Architectural Inspections, EI: Engineering Inspections, OI: Other Inspections
Sub-Groups   **SI:** [T] Testing Laboratory, [H] Hazardous Materials, [Y] Safety   **AI:** No Sub-Groups
EI: [V] Civil, [S] Structural, [M] Mechanical, [E] Electrical   **OI:** [O] Owner, [C] Contractor, [B] Subcontractor, [L] Legal, [G] Government, [F] Fire Protection, [P] Plumbing, [A] Acoustical

Construction Inspection Manual

SI	AI	EI	OI		
[ ]	[ ]	[S]	[B] [C]	28.	Verify curing methods are started as soon as possible. Verify curing compounds are as required and compatible with subsequent finishes.
[x]	[ ]	[S]	[B] [C]	29.	Verify finishing method provides evenness, smoothness, and levelness of surfaces within tolerance indicated. Verify slopes are provided to properly drain. Ensure marks left by tools are removed.
[ ]	[x]	[S]	[B] [C]	30.	Ensure that joints, edges, and corners are carefully finished and/or match.
[ ]	[ ]	[S]	[B] [C]	31.	Verify wet spray or moist curing method is adequately performed.
[ ]	[X]	[S]	[B] [C]	32.	Verify waterproof paper or similar covers are applied with sufficient lap and seal and adequately protected during curing period. (Colored surfaces may require special covers to avoid staining, etc.).
[ ]	[ ]	[S]	[C]	33.	Confirm loading and traffic are controlled over surfaces to protect concrete.
[ ]	[X]	[S]	[B] [C]	34.	Confirm methods of repairing defective areas, removal of fins, form marks, ties, etc., on formed surfaces are understood and provided as soon as possible upon removal of forms.

Legend: Upper Case Letter and <u>BOLD</u> = Primary Inspection; Lower Case = Secondary Inspection
Main Groups SI: Special Inspections, **AI:** Architectural Inspections, **EI:** Engineering Inspections, **OI:** Other Inspections
Sub-Groups <u>**SI:**</u> [T] Testing Laboratory, [H] Hazardous Materials, [Y] Safety   <u>**AI:**</u> No Sub-Groups
<u>**EI:**</u> [V] Civil, [S] Structural, [M] Mechanical, [E] Electrical   <u>**OI:**</u> [O] Owner, [C] Contractor, [B] Subcontractor, [L] Legal, [G] Government, [F] Fire Protection, [P] Plumbing, [A] Acoustical

Construction Inspection Manual

DIVISION 3 — CONCRETE — PRE-CAST CONCRETE 03400

STANDARDS: ACI (5) AISI (13) ANSI (16) ASTM (19) PCA (75) PCI (76)

SI   AI   EI   OI

[ ] [X] [s] [B]   1. Verify required samples have been submitted and approved as to color,
               [C]   texture, form, finish, treatment, anchors, and reinforcement. Mock-up is provided and approved if required.

[ ] [X] [s] [B]   2. Confirm special items and conditions are understood.
               [C]

[ ] [ ] [S] [B]   3. Verify conditions to receive precast material are as required. Verify
               [C]   structural support members are properly located, installed, and within tolerance. Confirm erection contractor accepts conditions.

[ ] [ ] [S] [C]   4. Verify masons and welders are qualified as required.

[t] [ ] [S] [B]   5. Verify testing procedure is scheduled and performed at fabrication source
               [C]   and in field.

[ ] [ ] [S] [B]   6. Verify connectors, anchors, and fasteners are of type, metal, finish, and
               [C]   size required and are installed as required.

[ ] [ ] [S] [B]   7. Verify shields to protect materials are provided during welding.
               [C]

[ ] [ ] [S] [B]   8. Confirm lead buttons or soft wood wedges are used to prevent crushing of
               [C]   mortar and are removed as required. Verify patching is performed.

[ ] [ ] [S] [C]   9. Verify cutting or drilling material to accommodate others' work is accurately and properly performed.

[ ] [X] [S] [C] 10. Verify flexible gaskets for joints are properly sealed.

[ ] [X] [ ] [B] 11. Confirm nonstaining back-up caulking, cement, or materials are
               [C]   provided and tests are performed or attested to in respect of non-staining qualities.

[ ] [X] [ ] [B] 12. Verify joint caulking, sealing, painting, etc., is performed under required
               [C]   conditions.

[ ] [X] [ ] [B] 13. Verify joint caulking, sealing, and painting, are of type, color, and
               [C]   method of installation required.

Legend: Upper Case Letter and **BOLD** = Primary Inspection; Lower Case = Secondary Inspection
Main Groups SI: Special Inspections, **AI:** Architectural Inspections, **EI:** Engineering Inspections, **OI:** Other Inspections
Sub-Groups   SI: [T] Testing Laboratory, [H] Hazardous Materials, [Y] Safety   **AI:** No Sub-Groups
EI: [V] Civil, [S] Structural, [M] Mechanical, [E] Electrical   **OI:** [O] Owner, [C] Contractor, [B] Subcontractor, [L] Legal, [G] Government, [F] Fire Protection, [P] Plumbing, [A] Acoustical

Construction Inspection Manual

SI	AI	EI	OI	
[x]	[x]	[S]	[B] [C]	14. Verify bolt or dowel holes in top of copings are grouted.
[ ]	[x]	[S]	[B] [C]	15. Verify expansion or control joints are located and provided as required.
[ ]	[X]	[ ]	[B] [C]	16. Verify built-in flashings are properly formed and located as required. Weep holes are provided.
[ ]	[x]	[S]	[B] [C]	17. Verify joint and panel alignment is within specified tolerances.
[ ]	[X]	[ ]	[B] [C]	18. Confirm water-repellent treatments are applied, using materials and methods required.
[ ]	[X]	[ ]	[B] [C]	19. Confirm protection is provided for pre-cast sills and projecting work.
[ ]	[X]	[ ]	[C]	20. Verify cleaning of work is performed as required, proceeding from top of building downward.
[ ]	[X]	[ ]	[C]	21. Verify adjacent metal and other materials are protected during cleaning operations.

Legend: Upper Case Letter and <u>BOLD</u> = Primary Inspection; Lower Case = Secondary Inspection
Main Groups SI: Special Inspections, AI: Architectural Inspections, EI: Engineering Inspections, OI: Other Inspections
Sub-Groups <u>SI</u>: [T] Testing Laboratory, [H] Hazardous Materials, [Y] Safety   <u>AI</u>: No Sub-Groups
<u>EI</u>: [V] Civil, [S] Structural, [M] Mechanical, [E] Electrical   <u>OI</u>: [O] Owner, [C] Contractor, [B] Subcontractor, [L] Legal, [G] Government, [F] Fire Protection, [P] Plumbing, [A] Acoustical

Construction Inspection Manual

DIVISION 4 — MASONRY — UNIT MASONRY 04200

STANDARDS: AISC (10) AINA (12) ANSI (16) ASTM (19) BIA (29) GA (45) MI (52) NCMA (58) PCA (75)

SI	AI	EI	OI	
[ ]	[x]	[S]	[B] [C]	1. Verify materials are suitably stored off the ground and covered with waterproof material.
[ ]	[X]	[s]	[B] [C]	2. Verify site materials match approved samples for color, texture, grade, and size and contain no defects such as chips, cracks, crazing, warps, kiln marks on face, and size differential, except for tolerances as allowed by the appropriate ASTM Standard. Verify required types and shapes are available and compatible with field materials.
[t]	[ ]	[S]	[C]	3. Confirm schedule of test and inspections is arranged before installation. Verify wall prisms, grout prisms, grout tests, mortar tests, type of mortar, mix and ingredients are as approved and required.
[ ]	[X]	[ ]	[B] [C]	4. Confirm sample panels have been provided and approved as required.
[ ]	[ ]	[S]	[B] [C]	5. Verify wetting of bricks is properly performed if required to assure that mortar will bond to brick. Concrete masonry units are not wet.
[ ]	[X]	[s]	[B] [C]	6. Verify mortar color is provided and approved if required.
[ ]	[x]	[S]	[B] [C]	7. Verify layout of work, coursing and dimensions are as required or indicated. Confirm story-pole is used if required.
[ ]	[X]	[s]	[B] [C]	8. Confirm joint size, type, tooling method, and equipment are understood and produced.
[ ]	[ ]	[S]	[B] [C]	9. Verify mortar is mixed as required, and methods and equipment are suitable to produce the approved mix.
[ ]	[X]	[s]	[B] [C]	10. Verify indicated bonding patterns are provided. Verify uniformity of laying.
[ ]	[ ]	[S]	[B] [C]	11. Generally observe mortar application to materials — full head and bed joints, shoving, and "buttering," Verify that complete filling of collar joints is as required in composite wall construction.

Legend: Upper Case Letter and **BOLD** = Primary Inspection; Lower Case = Secondary Inspection
Main Groups SI: Special Inspections, **AI:** Architectural Inspections, **EI:** Engineering Inspections, **OI:** Other Inspections
Sub-Groups **SI:** [T] Testing Laboratory, [H] Hazardous Materials, [Y] Safety   **AI:** No Sub-Groups
EI: [V] Civil, [S] Structural, [M] Mechanical, [E] Electrical   **OI:** [O] Owner, [C] Contractor, [B] Subcontractor, [L] Legal, [G] Government, [F] Fire Protection, [P] Plumbing, [A] Acoustical

Construction Inspection Manual

SI AI EI OI

[ ] [ ] [S] [B] [C]   12. Verify joints are tooled in such a manner as to provide a dense surface unless otherwise specified.

[ ] [x] [S] [B] [C]   13. Verify cutting of units is as required.

[ ] [ ] [S] [B] [C]   14. Verify cleanouts are provided as required.

[ ] [X] [s] [B] [C]   15. Verify spaces between wythes are of sizes required and kept free of excess droppings.

[ ] [ ] [S] [C]   16. Confirm provisions are adequate to protect work at least 48 hours from freezing or longer if required to properly cure. Consult with International Masonry Institute, All-Weather Council for cold weather requirements. Verify that acceptable cold weather precautions are provided when the temperature is less than 40°F.

[ ] [X] [s] [B] [C]   17. Confirm methods of cleaning are understood and performed as required. Ensure droppings and splatters on finished surfaces are cleaned as soon as possible.

[ ] [ ] [S] [B] [C]   18. Verify anchors and ties are of type of material and size required and are installed as required.

[ ] [ ] [S] [B] [C]   19. Verify reinforcement, is of type, size, splicing, and spacing required that it is properly doweled, tied, and installed. Confirm additional reinforcement is provided as required for corners, intersections, openings, and lintels. Refer also to 03200 "Concrete Reinforcement."

[ ] [ ] [S] [B] [C]   20. Do not allow bending rebar excessively to fit masonry cells. Verify approval has been obtained if required.

[ ] [ ] [S] [B] [C]   21. Verify bucks, anchors, forming, supports, and other embedded materials are available, secured, plumb, or level and otherwise properly installed.

[ ] [x] [X] [B] [C]   22. Confirm provision for flashings, cut-outs, and later installation of other items is made.

[ ] [X] [s] [B] [C]   23. Confirm provision for parging or treatment of backs of walls which are to receive backfill is performed as required.

Legend: Upper Case Letter and **BOLD** = Primary Inspection; Lower Case = Secondary Inspection
Main Groups **SI:** Special Inspections, **AI:** Architectural Inspections, **EI:** Engineering Inspections, **OI:** Other Inspections
Sub-Groups **SI:** [T] Testing Laboratory, [H] Hazardous Materials, [Y] Safety **AI:** No Sub-Groups
EI: [V] Civil, [S] Structural, [M] Mechanical, [E] Electrical **OI:** [O] Owner, [C] Contractor, [B] Subcontractor, [L] Legal, [G] Government, [F] Fire Protection, [P] Plumbing, [A] Acoustical

Construction Inspection Manual

SI AI EI OI

[ ] [ ] [S] [B]   24. Verify expansion joints for brick masonry and control joints for concrete
           [C]   masonry are located and provided as indicated or required.

[ ] [ ] [S] [B]   25. Verify weeps holes are provided if required.
           [C]

[ ] [ ] [S] [B]   26. Verify structural members to receive masonry are located, supported and
           [C]   anchored and have suitable attachments as required.

[ ] [ ] [S] [B]   27. Verify prior to grout pour, all wythes, reinforcement, etc., are properly
           [C]   cleaned. Ensure that precautions such as ties and braces are considered
                         so as to avoid "blow-outs."

[ ] [ ] [S] [B]   28. Verify pipes, conduits, sleeves, and boxes are located, secured,
           [C]   protected, insulated, and spaced as required.

[ ] [ ] [S] [C]   29. Verify measuring box is used for job mixed grout. Allow no shovel
                         measures.

[X] [ ] [S] [B]   30. Ensure agency inspection is performed before grouting, if required.
           [C]

[T] [ ] [s] [C]   31. Verify wall prisms are made at proper frequency and are properly stored,
                         delivered, cured and tested, all as required.

[t] [ ] [S] [B]   32. Verify lifts of grout are poured in a timely sequence and as required.
           [C]   Generally observe bond beam filling and compaction methods,
                         including consolidation. Note height limitations required for lifts.

[t] [ ] [S] [B]   33. Verify hollow metal frames are filled solid.
           [C]

[t] [ ] [S] [B]   34. Ensure backfilling is performed only after proper curing or support units
           [C]   are provided and as required.

[T] [ ] [ ] [C]   35. Confirm pointing, replacement of defective units, and repair of other
                         defects are promptly performed and as deemed necessary.

[ ] [X] [s] [C]   36. Verify water-proofing of walls is performed if required. Verify moisture
                         content prior to waterproofing is as required or specified.

[ ] [X] [s] [C]   37. Confirm methods of final cleaning are as required.

Legend: Upper Case Letter and BOLD = Primary Inspection; Lower Case = Secondary Inspection
    Main Groups SI: Special Inspections, **AI:** Architectural Inspections, **EI:** Engineering Inspections, **OI:** Other Inspections
    Sub-Groups   **SI:** [T] Testing Laboratory, [H] Hazardous Materials, [Y] Safety   **AI:** No Sub-Groups
    EI: [V] Civil, [S] Structural, [M] Mechanical, [E] Electrical   **OI:** [O] Owner, [C] Contractor, [B] Subcontractor,
      [L] Legal, [G] Government, [F] Fire Protection, [P] Plumbing, [A] Acoustical

Existing Unreinforced Masonry Retrofit

SI **AI** **EI** **OI**

[T] [ ] [s] [B] 38. Verify lay-up of existing walls, in particular full-length header courses,
           [C]   are in compliance with code requirements.

[T] [ ] [s] [c] 39. Test the quality of the existing mortar by performing in-place shear tests in accordance with code requirements.

[T] [ ] [s] [c] 40. Test in-place existing wall anchors in accordance with the code requirements.

[T] [ ] [s] [c] 41. Perform special inspection or testing of new shear bolts and combined tension and shear bolts in accordance to the code.

[T] [ ] [s] [c] 42. Verify archaic materials are investigated and tested in accordance with code requirements.

[ ] [ ] [S] [B] 43. Verify pointing of unreinforced masonry, when required due to
           [C]   deteriorated mortar points, is in accordance with the code.

Legend: Upper Case Letter and <u>BOLD</u> = Primary Inspection; Lower Case = Secondary Inspection
Main Groups SI: Special Inspections, **AI:** Architectural Inspections, **EI:** Engineering Inspections, **OI:** Other Inspections
Sub-Groups   **SI:** [T] Testing Laboratory, [H] Hazardous Materials, [Y] Safety   **AI:** No Sub-Groups
    EI: [V] Civil, [S] Structural, [M] Mechanical, [E] Electrical   **OI:** [O] Owner, [C] Contractor, [B] Subcontractor,
    [L] Legal, [G] Government, [F] Fire Protection, [P] Plumbing, [A] Acoustical

Construction Inspection Manual

DIVISION 5 — METALS — STRUCTURAL STEEL 05120

STANDARDS: AA (2) AHDGA (9) AISC (10) AISI (13) ASTM (19) AWS (23) FS (41) SSPC (89)

SI AI EI OI

[ ] [ ] [S] [B] [C] 1. Verify setting of foundation anchor bolts, size, and location are as required.

[ ] [ ] [S] [C] 2. Verify foundations are provided as required under all column locations.

[t] [ ] [S] [C] 3. Verify testing and inspection laboratory has observed shop fabrication as required or specified.

[ ] [ ] [S] [B] [C] 4. Verify delivered materials are of correct size, shape, and weight.

[ ] [ ] [S] [B] [C] 5. Ensure beams made up of welded plates are not substituted for specified rolled sections without approval.

[ ] [ ] [S] [B] [C] 6. Verify size and type of bolts, rivets, washers, and hole diameters are as required.

[t] [ ] [S] [c] 7. Verify welds on delivered materials are of size required.

[ ] [ ] [S] [C] 8. Verify shop painting is provided as required and items to be embedded are not shop coated unless required.

[ ] [ ] [S] [B] [C] 9. Verify delivered steel is new, undamaged, and free of distortions.

[ ] [ ] [S] [C] 10. Verify steel is suitably stored, blocked off ground, and covered, where prolonged storage occurs.

[ ] [ ] [S] [C] 11. Observe erection contractor's proposed methods, sequence of operations, and erection equipment. Ensure long trusses are to be erected by double choker or sling method to avoid overstress.

[ ] [ ] [S] [B] [C] 12. Verify column ends are milled and protected if required.

[ ] [ ] [S] [B] [C] 13. Observe setting of base and bearing plates. See that full engagement of nut occurs and that bending of anchor bolts or undue chipping of concrete does not occur. Verify clearances required for finish coverings or materials are provided.

Legend: Upper Case Letter and **BOLD** = Primary Inspection; Lower Case = Secondary Inspection
Main Groups SI: Special Inspections, **AI:** Architectural Inspections, **EI:** Engineering Inspections, **OI:** Other Inspections
Sub-Groups   SI: [T] Testing Laboratory, [H] Hazardous Materials, [Y] Safety   **AI:** No Sub-Groups
EI: [V] Civil, [S] Structural, [M] Mechanical, [E] Electrical   **OI:** [O] Owner, [C] Contractor, [B] Subcontractor, [L] Legal, [G] Government, [F] Fire Protection, [P] Plumbing, [A] Acoustical

Construction Inspection Manual

SI	AI	EI	OI	
[t]	[ ]	[S]	[B] [C]	14. Verify temporary connections to hold steel in place are provided and members are accurately wedged, shimmed, plumbed, fitted, aligned, or leveled before final rivets, bolts, or weld connections are made.
[ ]	[ ]	[S]	[B] [C]	15. See that concrete is cleaned and free of dirt and laitance, and grouting is properly performed. Space between concrete and bottom of bearing plate usually must not exceed 1/24 bearing plate width. See that dry pack mortar is properly rammed and cured.
[ ]	[ ]	[S]	[B] [C]	16. Verify shims are steel plates of varying thickness and not odd pieces of metal.
[ ]	[ ]	[S]	[B] [C]	17. See that temporary connections, guys, and braces are provided to hold work in place before permanent connections are completed. Verify work is installed plumb and to tolerances required.
[ ]	[ ]	[S]	[B] [C]	18. Verify drift pins are used to bring together the parts, but should not damage or distort metal.
[x]	[ ]	[S]	[B] [C]	19. Verify beam members are set with natural camber up. Verify camber is furnished where required.
[ ]	[ ]	[S]	[C]	20. Verify steel members have stud bolts, or other method of connection is provided to attach other materials suitably.
[x]	[x]	[X]	[C]	21. Verify steel members are not cut for passage of conduit, pipes, etc., unless so indicated on shop drawings.

Bolting

[x]	[ ]	[S]	[B] [C]	22. Ensure bolts and rivets are not allowed in same connection unless otherwise required.
[x]	[ ]	[S]	[B] [C]	23. Verify type, size, and length of bolt, size and type of washer, and size of hole are as required.
[x]	[ ]	[S]	[B] [C]	24. Verify all heads and nuts are resting squarely against metal; check to see that bolts have been drawn adequately tight.
[X]	[ ]	[S]	[B] [C]	25. Verify holes are properly aligned. Ensure burning to correct misalignment is not permitted.

Legend: Upper Case Letter and **BOLD** = Primary Inspection; Lower Case = Secondary Inspection
Main Groups **SI**: Special Inspections, **AI**: Architectural Inspections, **EI**: Engineering Inspections, **OI**: Other Inspections
Sub-Groups **SI**: [T] Testing Laboratory, [H] Hazardous Materials, [Y] Safety **AI**: No Sub-Groups
EI: [V] Civil, [S] Structural, [M] Mechanical, [E] Electrical **OI**: [O] Owner, [C] Contractor, [B] Subcontractor, [L] Legal, [G] Government, [F] Fire Protection, [P] Plumbing, [A] Acoustical

Construction Inspection Manual

SI   AI   EI   OI

[ ] [ ] [S] [B]   26. Verify use of ribbed bolts or turned bolts.
           [C]

High Strength Bolting

[T] [ ] [S] [B]   27. Ensure identification of bolt is made, and washer and nut are of proper type. (Testing may be required.)

[ ] [ ] [S] [B]   28. Verify whether paint is allowed on contact surface. Generally, all deleterious materials such as dirt, oil, or loose scale, or defects such as burrs or pits should not be present.
           [C]

[T] [ ] [S] [B]   29. Verify that impact wrenches are accurately calibrated and that frequency of calibration checks. Turn-of-the-nut method is allowed. Direct tension indicator devices are acceptable.
           [C]

[ ] [ ] [S] [B]   30. Verify slope of flanges (1:20); beveled washers as required.
           [C]

[ ] [ ] [S] [B]   31. Verify hardened washers are provided as required.
           [C]

Welding

[x] [ ] [S] [B]   32. Review mill certificate for metals to be joined.
           [C]

[T] [ ] [s] [B]   33. Review the Welding Procedure Specifications on-site and previously approved by the Engineer-of-Record for the specific project.
           [C]

[T] [ ] [s] [B]   34. Review welder qualifications for the specified welding.
           [C]

[T] [ ] [ ] [B]   35. Inspect weld preparation prior to welding for conformance with approved welding Procedure Specifications including:
           [C]

                       a. type of joint.
                       b. base metals.
                       c. filler metals, class and diameter.
                       d. shielding.
                       e. preheat.
                       f. welding positions.
                       g. electrical characteristic, current (type, polarity and amps) volts.
                       h. technique, travel speed
                       i. post weld heat treatment
                       j. process

[T] [ ] [ ] [C]   36. Inspect in-process welding.

[T] [ ] [s] [C]   37. See that non-destructive testing is performed as required.

Legend: Upper Case Letter and **BOLD** = Primary Inspection; Lower Case = Secondary Inspection
    Main Groups SI: Special Inspections, **AI:** Architectural Inspections, **EI:** Engineering Inspections, **OI:** Other Inspections
    Sub-Groups   **SI:** [T] Testing Laboratory, [H] Hazardous Materials, [Y] Safety   **AI:** No Sub-Groups
      EI: [V] Civil, [S] Structural, [M] Mechanical, [E] Electrical   **OI:** [O] Owner, [C] Contractor, [B] Subcontractor,
        [L] Legal, [G] Government, [F] Fire Protection, [P] Plumbing, [A] Acoustical

DIVISION 5 — METALS — METAL JOISTS 05200

STANDARDS: AISC (10) AISI (13) ASTM (19) AWS (23) SJI (88) SSPC (89)

SI	AI	EI	OI	
[ ]	[ ]	[S]	[B] [C]	1. Verify proper equipment is used for unloading and handling.
[ ]	[ ]	[S]	[B] [C]	2. Verify joists are stored with top or bottom chords up.
[ ]	[ ]	[S]	[B] [C]	3. Verify joists are coated with type of paint and number of coats required.
[T]	[ ]	[S]	[C]	4. Visually inspect welds for length and size.
[ ]	[ ]	[S]	[B] [C]	5. Verify nailer on top and/or bottom chord is provided if required.
[ ]	[ ]	[S]	[B] [C]	6. Verify holes in bearing plate at one end have been slotted if required.
[ ]	[ ]	[S]	[B] [C]	7. Verify joists are accurately spaced and have proper bearing and anchorage.
[x]	[ ]	[S]	[B] [C]	8. Verify installation and connections are as required.
[ ]	[ ]	[S]	[B] [C]	9. Verify ceiling extensions are provided where required.
[ ]	[ ]	[S]	[B] [C]	10. Verify bridging and anchoring are installed as soon as joists are placed and before application of loads.
[ ]	[ ]	[S]	[B]	11. Verify ends of bridging lines terminating at walls or beams are anchored at plane of top and bottom chords as required.
[ ]	[ ]	[S]	[C]	12. Allow no cutting or drilling of web or chord members.
[ ]	[ ]	[S]	[C]	13. Do not allow excessive concentrated loads of heavy building materials or moving of heavy equipment over joists.
[ ]	[X]	[s]	[C]	14. Verify all rust, scale, slag, and splatter are removed and joist is clean before it is painted.

Legend: Upper Case Letter and **BOLD** = Primary Inspection; Lower Case = Secondary Inspection
Main Groups **SI**: Special Inspections, **AI**: Architectural Inspections, **EI**: Engineering Inspections, **OI**: Other Inspections
Sub-Groups **SI**: [T] Testing Laboratory, [H] Hazardous Materials, [Y] Safety   **AI**: No Sub-Groups
EI: [V] Civil, [S] Structural, [M] Mechanical, [E] Electrical   **OI**: [O] Owner, [C] Contractor, [B] Subcontractor, [L] Legal, [G] Government, [F] Fire Protection, [P] Plumbing, [A] Acoustical

Construction Inspection Manual

DIVISION 5 — METALS — METAL DECK 05300

STANDARDS: AHDGA (9) AISC (10) AISI (13) ASTAM (19) AWS (23) FMU (40) SDI (86)

<u>SI</u> <u>AI</u> <u>EI</u> <u>OI</u>

[ ] [ ] [S] [B] [C]   1. Verify material is of approved type, material, finish, shapes, gauge, and size. Verify approved samples are on job if required. Review approved decking layout submissions.

[ ] [x] [S] [B] [C]   2. Verify material is properly stored on site and protected.

[ ] [ ] [S] [B] [C]   3. Verify all accessory items are furnished and approved type and sequence of fastening is performed.

[ ] [ ] [S] [B] [C]   4. Verify closures at edges, and over walls are provided as required.

[t] [ ] [S] [B] [C]   5. Verify welders are certified if required. Verify that a welding inspection by a testing laboratory is provided if required. Make sure decking is in contact with beams and joists to insure proper tack welds.

[ ] [ ] [X] [C]   6. Verify provisions such as tabs, hangers, and supports are met in regard to mechanical and electrical equipment, and hung ceilings.

[ ] [ ] [X] [C]   7. Verify coordination is performed with related trades: sheet metal, roofing, insulation, and electrical.

[T] [ ] [s] [B] [C]   8. Verify decking is continuous over supports if required, and welded connections and spacings are as required. Observe panel to panel seams for tack weld or, button punch connections. Check seam welding for burn-outs.

[ ] [ ] [S] [B] [C]   9. Verify reinforcement around columns and at penetrations is provided as required.

[x] [ ] [S] [B] [C]   10. Verify reinforcement at locations of major concentrated loads is as required.

[ ] [ ] [S] [B] [C]   11. Verify seams are aligned. Verify type and spacing of seam connections are as required.

[ ] [ ] [S] [B] [C]   12. Verify electric cell units are aligned, and no rough or dented edges occur so that insulation is not damaged when wire is pulled through cell. Observe that butt ends are taped to keep concrete fill out of cell.

[ ] [ ] [S] [C]   13. Ensure no concentrated loads are placed on decking during construction.

Legend: Upper Case Letter and <u>BOLD</u> = Primary Inspection; Lower Case = Secondary Inspection
  Main Groups SI: Special Inspections, **AI:** Architectural Inspections, **EI:** Engineering Inspections, **OI:** Other Inspections
  Sub-Groups   <u>SI</u>: [T] Testing Laboratory, [H] Hazardous Materials, [Y] Safety   <u>AI</u>: No Sub-Groups
    <u>EI</u>: [V] Civil, [S] Structural, [M] Mechanical, [E] Electrical   <u>OI</u>: [O] Owner, [C] Contractor, [B] Subcontractor,
       [L] Legal, [G] Government, [F] Fire Protection, [P] Plumbing, [A] Acoustical

SI	AI	EI	OI	
[ ]	[ ]	[S]	[B] [C]	14. If decking is to receive concrete topping, verify if shoring is required.
[ ]	[ ]	[S]	[B] [C]	15. If decking is to receive a topping material, observe that decking is free of loose dirt, oil, or debris.
[ ]	[ ]	[S]	[B] [C]	16. On roof decking, verify deslag welds and paint out with approved zinc preparation as required. Point out all abrasions.
[ ]	[ ]	[S]	[B] [C]	17. Verify if U.L. labels are required.
[ ]	[X]	[s]	[B] [C]	18. Deformation of ¼" or more across any three ribs is unacceptable if roof decking is to receive insulation and built-up-roofing conforming to FM standards. Verify compliance.
[ ]	[X]	[s]	[ ]	19. Verify roof ventilation provisions are met.
[ ]	[X]	[s]	[C]	20. Inspect exterior siding or roofing during installation, especially factory finished surfaces requiring sequence installation.
[ ]	[ ]	[S]	[B] [C]	21. Verify touch up of exposed and cut galvanized metal is performed.
[ ]	[ ]	[S]	[B]	22. Report all damaged panels to contractor and architect.
[T]	[ ]	[s]	[B] [C]	23. Check weld washers used, if required (check gauge).

Legend: Upper Case Letter and **BOLD** = Primary Inspection; Lower Case = Secondary Inspection
Main Groups **SI:** Special Inspections, **AI:** Architectural Inspections, **EI:** Engineering Inspections, **OI:** Other Inspections
Sub-Groups **SI:** [T] Testing Laboratory, [H] Hazardous Materials, [Y] Safety   **AI:** No Sub-Groups
EI: [V] Civil, [S] Structural, [M] Mechanical, [E] Electrical   **OI:** [O] Owner, [C] Contractor, [B] Subcontractor, [L] Legal, [G] Government, [F] Fire Protection, [P] Plumbing, [A] Acoustical

Construction Inspection Manual

DIVISION 5 — METALS — METAL FABRICATIONS 05500

STANDARDS: AA (2) AAMA (26) AHDGA (9) AISC (10) AWS (23) CDA (37) NAAMM (54) SMUACNA (82) SSPC (89)

<u>SI</u> <u>AI</u> <u>EI</u> <u>OI</u>

[ ] [x] [S] [B] 1. Verify delivered materials are fabricated from approved shop drawings, if
          [C]   required, and meet specifications of steel fabrication, riveting, welding,
                    galvanizing, sheet metal, or other metals required.

[ ] [ ] [S] [B] 2. Verify templates are furnished for proper placement and anchorage if
          [C]   required.

[ ] [ ] [S] [B] 3. Ensure prior provisions are made for adequate bracing, blocking, and
          [C]   anchorage of items.

[ ] [ ] [S] [B] 4. Verify sleeves, bolts, cut-outs, holes, and connectors are located and
          [C]   provided as required.

[ ] [X] [s] [C] 5. Verify materials are protected from damage before and after installation.

[ ] [X] [s] [B] 6. Verify opening bucks, angles, and thresholds are adequately braced,
          [C]   anchored, and aligned and bear labels if required.

[ ] [X] [s] [B] 7. On railings, see that heights, vertical spacing, returns, and anchorage will
          [C]   meet code requirements after installation. Notify architect of
                    discrepancies.

[ ] [X] [s] [B] 8. On metal stairs, verify bearing of supports, levelness, variation of risers,
          [C]   coordination with adjacent finish surfaces, spacings, and tolerances
                    required are adequate. Verify nosings of pan treads are protected.

[ ] [X] [s] [B] 9. Observe installation techniques and workmanship: smooth and ground
          [C]   welds, touch up, caulking, galvanizing, and touch-up shop coats.

[ ] [X] [s] [B] 10. Verify closed risers, smooth nosings, railing extensions for stairs and
          [C]   ramps, marking of stairs for visually impaired and other accessibility
                     compliance items are observed.

Legend: Upper Case Letter and **BOLD** = Primary Inspection; Lower Case = Secondary Inspection
Main Groups SI: Special Inspections, **AI:** Architectural Inspections, **EI:** Engineering Inspections, **OI:** Other Inspections
Sub-Groups <u>**SI**</u>: [T] Testing Laboratory, [H] Hazardous Materials, [Y] Safety   <u>**AI**</u>: No Sub-Groups
  <u>**EI**</u>: [V] Civil, [S] Structural, [M] Mechanical, [E] Electrical   <u>**OI**</u>: [O] Owner, [C] Contractor, [B] Subcontractor,
  [L] Legal, [G] Government, [F] Fire Protection, [P] Plumbing, [A] Acoustical

DIVISION 6 — WOOD AND PLASTICS — ROUGH CARPENTRY 06100

STANDARDS: AITC (11) ANSI (16) APA (17) AWPA (25) CRA (30) CS (35) DFPA (38) FPL (43) FS (41) NFoPa (64) NPA (67) PS (77) RIS (79) RIS (79) SPR (83) SPA (85) TPI (92) UL (93) WCLIB (95) WRCLA (96) WWPA (97)

SI   AI   EI   OI

[X] [ ] [s] [c]   1. Verify delivered lumber is of proper species and grade and has treatment required.

[X] [ ] [s] [c]   2. Verify framing lumber is grade-stamped or suitably identified.

[X] [ ] [s] [c]   3. Generally spot check for splits, shake, decay, pockets, wane, crook, bow, cup, loose knots, or other defects not in compliance with specified grade.

[X] [ ] [s] [c]   4. Verify lumber is suitable stored off the ground, stacked to prevent warp, and protected to prevent increase in moisture content.

[X] [ ] [s] [c]   5. Verify grade stamp indicates that moisture content is as specified.

[X] [ ] [s] [c]   6. Preservative treatment is as required and affidavits are supplied if required.

[X] [ ] [s] [c]   7. Verify materials in contact with concrete or masonry or near earth are treated or of suitably graded species of lumber for these conditions.

[X] [x] [s] [c]   8. Verify surfaces to be painted are treated with proper preservatives.

[X] [ ] [s] [c]   9. Verify framing is in alignment, plumb and level, and temporary bracing is provided during construction.

[X] [ ] [s] [c]   10. Verify nails, bolts, and connectors are as required. Observe usage of box and common nails. Observe spacing penetration of nails. Verify no coatings are provided on nails such as to reduce friction.

[X] [ ] [s] [c]   11. Verify predrill for nails if required.

[X] [ ] [s] [c]   12. Verify joists are set with crown up, have adequate bearing, and are properly fire cut when bearing in masonry walls.

[X] [ ] [s] [c]   13. Verify allowance is made for expansion or contraction of lumber, concrete, masonry, and steel.

[X] [ ] [s] [c]   14. Make sure that unspecified notching, drilling or cutting of framing members have been reviewed for structural adequacy.

Legend: Upper Case Letter and **BOLD** = Primary Inspection; Lower Case = Secondary Inspection
Main Groups SI: Special Inspections, **AI:** Architectural Inspections, **EI:** Engineering Inspections, **OI:** Other Inspections
Sub-Groups   SI: [T] Testing Laboratory, [H] Hazardous Materials, [Y] Safety   **AI:** No Sub-Groups
EI: [V] Civil, [S] Structural, [M] Mechanical, [E] Electrical   **OI:** [O] Owner, [C] Contractor, [B] Subcontractor, [L] Legal, [G] Government, [F] Fire Protection, [P] Plumbing, [A] Acoustical

Construction Inspection Manual

SI AI EI OI

[**X**] [ ] [s] [c]   15. Verify bridging, blocks and bracing are provided as required and in suitable manner. Verify fire blocking is provided as required.

[x] [ ] [s] [**C**]   16. Verify blocking is provided for, equipment and other features to be attached.

[**X**] [ ] [s] [c]   17. Verify plates are lapped and properly connected.

[x] [ ] [s] [**C**]   18. Verify metal connectors will not protrude or interfere with finish surfaces.

[**X**] [ ] [s] [c]   19. Verify metal connectors are adequately nailed or fastened.

[**X**] [ ] [s] [c]   20. Verify connections to metal are as required.

[**X**] [ ] [s] [c]   21. Verify framing members are doubled where required.

[**X**] [ ] [s] [c]   22. Verify headers are of size required, have proper bearing, and are suitably connected.

[**X**] [x] [s] [c]   23. Verify attention is given to ventilation of lumber and enclosed spaces.

[**X**] [ ] [s] [c]   24. Verify plywood sheathing is applied as specified: grade, dimension, staggering, nailing, blocking, number of plies and no overdriven nails.

[**X**] [x] [s] [c]   25. Verify clearances are provided, such as for chimneys and flues; or other spacing requirements are indicated.

[**X**] [x] [s] [c]   26. Verify furring and grounds are as required, properly aligned and plumb.

[**X**] [ ] [s] [c]   27. Verify all bolts are tight or retightened before closing up.

[**X**] [x] [ ] [c]   28. Verify sealing, especially for acoustical or waterproofing purposes, is provided where required.

[**X**] [x] [ ] [c]   29. Verify sheathing paper air infiltration or vapor barrier is provided as required, installed properly and not damaged.

[**X**] [x] [s] [c]   30. Verify seasoned, preservative-treated, or fire-resistive lumber is identified and is provided where required.

[**X**] [x] [s] [c]   31. Verify termite prevention, such as shields or spacing from earth, or other treatment are provided as required.

[x] [x] [s] [**C**]   32. Ensure agency inspection is provided before closing-up if required.

Legend: Upper Case Letter and **BOLD** = Primary Inspection; Lower Case = Secondary Inspection
  Main Groups SI: Special Inspections, **AI:** Architectural Inspections, **EI:** Engineering Inspections, **OI:** Other Inspections
  Sub-Groups   **SI:** [T] Testing Laboratory, [H] Hazardous Materials, [Y] Safety   **AI:** No Sub-Groups
  EI: [V] Civil, [S] Structural, [M] Mechanical, [E] Electrical   **OI:** [O] Owner, [C] Contractor, [B] Subcontractor, [L] Legal, [G] Government, [F] Fire Protection, [P] Plumbing, [A] Acoustical

DIVISION 6 — WOOD AND PLASTICS — CARPENTRY PREFABRICATED STRUCTURAL WOOD 06170

STANDARDS: AITC (11) CRA (30) NFoPA (64) SPA (85) TPI (92) WCLIB (95) WWPA (97)

<u>SI</u> <u>AI</u> <u>EI</u> <u>OI</u>

[**X**] [ ] [s] [c] 1. Ensure testing laboratory and inspection report are provided before erection if required.

[x] [ ] [s] [**C**] 2. Verify material is handled on site using proper equipment and is suitably stored and protected.

[**X**] [ ] [s] [c] 3. Verify protective covering is provided if required and is not damaged.

[**X**] [ ] [s] [c] 4. Verify type, species, grade and finish are as specified for glue-laminated beams and wood trusses.

[**X**] [ ] [s] [c] 5. Verify water resistant adhesive is used for exterior exposure.

[**X**] [ ] [s] [c] 6. Verify cutting, notching, drilling, and fitting are performed in an acceptable manner.

[**X**] [ ] [s] [c] 7. Verify fastenings and connections are provided as required.

[**X**] [ ] [s] [c] 8. Verify field splices and connections are observed.

[**X**] [ ] [s] [c] 9. Allow no unscheduled drilling or notching without consultant's approval.

[**X**] [ ] [s] [c] 10. Verify all cuts are sealed as required.

[**X**] [ ] [s] [c] 11. Verify exposed work is protected from weather and humidity.

[**Y**] [ ] [s] [c] 12. Verify bracing schedule is provided and adhered to during erection sequence.

[**X**] [x] [s] [c] 13. Verify metal end fittings are furnished and fitted by fabricator of structural laminated timber if required.

Legend: Upper Case Letter and **BOLD** = Primary Inspection; Lower Case = Secondary Inspection
Main Groups SI: Special Inspections, **AI:** Architectural Inspections, **EI:** Engineering Inspections, **OI:** Other Inspections
Sub-Groups   SI: [T] Testing Laboratory, [H] Hazardous Materials, [Y] Safety   **AI:** No Sub-Groups
   EI: [V] Civil, [S] Structural, [M] Mechanical, [E] Electrical   **OI:** [O] Owner, [C] Contractor, [B] Subcontractor,
   [L] Legal, [G] Government, [F] Fire Protection, [P] Plumbing, [A] Acoustical

DIVISION 6 — WOOD AND PLASTICS — FINISH CARPENTRY 06200

STANDARDS: APA (17) AWI (27) AWPA (25) CRA (30) CS (35) DFPA (38) HPMA (46) NEMA (60) VMWA (71) PS (77) RIS (79) SPA (85) SPI (84) SPR (83) UL (93) WIC (99) WRCLA (96) WWPA (97)

SI   AI   EI   OI

[**X**] [ ] [ ] [c]   1. Verify materials delivered to site are of grade, species, type, and sizes approved or specified and are suitably stored. Verify kiln dried materials are not exposed to conditions that affect moisture content.

[**X**] [x] [ ] [c]   2. Inspect for condition of materials — warps, splits, graining and finishing.

[**X**] [x] [s] [c]   3. Verify preservative treatment or backpriming are performed if required. (All wood resting against masonry or concrete usually requires treatment).

[**X**] [x] [s] [c]   4. Observe workmanship — sawing, fitting, splicing, coping, shouldering, and mitering.

[**X**] [x] [s] [c]   5. Observe fastening and that types and methods are understood.

[**X**] [x] [s] [c]   6. Fastenings such as bolts and nails on exterior work are usually required to be corrosion resistant. Verify with requirements.

[**X**] [x] [s] [c]   7. Verify type of nail head, set of nails, exposure of nails, pattern of nails, and puttying are provided as required. Ensure excessive nailing is avoided.

[**X**] [x] [s] [c]   8. Verify material is of length to provide indicated joints, splicing is staggered and avoidance of excessive splicing is observed.

[**X**] [x] [s] [c]   9. Verify on exterior work, metal installation and accessories should be observed for weather tightness.

[**X**] [x] [s] [c]   10. Verify grounds and anchorage provisions are adequate for finish attachment.

[**X**] [x] [s] [c]   11. Verify kerfing and hollow backs are provided if required or indicated.

[**X**] [x] [ ] [c]   12. Verify work is sanded smooth and edges are eased as required.

[**X**] [x] [s] [c]   13. Verify gluing and other means of fastenings are of types or methods required.

Legend: Upper Case Letter and **BOLD** = Primary Inspection; Lower Case = Secondary Inspection
Main Groups SI: Special Inspections, AI: Architectural Inspections, EI: Engineering Inspections, OI: Other Inspections
Sub-Groups   **SI**: [T] Testing Laboratory, [H] Hazardous Materials, [Y] Safety   **AI**: No Sub-Groups
EI: [V] Civil, [S] Structural, [M] Mechanical, [E] Electrical   **OI**: [O] Owner, [C] Contractor, [B] Subcontractor, [L] Legal, [G] Government, [F] Fire Protection, [P] Plumbing, [A] Acoustical

Construction Inspection Manual

SI   **AI**   **EI**   **OI**

[**X**] [x] [s] [c]   14. Splits due to nailing should be pointed out during installation and corrected. So should hammer marks and dents, poor quality material, etc.

[**X**] [x] [ ] [c]   15. Verify sequence of setting base to floor is understood. If shoe mold is required, nailing is to base only.

[**X**] [x] [ ] [c]   16. Verify doors are installed with necessary beveling and uniform tolerances as required for proper operation and good practice. Ensure warped doors are not installed.

[**X**] [x] [ ] [c]   17. Verify top and bottom edges of doors are immediately sealed, stained, painted or otherwise protected.

[**X**] [x] [ ] [c]   18. Verify clearances required for thresholds, carpeting, and weather-stripping, etc., on doors are understood.

[**X**] [x] [ ] [c]   19. Verify scribing or scribe strips are provided as required.

[**X**] [x] [ ] [c]   20. Verify installation of equipment furnished by others is done in accordance with requirements.

Legend: Upper Case Letter and **BOLD** = Primary Inspection; Lower Case = Secondary Inspection
Main Groups SI: Special Inspections, **AI:** Architectural Inspections, **EI:** Engineering Inspections, **OI:** Other Inspections
Sub-Groups   **SI:** [T] Testing Laboratory, [H] Hazardous Materials, [Y] Safety   **AI:** No Sub-Groups
     EI: [V] Civil, [S] Structural, [M] Mechanical, [E] Electrical   **OI:** [O] Owner, [C] Contractor, [B] Subcontractor,
     [L] Legal, [G] Government, [F] Fire Protection, [P] Plumbing, [A] Acoustical

Construction Inspection Manual

DIVISION 6 — CARPENTRY — ARCHITECTURAL WOODWORK 06400

STANDARDS: APA (17) AWI (27) AWPA (25) CRA (30) CS (35) HPMA (46) NPA (67) NWMA (71) UL (9) WIC (99) WWPA (97)

SI AI EI OI

[**X**] [x] [ ] [c] 1. Refer to Section 12300 "manufactured casework" where applicable to this section.

[**X**] [x] [ ] [c] 2. Refer to Section 06200 "Finish Carpentry" where applicable to this section.

[**X**] [x] [ ] [c] 3. Refer to Section 09900 "Paints" and coatings where applicable to this section.

[**X**] [x] [ ] [c] 4. Verify furring and blocking are provided and installed as required to receive materials.

[**X**] [x] [ ] [c] 5. Verify subsurfaces to receive finish materials are as required. Verify moisture is checked as required.

[**X**] [x] [ ] [c] 6. Verify finish materials are not delivered before closing-in building, and interior conditions such as temperature, humidity, and sequence are as required.

[**X**] [x] [ ] [c] 7. Verify materials are of grade, species, treatment, construction, thickness, pattern, finish, matching, and appearance required.

[**X**] [x] [ ] [c] 8. Verify methods of installation and connection are as required. Verify workmanship is adequate, with no tool marks, open joints, or other defects.

[**X**] [x] [ ] [c] 9. Verify surfaces are thoroughly cleaned and finished as required.

[**X**] [x] [ ] [c] 10. Verify surfaces are protected as required.

[**X**] [x] [ ] [c] 11. Verify requirements for access to kitchen counters and lavatory countertops for persons with physical disabilities.

Legend: Upper Case Letter and <u>BOLD</u> = Primary Inspection; Lower Case = Secondary Inspection
Main Groups **SI:** Special Inspections, **AI:** Architectural Inspections, **EI:** Engineering Inspections, **OI:** Other Inspections
Sub-Groups   **SI:** [T] Testing Laboratory, [H] Hazardous Materials, [Y] Safety   **AI:** No Sub-Groups
  EI: [V] Civil, [S] Structural, [M] Mechanical, [E] Electrical   **OI:** [O] Owner, [C] Contractor, [B] Subcontractor,
  [L] Legal, [G] Government, [F] Fire Protection, [P] Plumbing, [A] Acoustical

DIVISION 7 — THERMAL AND MOISTURE PROTECTION — WATERPROOFING
07100

STANDARDS: FS (41)

SI AI EI OI

[ ] **[X]** [ ] **[C]**   1. Where required by contract documents or warranties, verify that waterproofing is installed by manufacturer's approved installers.

[x] [ ] [ ] **[C]**   2. Before waterproofing or dampproofing subcontractor is allowed to commence work, see that:

   a. Surfaces are free from foreign material.
   b. Excess mortar or concrete is removed; all holes, joints and cracks are pointed, and rough or high spots are ground smooth.
   c. Wood nailers or other attachment conditions are adequate.
   d. Surfaces are dry to receive heated asphalt, coal tar or other membrane. Check for dampness if necessary.
   e. Special conditions are provided as required at corners, intersections, and connections to existing works.

[x] [ ] [ ] **[C]**   3. Installation should not proceed if temperature is below 40°F or weather is damp or foggy.

[x] [ ] [ ] **[C]**   4. Hot-applied bituminous materials must comply with manufacturer recommended EVT. If asphalt is being used, heated requirements is EVT, plus or minus 25°F, at point of application. (EVT [Equiviscous Temperature] is the temperature at which asphalt will attain a viscosity of 125 centistokes which is the practical and optimum temperature for wetting and fusion at the point of application). In the event EVT information is not furnished by the manufacturer, the following maximum heating temperatures may be used as guidelines:

   Dead Level Asphalt .... Type I 475°F maximum
   Flat Grade Asphalt .... Type II 500°F maximum
   Steep Grade Asphalt .... Type III 525°F maximum
   Special Steep Asphalt ...... Type IV 525°F maximum

   In no case should kettle or tanker be heated above flash point. Observe kettle temperature. Verify method and sequence of transporting and application of bituminous materials are appropriate.

[ ] [ ] [ ] **[C]**   5. Verify surfaces are primed, if required.

Legend: Upper Case Letter and **BOLD** = Primary Inspection; Lower Case = Secondary Inspection
  Main Groups SI: Special Inspections, **AI:** Architectural Inspections, **EI:** Engineering Inspections, **OI:** Other Inspections
  Sub-Groups   **SI:** [T] Testing Laboratory, [H] Hazardous Materials, [Y] Safety   **AI:** No Sub-Groups
    EI: [V] Civil, [S] Structural, [M] Mechanical, [E] Electrical   **OI:** [O] Owner, [C] Contractor, [B] Subcontractor,
    [L] Legal, [G] Government, [F] Fire Protection, [P] Plumbing, [A] Acoustical

SI AI EI OI

[x] [ ] [ ] [C]   6. Ensure pipes, ducts, conduits, and other items penetrating membrane are watertight.

[x] [ ] [ ] [C]   7. Observe appropriateness of changes in planes membrane lap inside and outside corners and that perimeter secured and flashed as required.

[ ] [ ] [ ] [C]   8. Verify joinery between each day's work is adequate.

[ ] [ ] [ ] [C]   9. Verify that interstitial moisture is not present .

[x] [ ] [ ] [C]   10. Verify stored materials are protected against moisture.

[ ] [ ] [ ] [C]   11. Verify proper nails, adhesives and fastenings are used.

[**X**] [ ] [ ] [ ]   12. Verify proper coverage and quantities of materials such as mil thickness.

[ ] [ ] [ ] [C]   13. Verify membrane is applied smooth with no "fish mouths" or buckles.

[ ] [ ] [ ] [C]   14. Verify protective covering is provided and installed as required and backfilling takes place immediately and the covering remains in place during backfilling.

[ ] [ ] [ ] [C]   15. Ensure installation is protected from damage by other trades or by general contractor during installation and following completion. If subject to heavy traffic, movement of equipment, plywood sheets or other protection is provided.

Legend: Upper Case Letter and BOLD = Primary Inspection;   Lower Case = Secondary Inspection
Main Groups SI: Special Inspections, **AI:** Architectural Inspections, **EI:** Engineering Inspections, **OI:** Other Inspections
Sub-Groups   SI: [T] Testing Laboratory, [H] Hazardous Materials, [Y] Safety   **AI:** No Sub-Groups
EI: [V] Civil, [S] Structural, [M] Mechanical, [E] Electrical   **OI:** [O] Owner, [C] Contractor, [B] Subcontractor,
[L] Legal, [G] Government, [F] Fire Protection, [P] Plumbing, [A] Acoustical

DIVISION 7 — THERMAL AND MOISTURE PROTECTION — INSULATION 07200

STANDARDS: FM (40) NFPA (63) PI (72) UL (93) VI (94)

<u>SI</u> <u>AI</u> <u>EI</u> <u>OI</u>

[ ] [x] [ ] [C] 1. Verify materials are of type required (surfaces, treatment, ratings, sizes, thickness, etc.)

[x] [x] [ ] [C] 2. Ensure materials are stored to prevent moisture infiltration are installed dry, and remain dry until covered by rooting.

Rigid Insulation

[ ] [ ] [ ] [C] 3. Verify wood nailers/stops are provided as required.

[ ] [ ] [ ] [C] 4. Ensure vapor barrier is provided if required. Check that vapor barrier seals insulation at gravel stops, walls, and openings.

[ ] [ ] [ ] [C] 5. Verify method of installation.

[x] [ ] [ ] [C] 6. Verify fasteners, when specified, are of proper type and spaced as required.

[x] [ ] [ ] [C] 7. Verify joints are staggered, except when joints are to be taped. When two layers are being installed, vertical joints should be offset. Do not allow joints over flute openings in steel deck.

[ ] [ ] [ ] [C] 8. Insure insulation is covered by roofing each day.

[ ] [ ] [ ] [C] 9. Ensure the components of fire-rated assemblies are in compliance with fire ratings.

[x] [ ] [ ] [C] 10. Verify expansion provisions are observed.

Fibrous and Reflective Insulation

[x] [ ] [ ] [C] 11. Verify batt insulation is of adequate width to fit between framing members and is of specified thickness or "R" value.

[x] [ ] [ ] [C] 12. Verify facing/vapor barrier is provided where required.

[ ] [ ] [ ] [C] 13. Verify insulation is properly secured.

Legend: Upper Case Letter and <u>BOLD</u> = Primary Inspection; Lower Case = Secondary Inspection
Main Groups SI: Special Inspections, **AI:** Architectural Inspections, **EI:** Engineering Inspections, **OI:** Other Inspections
Sub-Groups   **SI:** [T] Testing Laboratory, [H] Hazardous Materials, [Y] Safety   **AI:** No Sub-Groups
EI: [V] Civil, [S] Structural, [M] Mechanical, [E] Electrical   **OI:** [O] Owner, [C] Contractor, [B] Subcontractor, [L] Legal, [G] Government, [F] Fire Protection, [P] Plumbing, [A] Acoustical

SI   AI   EI   OI

[x] [ ] [ ] [C]   14. Verify fitting and cutting around penetrations is tight and surfaces are not torn or damaged.

[x] [ ] [ ] [C]   15. Verify ventilation and airspace are maintained open and free.

[ ] [ ] [ ] [C]   16. Verify reflective vapor barrier material is on correct warm side (typically inside except at refrigerated spaces).

Loose Fill Insulation

[x] [ ] [ ] [C]   17. Verify there is no settlement. Verify there are no voids.

[x] [ ] [ ] [C]   18. Verify horizontal surfaces are filled to required depth.

Legend: Upper Case Letter and <u>BOLD</u> = Primary Inspection;   Lower Case = Secondary Inspection
   Main Groups SI: Special Inspections, **AI:** Architectural Inspections,   **EI:** Engineering Inspections,   **OI:** Other Inspections
   Sub-Groups   SI: [T] Testing Laboratory, [H] Hazardous Materials, [Y] Safety   **AI:** No Sub-Groups
      EI: [V] Civil, [S] Structural, [M] Mechanical, [E] Electrical   **OI:** [O] Owner, [C] Contractor, [B] Subcontractor,
      [L] Legal, [G] Government, [F] Fire Protection, [P] Plumbing, [A] Acoustical

DIVISION 7 — THERMAL AND MOISTURE PROTECTION — SHINGLES AND ROOFING TILES 07300

STANDARDS: RCSHSB (78) UL (93)

SI AI EI OI

[ ] [ ] [ ] **[C]** 1. Delivered materials are of grades, types, shapes, sizes, colors, fire-rated classification, and pattern texture required as applicable.

[ ] [ ] [ ] **[B]** 2. Carefully check subsurfaces and decking before installation and have corrected as required. Knotholes and splits should be covered.

Wood Shingles and Shakes

[ ] [x] [ ] **[B]** 3. Verify underpayment where required is provided and installed with proper lapping shingle style.

[ ] [x] [ ] **[C]** 4. Verify exposure and fastening as required, layout to provide full shingle courses, and overall alignment is controlled. Review pattern is indicated.

[ ] [ ] [ ] **[B]** [c] 5. Verify starter courses provided as required. Verify gutter installed if of type attached to decking.

[ ] [ ] [ ] **[B]** [c] 6. Verify construction methods: valley construction — open or closed; hip, ridge and venting construction.

[ ] [x] [ ] **[C]** 7. Check flashings to be installed in conjunction with shingling.

[ ] [x] [ ] **[C]** 8. Shingles narrower than 3 inches, shakes narrower than 4 inches, and shingle and shakes wider than 14 inches should not be installed unless permitted by code. Verify compliance.

[ ] [x] [ ] **[C]** 9. Verify that all jointing is as required; nailing is as required. Observe that nails do not penetrate underside of exposed decking.

[ ] [ ] [ ] **[C]** [b] 10. Verify slope is as required. If roof surfaces do not have sufficient slope, contact architect before installation of roofing starts.

[ ] [x] [ ] **[C]** 11. Verify interlay is performed for shake installation as required.

[ ] [x] [ ] **[C]** 12. Verify details of fire treatment or fire retardant spray coatings are appropriate for fire-resistive assemblies.

Legend: Upper Case Letter and **BOLD** = Primary Inspection; Lower Case = Secondary Inspection
Main Groups **SI:** Special Inspections, **AI:** Architectural Inspections, **EI:** Engineering Inspections, **OI:** Other Inspections
Sub-Groups **SI:** [T] Testing Laboratory, [H] Hazardous Materials, [Y] Safety   **AI:** No Sub-Groups
EI: [V] Civil, [S] Structural, [M] Mechanical, [E] Electrical   **OI:** [O] Owner, [C] Contractor, [B] Subcontractor, [L] Legal, [G] Government, [F] Fire Protection, [P] Plumbing, [A] Acoustical

Construction Inspection Manual

Asphalt Shingles

SI AI EI OI

[ ] [x] [ ] [ ] [C]   13. Verify underlayment or membrane is provided if required.

[ ] [ ] [ ] [ ] [C]
                [b]   14. Verify slope is as required. If roof surfaces do not have sufficient slope, contact architect before installation of roofing starts.

[ ] [ ] [ ] [ ] [B]   15. Verify temperature is over 40°F.

[ ] [ ] [ ] [ ] [C]
                [b]   16. Verify flashings are installed or on hand to be installed in conjunction with roofing.

[ ] [ ] [ ] [ ] [C]
                [b]   17. Verify starter courses are provided as required. Verify gutter is installed if of type attached to decking.

[ ] [ ] [ ] [ ] [C]
                [b]   18. Verify alignment of layers and rows is maintained. Verify overall layout is made to provide full shingle courses.

[ ] [x] [ ] [ ] [C]   19. Verify exposure required is provided and nailing is as required.

[ ] [ ] [ ] [ ] [B]   20. Verify tabbing of shingles is performed as required. Verify self-sealing shingles are installed as required.

Slate Shingles

[ ] [x] [ ] [ ] [B]
                [c]   21. See items above where applicable.

[ ] [x] [ ] [ ] [C]   22. Verify bedding of slaters is performed at valleys, ridges, chimneys, and dormers as required.

Concrete and Clay Tile

[ ] [x] [ ] [ ] [B]   23. Verify underlayment where required is provided and installed with proper lapping shingle style.

[ ] [x] [ ] [ ] [C]   24. Verify exposure and fastening as required. Layout to provide tile courses, and overall alignment is controlled. Review patterns if indicated.

[ ] [ ] [ ] [ ] [C]
                [b]   25. Verify starter courses provided as required. Verify gutter installed if of type attached to decking.

Legend: Upper Case Letter and <u>BOLD</u> = Primary Inspection; Lower Case = Secondary Inspection
Main Groups SI: Special Inspections, **AI:** Architectural Inspections, **EI:** Engineering Inspections, **OI:** Other Inspections
Sub-Groups   **SI:** [T] Testing Laboratory, [H] Hazardous Materials, [Y] Safety   **AI:** No Sub-Groups
    EI: [V] Civil, [S] Structural, [M] Mechanical, [E] Electrical   **OI:** [O] Owner, [C] Contractor, [B] Subcontractor,
    [L] Legal, [G] Government, [F] Fire Protection, [P] Plumbing, [A] Acoustical

SI AI EI OI

[ ] [ ] [ ] [C] 26. Verify construction method is appropriate and as specified Valley
        [b]   construction is open or closed; hip and ridge construction. Check manufacturer's specification or code enforcement reports for requirements for mortar at hips/ridges. If required, check for cement/sand ratio and color requirement.

[ ] [x] [ ] [C] 27. Verify flashings to be installed in conjunction with laying of the tile. Some tile configurations may require lead flashings or flashings made from other flexible material.

[ ] [x] [ ] [C] 28. Verify all jointing as required; fastening is as required. Observe that nails do not penetrate underside of exposed decking. Exposed decking should be minimum of 1 inch (net) in thickness to provide adequate base to receive nails. Code requires nails to be driven into deck minimum ¾ inch, or through the sheathing if less than ¾ inch thick, verify with manufacturer's specifications or code reports.

[ ] [ ] [ ] [C] 29. Verify slope is as required. If roof surfaces do not have sufficient slope,
        [b]   contact architect before installation of roofing starts.

[ ] [ ] [ ] [C] 30. If wood stripping is to be installed over sheathing for purposes of
        [b]   anchoring tile, check for type (i.e. redwood, D.F., treated D.F., etc.), securement to sheathing and spacing which controls tile exposure.

[ ] [ ] [ ] [c] 31. Before roofing contractor is allowed to commence work
        [b]

           a. If tile to be installed over open-spaced sheathing, check spacing; and
           b. If nailers required at hips, ridges, eaves, check size and attachment.

[ ] [x] [ ] [C] 32. Verify proper fastener is provided for the tile being used.

[ ] [x] [ ] [O] 33. Inspect for broken or missing tile promptly as soon as roofing
        [b]   installation is completed. Report to general contractor and roofing
        [c]   contractor so that final repairs can be made before roofing contractor leaves jobsite. Protection of finished roof is the responsibility of the general contractor.

Legend: Upper Case Letter and **BOLD** = Primary Inspection; Lower Case = Secondary Inspection
Main Groups SI: Special Inspections, **AI:** Architectural Inspections, **EI:** Engineering Inspections, **OI:** Other Inspections
Sub-Groups SI: [T] Testing Laboratory, [H] Hazardous Materials, [Y] Safety   **AI:** No Sub-Groups
    EI: [V] Civil, [S] Structural, [M] Mechanical, [E] Electrical   **OI:** [O] Owner, [C] Contractor, [B] Subcontractor,
    [L] Legal, [G] Government, [F] Fire Protection, [P] Plumbing, [A] Acoustical

Construction Inspection Manual

125

DIVISION 7 — THERMAL AND MOISTURE PROTECTION — MEMBRANE ROOFING 07500

STANDARDS: ASTM (19) FS (41) FM (40)

SI AI EI OI

[ ] [ ] [ ] [ ] 1. Refer to Section 07300 shingles, roof tiles, and roof covering and Section 07600 "Flashing and Sheet Metal."

[ ] [x] [ ] **[B]** **[C]** 2. Before roofing contractor is allowed to commence work verify that:
    a. Surfaces are free from foreign material;
    b. Excess mortar or concrete is removed; all holes, joints and cracks are pointed, and rough or high spots are ground smooth;
    c. Wood nailers or other attachment conditions are adequate;
    d. Surfaces are dry to receive membrane heated asphalt, coal tar, and petroleum solvent asphalt mastics. Surfaces are tested for dampness if necessary;
    e. Slope is as required. If roof surface does not have sufficient slope, contact architect;
    f. Pipes, conduits and other items penetrating the membrane are in place and ready to receive flashings; and
    g. All sheet metal and roof accessories are in place or on hand to be installed in conjunction with roofing as required.

[ ] [ ] [ ] **[C]** [b] 3. Verify materials of types required are provided. Verify softening point of bitumen is as required. Verify materials are identifiable and comply with ASTM or FS standards. Verify roll roofing is stood on end and kept free of contact with earth or moisture. Verify protective coverings of stored roll roofing is vented so condensation will not occur.

[ ] [ ] [ ] **[B]** [c] 4. Verify nails and fasteners are of length, shank, head, and coating as required.

[ ] [ ] [ ] **[C]** 5. Verify felts for use with asphalt are asphalt-saturated; felts for use with coal tar pitch are coal-tar saturated.

[ ] [ ] [ ] **[B]** [c] 6. Verify surface to receive roofing is primed or other wise prepared if required.

Legend: Upper Case Letter and <u>BOLD</u> = Primary Inspection; Lower Case = Secondary Inspection
Main Groups SI: Special Inspections, **AI:** Architectural Inspections, **EI:** Engineering Inspections, **OI:** Other Inspections
Sub-Groups **SI:** [T] Testing Laboratory, [H] Hazardous Materials, [Y] Safety   **AI:** No Sub-Groups
    EI: [V] Civil, [S] Structural, [M] Mechanical, [E] Electrical   **OI:** [O] Owner, [C] Contractor, [B] Subcontractor,
        [L] Legal, [G] Government, [F] Fire Protection, [P] Plumbing, [A] Acoustical

SI AI EI OI

[ ] [ ] [ ] **[B]** 7: Verify asphalt or pitch is not overheated. Check kettle thermometer.
        [c]   Verify methods to transport heated material are provided to avoid overcooling. Measure installation temperature. If asphalt is being used, heated requirements is EVT, plus or minus 25°F, at point of application. (EVT [Equiviscous Temperature] is the temperature, at which asphalt will attain a viscosity of 125 centistokes which is the practical and optimum temperature for wetting and fusion at the point of application). In the event EVT information is not furnished by the manufacturer, the following maximum heating temperatures should be used as guidelines:

    Dead Level Asphalt .... Type I 475°F maximum
    Flat Grade Asphalt .... Type II 500°F maximum
    Steep Grade Asphalt .... Tine III 525°F maximum
    Special Steep Asphalt ...... Type IV 525°F maximum

In no case should kettle or tanker be heated above flash point. Final blowing temperature should not be exceeded for more than four hours.

[ ] [ ] [ ] **[B]** 8. Roofing materials should not be applied unless correct bitumen application temperatures can be maintained. Ensure that the correct temperatures are maintained.
        [c]

[ ] [ ] [ ] **[B]** 9. Observe lap, mailing, and quantity of pitch or asphalt applied. In no case should felt touch felt; no bare spots.
        [c]

[ ] [ ] [ ] **[B]** 10. See that felts are laid so that each layer is free of air pockets, wrinkles, and buckles. Brooming may be required. Glass fiber felts should not be broomed. Do not allow "flopping" of roofing felts, except in the application of cap sheets. See that no felt touches felt. Verify mopping is full to ply lines.
        [c]

[ ] [ ] [ ] **[B]** 11. Verify all surfaces are kept moisture-free. Under no condition allow exposure of insulation or felts overnight without a mopping. Verify stored material from moisture.
        [c]

[ ] [x] [ ] **[B]** 12. When felt layer equipment is used, observe that jets are clear and an adequate and uniform layer of bitumen is deposited.
        [c]

[ ] [x] [ ] **[B]** 13. Observe installation of roofing at cant strips, vertical surfaces, reglets, and penetration. Observe sealing of roofing membrane envelopes where use of envelope is required.
        [c]

Legend: Upper Case Letter and **BOLD** = Primary Inspection; Lower Case = Secondary Inspection
Main Groups SI: Special Inspections, **AI:** Architectural Inspections, **EI:** Engineering Inspections, **OI:** Other Inspections
Sub-Groups SI: [T] Testing Laboratory, [H] Hazardous Materials, [Y] Safety   **AI:** No Sub-Groups
EI: [V] Civil, [S] Structural, [M] Mechanical, [E] Electrical   **OI:** [O] Owner, [C] Contractor, [B] Subcontractor, [L] Legal, [G] Government, [F] Fire Protection, [P] Plumbing, [A] Acoustical

Construction Inspection Manual

SI AI EI OI

[ ] [x] [ ] [C] 14. Verify concrete walls to receive roofing are primed. Verify wall membranes are properly prepared and attached or fastened as specified.

[ ] [ ] [ ] [B] 15. Observe aggregates for surfacing of type, color and size, specified, clean and dry.

[ ] [ ] [ ] [B] 16. Verify aggregates in quantity required is spread over flood coat while bitumen is hot.

[ ] [x] [ ] [B] [c] 17. Verify roll roofing or cap sheet, if utilized, is of weight, selvage, finish, color, as required. Verify cap sheet installed as required.

[ ] [x] [ ] [C] [b] 18. Verify operations are performed in a manner to avoid plugging of drains, and weeps and do not damage or interfere with adjoining surfaces.

[ ] [x] [ ] [C] 19. Observe that roof drains are set to permit proper drainage.

[ ] [ ] [ ] [C] [b] 20. Verify roofing plies are mopped into clamping ring. Verify lead collar flashing is installed and stripped in, if required.

[ ] [x] [ ] [C] [b] 21. Verify roofing is protected from damage by other trades or by general contractor during installation and following completion. If subject to heavy traffic, movement of equipment, storage or materials, or used as a work surface, verify that runways, plywood sheets or other protection is provided.

[ ] [x] [ ] [B] [c] 22. Observe and/or cut samples if required. Verify patching is properly performed where samples are cut. Samples are to be taken before finish surface (aggregate, cap sheet, emulsion) is applied.

[ ] [x] [ ] [C] 23. Verify clean-up is provided after installation, drains cleared and debris is removed from site.

Legend: Upper Case Letter and **BOLD** = Primary Inspection; Lower Case = Secondary Inspection
Main Groups SI: Special Inspections, **AI:** Architectural Inspections, **EI:** Engineering Inspections, **OI:** Other Inspections
Sub-Groups **SI:** [T] Testing Laboratory, [H] Hazardous Materials, [Y] Safety   **AI:** No Sub-Groups
EI: [V] Civil, [S] Structural, [M] Mechanical, [E] Electrical   **OI:** [O] Owner, [C] Contractor, [B] Subcontractor, [L] Legal, [G] Government, [F] Fire Protection, [P] Plumbing, [A] Acoustical

DIVISION 7 — THERMAL AND MOISTURE PROTECTION — FLASHING AND SHEET METAL 07600

STANDARDS: ASTM (19) CDA (37) SMACNA (82)

SI	AI	EI	OI	
[ ]	[ ]	[ ]	[C] [b]	1. Verify delivered material is of approved type, shape, gauge, metal, fabrication, and priming, as required, and all accessories are provided.
[ ]	[x]	[ ]	[C] [b]	2. Verify isolation provisions are made for dissimilar metals. Do not allow copper and aluminum flashings to be in contact with each other or with ferrous metal. Copper or aluminum flashings are to be fastened with non-ferrous nails or screws. Ferrous equipment bases are not to be set on copper flashings. Verify flanges embedded in plastic cement or asphalt are asphalt primed.
[ ]	[x]	[ ]	[C] [b]	3. Verify expansion joints are provided and installed as required or as specified. Note location of joints with respect to drains, downspouts, scuppers, corners, and other outlets.
[X]	[x]	[ ]	[ ]	4. Observe methods of installation — nailing and cleating types for spacing and location; also soldering, welding, bolting, and riveting.
[ ]	[x]	[ ]	[C]	5. Verify flashing does not interfere with structural requirements.
[x]	[x]	[ ]	[B]	6. Generally see that edge metal is lapped a minimum of 4 inches with 12 inches staggered nailing or fastening through the back flange unless otherwise required.
[ ]	[x]	[ ]	[B] [c]	7. Verify all edge metal laps are coated with sealant on horizontal flange and vertical rise. Verify coating covers entire lap and is sandwiched between.
[ ]	[x]	[ ]	[C] [b]	8. Verify lengths are as long as practical or specified.
[ ]	[x]	[ ]	[C]	9. Verify installation is coordinated with roofing and/or siding installation.
[ ]	[ ]	[ ]	[C] [b]	10. Verify nailer or cant strip is provided for fastening flashing to roof deck; is of proper material, well secured, and allows venting if required or specified.
[ ]	[x]	[ ]	[C] [b]	11. Verify flashing is embedded and installed over roof membrane assembly with additional roofing membrane material.
[x]	[ ]	[ ]	[C]	12. Verify method of anchoring lower edge of fascia is as required. Observe alignment, and stiffness.

Legend: Upper Case Letter and **BOLD** = Primary Inspection; Lower Case = Secondary Inspection
Main Groups SI: Special Inspections, **AI:** Architectural Inspections, **EI:** Engineering Inspections, **OI:** Other Inspections
Sub-Groups **SI:** [T] Testing Laboratory, [H] Hazardous Materials, [Y] Safety   **AI:** No Sub-Groups
EI: [V] Civil, [S] Structural, [M] Mechanical, [E] Electrical   **OI:** [O] Owner, [C] Contractor, [B] Subcontractor, [L] Legal, [G] Government, [F] Fire Protection, [P] Plumbing, [A] Acoustical

Construction Inspection Manual

SI   AI   EI   OI

[ ] [x] [ ] [C]   13. Verify gravel stops are flush with deck unless otherwise required.

Gutters

[x] [x] [ ] **[B]**
            [c]   14. Verify gutters are adequately supported and allow for movement. Observe attachment size, type, location, and spacing of hangers and supports.

[ ] [x] [ ] [C]   15. Verify gutters are pitched if required and provide for drainage to outlets. Factors of settlement of wood and concrete cantilevered overhangs sometimes affect drainage pattern.

[x] [x] [ ] [C]   16. Verify gutter joints are lapped in direction of flow.

[x] [x] [ ] [C]   17. Verify expansion joints, concealed or standing, are provided midway between outlets or downspouts and/or as required.

[x] [x] [ ] [C]   18. Verify scuppers are installed low enough not to dam water on roof. Verify overflow drains and scuppers if indicated or required by code are provided, located properly, i.e., low point of roof, are of size required and have correct inlet flow elevation.

[ ] [x] [ ] [C]
            [b]   19. Verify accessories are provided if required — basket strainer, bird screens, and covers.

Downspouts

[ ] [x] [ ] [C]   20. Verify lengths are as long as practical and in accordance with specifications.

[ ] [x] [ ] [C]   21. Verify slip joints in direction of flow or allowance for movement is provided.

[ ] [x] [ ] [C]
            [b]   22. Verify hangers or straps as required are provided. Verify spacing and location are as required or specified, and each section is supported. Connection of hangers does not damage finish wall material.

[ ] [x] [ ] [C]
            [b]   23. Verify contact is not made with wall surfaces except for supports.

[ ] [x] [ ] [C]   24. Verify downspouts are installed plumb, without excessive lateral or angled joints, unless indicated or if required to conduct drainage.

[ ] [x] [ ] [C]   25. Verify special items are furnished: heads, scuppers, and linings.

Legend: Upper Case Letter and **BOLD** = Primary Inspection; Lower Case = Secondary Inspection
  Main Groups   **SI:** Special Inspections, **AI:** Architectural Inspections, **EI:** Engineering Inspections, **OI:** Other Inspections
  Sub-Groups   **SI:** [T] Testing Laboratory, [H] Hazardous Materials, [Y] Safety   **AI:** No Sub-Groups
    EI: [V] Civil, [S] Structural, [M] Mechanical, [E] Electrical   **OI:** [O] Owner, [C] Contractor, [B] Subcontractor,
    [L] Legal, [G] Government, [F] Fire Protection, [P] Plumbing, [A] Acoustical

SI AI EI OI

[ ] [x] [ ] [B] 26. Verify downspouts that are indicated to terminate in drainage lines are nearly fitted and are cleaned and free of building debris or other materials.

Base and Cap Flashings

[ ] [x] [ ] [C] 27. Verify flashing is provided to suit conditions; cant, size, gauge, and fabrication.

[ ] [x] [ ] [C]
           [b] 28. Verify base flashing extends up sufficiently; flange is properly secured and embedded at least 4 inches in roofing membrane and is installed similarly to gravel stops. Verify mopped felt or suitable membrane covering flashings or cleats is provided. It is good practice to cover as much metal as practical to avoid movement from temperature variations.

[ ] [x] [ ] [C] 29. Verify seams are lapped, locked, and soldered as required.

[ ] [x] [ ] [C]
           [b] 30. Verify secure anchorage is provided for size, spacing, and fixing of cleats or other equipment mountings.

[ ] [x] [ ] [C]
           [b] 31. Verify cap flashings are of shapes, sizes, and gauges required and are installed to provide secure anchorage, allow movement, and have sufficient laps and spacing.

[ ] [x] [ ] [B]
           [b] 32. Verify counter flashing is extended sufficiently into masonry walls or into reglet and is securely anchored and caulked, if necessary.

Other Roof Flashing

[ ] [x] [ ] [C]
           [b] 33. Verify hip and ridge flashing and venting is provided as required. Check fabrication, size, gauge, anchorage, and lap. Observe caulking and painting procedures.

[ ] [x] [ ] [C]
           [b] 34. Verify valley flashing is provided as required: open or closed, width, gauge, anchorage, and lap.

[ ] [x] [ ] [C] 35. Verify stepped flashing is provided as required: depth of insertion into wall, and length of material attached to deck and lap. Verify plastic cement or approved material is filled into joints between edges of shingles and flashing as required.

Legend: Upper Case Letter and **BOLD** = Primary Inspection; Lower Case = Secondary Inspection
Main Groups SI: Special Inspections, AI: Architectural Inspections, EI: Engineering Inspections, OI: Other Inspections
Sub-Groups   **SI**: [T] Testing Laboratory, [H] Hazardous Materials, [Y] Safety   **AI**: No Sub-Groups
    EI: [V] Civil, [S] Structural, [M] Mechanical, [E] Electrical   **OI**: [O] Owner, [C] Contractor, [B] Subcontractor,
       [L] Legal, [G] Government, [F] Fire Protection, [P] Plumbing, [A] Acoustical

Construction Inspection Manual

SI AI EI OI

[ ] [x] [ ] [C] 36. Verify reglets are provided at required areas: observe the setting in concrete or masonry to assure firm anchorage. Verify reglets are protected to prevent deformation or filling during installation.

[ ] [x] [ ] [C] 37. Observe installation of sheet metal into reglets for tightness, weatherproofness, caulking, and lap.

Wall and Through Wall Flashing

[ ] [x] [ ] [B]
           [c] 38. Verify locations for flashings fabrication and design with contractor.

[ ] [x] [ ] [C] 39. Verify lap, turn up, location in wall, depth in masonry, length, etc., are as required.

[ ] [x] [ ] [C]
           [b] 40. Verify sill flashing and pans extend full depth, are turned up, extend beyond horns or 4 inches, and are installed for proper drainage.

Miscellaneous

[ ] [x] [ ] [C] 41. Verify louvers and vents have adequate flanges and connections for anchorage and flashings are watertight against driving rains after installation. Verify insect screen, bird screen, and shutters are provided as required.

[ ] [ ] [ ] [C]
           [b] 42. Review drawings and specifications for sheet metal items.

[ ] [x] [ ] [C]
           [b] 43. Items such as skylight, roof, hatches, and fans may be suited for installation with or without flashing. Verify installation meets manufacturers' specifications when required.

[ ] [x] [ ] [C]
           [b] 44. Verify plastic flashing is of type required and is installed in accordance with requirements.

[ ] [x] [ ] [C] 45. Verify sheet metal termite shields are provided as required.

Legend: Upper Case Letter and **BOLD** = Primary Inspection; Lower Case = Secondary Inspection
Main Groups SI: Special Inspections, **AI:** Architectural Inspections, **EI:** Engineering Inspections, **OI:** Other Inspections
Sub-Groups SI: [T] Testing Laboratory, [H] Hazardous Materials, [Y] Safety   **AI:** No Sub-Groups
    EI: [V] Civil, [S] Structural, [M] Mechanical, [E] Electrical   **OI:** [O] Owner, [C] Contractor, [B] Subcontractor,
    [L] Legal, [G] Government, [F] Fire Protection, [P] Plumbing, [A] Acoustical

DIVISION 7 — THERMAL AND MOISTURE PROTECTION — CEMENTITIOUS FIREPROOFING 07810

STANDARDS: UBC, ASTM (19), AWCI (26-A)

SI AI EI OI

[ ] [X] [ ] [C] 1. Verify installation is by a firm approved by the sprayed fireproofing material manufacturer.

[T] [ ] [ ] [C] 2. Verify fireproofing provides fire-resistance ratings as tested by appropriate laboratories.

[ ] [ ] [ ] [C] 3. Verify manufacturers' data and instructions for proper application of sprayed fireproofing are on file.

[T] [ ] [ ] [C] 4. Verify laboratory test results for fireproofing are on file for performance criteria such as:
   A. Bond Strength.
   B. Compressive Strength.
   C. Deflection.
   D. Bond Impact.
   E. Air Erosion.
   F. Corrosion Resistance.

[ ] [ ] [ ] [C] 5. Verify material is delivered in original unopened packages, fully identified as to manufacturer, brand or other identifying data, and bears the proper Underwriters' Laboratories, Inc. labels for fire hazards and fire-resistance classification.

[ ] [ ] [ ] [C] 6. Verify material is stored (above ground), under cover and in a dry location until ready for use. Ensure bags exposed to water before use are discarded as unsuitable.

[ ] [ ] [ ] [C] 7. Verify temperature range is within manufacturer's recommended range.

[ ] [ ] [ ] [C] 8. Verify ventilation for proper drying of the fireproofing.

[ ] [ ] [ ] [B] 9. Verify surfaces to receive sprayed fireproofing so that they are free of oil, grease, paints/primers, loose mill scale, dirt or other foreign substances which may impair proper adhesion of the fireproofing to the substrate.

[ ] [ ] [ ] [C] 10. Ensure prior to application of fireproofing, clips, hangers, support sleeves and other attachments required to penetrate the fireproofing are in place.

Legend: Upper Case Letter and **BOLD** = Primary Inspection; Lower Case = Secondary Inspection
Main Groups SI: Special Inspections, **AI:** Architectural Inspections, **EI:** Engineering Inspections, **OI:** Other Inspections
Sub-Groups   **SI:** [T] Testing Laboratory, [H] Hazardous Materials, [Y] Safety   **AI:** No Sub-Groups
EI: [V] Civil, [S] Structural, [M] Mechanical, [E] Electrical   **OI:** [O] Owner, [C] Contractor, [B] Subcontractor, [L] Legal, [G] Government, [F] Fire Protection, [P] Plumbing, [A] Acoustical

Construction Inspection Manual

SI AI EI OI

[ ] [ ] [ ] **[C]** 11. Verify that prior to application of the fireproofing to the underside of metal roof decks, roofing applications are completed. Ensure roof traffic is prohibited upon commencement of the fireproofing application and until the fireproofing material is cured and fully dried.

[T] [ ] [ ] [ ] 12. Ensure an independent testing laboratory is to sampling and verifying the thickness and density of the fireproofing.

[x] [ ] [ ] **[C]** 13. Check final condition of fireproofing after the work of other construction trades is completed.

Legend: Upper Case Letter and <u>BOLD</u> = Primary Inspection; Lower Case = Secondary Inspection
Main Groups SI: Special Inspections, **AI:** Architectural Inspections, **EI:** Engineering Inspections, **OI:** Other Inspections
Sub-Groups **SI:** [T] Testing Laboratory, [H] Hazardous Materials, [Y] Safety   **AI:** No Sub-Groups
EI: [V] Civil, [S] Structural, [M] Mechanical, [E] Electrical   **OI:** [O] Owner, [C] Contractor, [B] Subcontractor,
[L] Legal, [G] Government, [F] Fire Protection, [P] Plumbing, [A] Acoustical

DIVISION 8 — DOORS AND WINDOWS — METAL DOORS AND FRAMES 08100

STANDARDS: AA (2) AAMA (26) AInA (12) ANSI (16) ASTM (19) CS (35) FM (40) NAAMA (54) NFoPA (64) SDI (87) SSPC (89) UL (93)

Doors

<u>SI</u> <u>AI</u> <u>EI</u> <u>OI</u>

[ ] [x] [ ] [C] 1. Verify doors are as approved — size, type design, panel, lights, louvers, and features — and have no defects such as dents, buckles, or warps.

[ ] [x] [ ] [C] 2. Verify fabrication, construction, and workmanship: look for smooth edges, and joints, finish, and straightness.

[ ] [x] [ ] [C] 3. Verify additional reinforcement is provided for hardware. Observe backing plates during drilling operations. Observe that closure channels are provided as required.

[ ] [x] [ ] [C] 4. Verify provisions to receive hardware required are adequate. Observe type and installation of factory-applied hardware. Verify backset is matched to finish hardware.

[ ] [x] [ ] [C] 5. Verify type of stile edges, and astragals required for pairs of doors.

[ ] [x] [ ] [C] 6. Verify fire-rated doors have labels and proper identification. Verify rated glass and louvers are provided as required.

[ ] [x] [ ] [C] 7. Observe installation and verify proper clearances provided. Verify doors are hung straight, level, and plumb.

[ ] [x] [ ] [C] 8. Verify doors function smoothly and easily, and hardware is properly adjusted.

[ ] [x] [ ] [C] 9. Observe glazing operation.

[ ] [x] [ ] [C] 10. Verify factory prime is retouched; factory finish is not damaged and surfaces are adequate to receive applied finish.

[ ] [x] [ ] [C] 11. Verify doors are cleaned. Report doors that cannot be properly cleaned. Verify protection is provided as required to avoid marring and other damage.

Legend: Upper Case Letter and <u>BOLD</u> = Primary Inspection; Lower Case = Secondary Inspection
Main Groups SI: Special Inspections, **AI:** Architectural Inspections, **EI:** Engineering Inspections, **OI:** Other Inspections
Sub-Groups **SI:** [T] Testing Laboratory, [H] Hazardous Materials, [Y] Safety   **AI:** No Sub-Groups
EI: [V] Civil, [S] Structural, [M] Mechanical, [E] Electrical   **OI:** [O] Owner, [C] Contractor, [B] Subcontractor, [L] Legal, [G] Government, [F] Fire Protection, [P] Plumbing, [A] Acoustical

Construction Inspection Manual

SI AI EI OI

[ ] [x] [ ] [C]   12. Verify that raised and recessed door trim and molding do not interfere with opening doors for wheelchair passage.

[ ] [x] [ ] [C]   13. Verify doors and space at doors allow for wheelchair passage.

Frames

[ ] [x] [ ] [C]   14. Verify fabrication and construction of frame is as required. Verify smooth joints, welded if required; gauge, size, straightness, and features are as required. Verify frames are prefabricated if required.

[ ] [x] [ ] [C]   15. Verify additional reinforcement is provided at head, corners, and hardware locations as required.

[ ] [x] [ ] [C]   16. Verify adequate provisions are made to receive hardware.

[ ] [x] [ ] [C]   17. Verify proper type and number of anchors are provided. Verify if adequate anchorage is made during installation.

[ ] [x] [ ] [C]   18. Verify sound-deadening treatment is provided if required.

[ ] [x] [ ] [C]   19. Verify fire-rated frames have labels and proper identification.

[ ] [x] [ ] [C]   20. Verify special light-proof, sound-proof, and lead-lined frames are provided and function as required.

[ ] [x] [ ] [C]   21. Verify frames are provided with special features such as silencer holes if required.

[ ] [x] [ ] [C]   22. Verify frame is grouted during installation if required. Frame is caulked if required.

[ ] [x] [ ] [C]   23. Verify frames are installed straight, level, and plumb and adequately braced where "built in."

[ ] [x] [ ] [C]   24. Verify priming or factory finish is provided. See also "Doors" above.

Legend: Upper Case Letter and **BOLD** = Primary Inspection; Lower Case = Secondary Inspection
Main Groups SI: Special Inspections, **AI:** Architectural Inspections, **EI:** Engineering Inspections, **OI:** Other Inspections
Sub-Groups   **SI:** [T] Testing Laboratory, [H] Hazardous Materials, [Y] Safety   **AI:** No Sub-Groups
EI: [V] Civil, [S] Structural, [M] Mechanical, [E] Electrical   **OI:** [O] Owner, [C] Contractor, [B] Subcontractor, [L] Legal, [G] Government, [F] Fire Protection, [P] Plumbing, [A] Acoustical

DIVISION 8 — DOORS AND WINDOWS — WOOD AND PLASTIC DOORS 08200

STANDARDS: AWI (27) CS (35) NEMA (60) NWMA (80) WIC (99)

<u>SI</u> <u>AI</u> <u>EI</u> <u>OI</u>

[ ] [x] [ ] [C] 1. Verify rough framing for doors and jambs is as required.

[ ] [x] [ ] [C] 2. Verify doors are stored and handled properly. Ensure doors are stored flat and in clean, dry surroundings. They should be protected from dirt, water, and abuse. If stored for long periods, doors should be sealed with a non-water based sealer or primer. Doors should not be exposed to excessive moisture, heat, dryness, or direct sunlight. Doors should always be handled with clean hands or while wearing clean gloves. Doors should be lifted and carried, not dragged across one another. Verify doors are as approved and have identification, and/or certification.

[ ] [x] [ ] [C] 3. Verify jambs for doors are blocked as required for hinges, and securely anchored to rough bucks. Verify they are installed plumb and square.

[ ] [x] [ ] [C] 4. Verify doors are as required: check grade, species, veneer cut, number of piles, match, and edge banded. Observe manufacturer's instruction or standards.

[ ] [x] [ ] [C] 5. Verify core type is as required.

[ ] [x] [ ] [C] 6. Verify wood doors are sealed with a non-based sealer or primer. Trop and bottoms are sealed, after fitting.

[ ] [x] [ ] [C] 7. Verify doors are factory pre-drilled, routed, cut, or otherwise prefabricated when required and in accordance with hardware templates and other requirements.

[ ] [x] [ ] [C] 8. Verify doors are installed with uniform and required tolerance. Verify doors are beveled as required and suit astragal requirements for pairs of doors.

[ ] [x] [ ] [C] 9. Verify stiles and rails are cut a minimum amount so as not to impair integrity of door. Verify that backsets match finish hardware and stiles are not weakened.

[ ] [x] [ ] [C] 10. Verify hardware installation does not impair integrity of door. Observe that lock blocks and other blocking is provided.

Legend: Upper Case Letter and <u>BOLD</u> = Primary Inspection; Lower Case = Secondary Inspection
Main Groups SI: Special Inspections, **AI:** Architectural Inspections, **EI:** Engineering Inspections, **OI:** Other Inspections
Sub-Groups   **SI:** [T] Testing Laboratory, [H] Hazardous Materials, [Y] Safety   **AI:** No Sub-Groups
EI: [V] Civil, [S] Structural, [M] Mechanical, [E] Electrical   **OI:** [O] Owner, [C] Contractor, [B] Subcontractor, [L] Legal, [G] Government, [F] Fire Protection, [P] Plumbing, [A] Acoustical

SI AI EI OI

[ ] [x] [ ] [C] 11. Verify cut-outs for glazed lights, and vents are within allowed standards of door manufacturers. Ensure edges are resealed immediately after cutting.

[ ] [x] [ ] [C] 12. Verify clearances for acceptance of thresholds, weatherstripping, gasketing, and carpeting are made.

[ ] [x] [ ] [C] 13. Verify fire-rated doors and assemblies are provided as required. Labels and identification are provided on door and assembly. Ensure fire-rated doors are not cut out or trimmed in any manner that would void their rating. Verify openings meet rating requirements. Verify ball-bearing type hinges or other approved alternates are used.

[ ] [x] [ ] [C] 14. Verify plastic doors are of approved fabrication and are undamaged and protected.

[ ] [x] [ ] [C] 15. Verify face sheets on plastic doors are lapped over stile pieces.

[ ] [x] [ ] [C] 16. Verify doors, openings and space at doors allow for accessibility requirements.

[ ] [x] [ ] [C] 17. Verify provisions for cap flashing of weather exposed areas are provided.

Legend: Upper Case Letter and <u>BOLD</u> = Primary Inspection;   Lower Case = Secondary Inspection
Main Groups SI: Special Inspections, **AI:** Architectural Inspections, **EI:** Engineering Inspections, **OI:** Other Inspections
Sub-Groups   SI: [T] Testing Laboratory, [H] Hazardous Materials, [Y] Safety   **AI:** No Sub-Groups
  EI: [V] Civil, [S] Structural, [M] Mechanical, [E] Electrical   **OI:** [O] Owner, [C] Contractor, [B] Subcontractor,
    [L] Legal, [G] Government, [F] Fire Protection, [P] Plumbing, [A] Acoustical

DIVISION 8 — DOORS AND WINDOWS — ENTRANCES AND STOREFRONTS 08400

STANDARDS: AA (2) NAAMM (54)

SI   AI   EI   OI

[ ] [**X**] [ ] **[B]** 1. Refer to Division 8 "Hardware 08700," Division 8 "Glazing 08800" and
          [C]   Division 8 "Glazed Curtain Wall 08900."

[ ] [x] [ ] **[C]** 2. Verify components or pre-assembled panels are checked for shipping damage after uncrating; and size, shape, thickness of metal extrusions or parts match full size details when available. Check that gauges, patterns and colors are as approved and match samples.

[ ] [x] [ ] **[C]** 3. Verify protective coating and/or lacquers are provided to proper thickness.

[ ] [x] [ ] **[C]** 4. Verify shop-applied sealant is provided at shop-assembled joints as required.

[ ] [x] [ ] **[C]** 5. Verify field-applied sealant is provided as required.

[ ] [x] [ ] **[C]** 6. Verify color matches between panels and parts are within specified range.

[ ] [x] [ ] **[C]** 7. Verify dissimilar metals and materials are isolated; for example, aluminum, in contact with other metals and cementitious surfaces, may require nylon, polystyrene or pressure tape, separators or stainless steel bolts.

[ ] [x] [ ] **[C]** 8. Verify field-applied sealant is of proper type and color and applied where required.

[ ] [x] [ ] **[C]** 9. Verify expansion joints are provided between units as required.

[ ] [x] [ ] **[C]** 10. Verify weep holes and drainage systems are provided and are clean before and after erection.

[ ] [x] [ ] **[C]** 11. Verify erection tolerances are maintained regarding horizontal and vertical alignment and plumbness.

[ ] [x] [ ] **[C]** 12. Verify reveals are of consistent size and align.

[ ] [x] [ ] **[C]** 13. Verify anchorage to structure is secure for transfer of wind load and is required and permanently tightened after alignment.

[ ] [x] [ ] **[C]** 14. Verify hardware provisions have been coordinated.

Legend: Upper Case Letter and **BOLD** = Primary Inspection; Lower Case = Secondary Inspection
   Main Groups SI: Special Inspections, **AI:** Architectural Inspections, **EI:** Engineering Inspections, **OI:** Other Inspections
     Sub-Groups   **SI:** [T] Testing Laboratory, [H] Hazardous Materials, [Y] Safety   **AI:** No Sub-Groups
       EI: [V] Civil, [S] Structural, [M] Mechanical, [E] Electrical   **OI:** [O] Owner, [C] Contractor, [B] Subcontractor,
        [L] Legal, [G] Government, [F] Fire Protection, [P] Plumbing, [A] Acoustical

SI	AI	EI	OI	
[ ]	[x]	[ ]	[C]	15. Verify electric or pneumatic outlets and locations, if required, are provided.
[ ]	[x]	[ ]	[C]	16. Verify exterior is maintained reasonably clean after erection, especially free from cementitious materials.
[ ]	[x]	[ ]	[C]	17. Verify final cleaning is performed as required.
[ ]	[x]	[ ]	[C]	18. Verify doors, openings and space at doors allow for accessibility requirements.

DIVISION 8 — DOORS AND WINDOWS — METAL WINDOWS 08500

STANDARDS: AA (2) AAMA (26) ANSI (16) ASTM (19) NAAMM (54) SWI (90) .

SI   AI   EI   OI

[ ] [X] [ ] [B] 1. Refer to Division 8 "Glazing 08800" for applicable items.
          [C]

[ ] [x] [ ] [C] 2. Verify delivered windows are of type, size, finish, and operation as approved.

[ ] [x] [ ] [C] 3. Verify windows are properly stored and clipped shut until hardware is installed.

[ ] [x] [ ] [C] 4. Verify hardware is of required type, metal, finish, and function.

[ ] [x] [ ] [C] 5. Verify special items are furnished, such as window cleaner's bolts, pull-down hooks, poles, special mullions, and trim.

[ ] [x] [ ] [C] 6. Verify required type of glazing beads or stops are provided, and are suitable to receive glass and glazing thicknesses. Verify method of Fastening is as required.

[ ] [x] [ ] [C] 7. Verify windows are set plumb, square, and level in alignment and at proper location and elevation.

[ ] [x] [ ] [C] 8. Verify windows have provision for suitable anchorage, and it is provided during installation. Verify windows are adequately braced where "built in."

[ ] [x] [ ] [C] 9. Verify windows are sealed as required for metal-to-metal surfaces and other surfaces. Observe that solid grouting, caulking and backup are provided if required.

[ ] [x] [ ] [C] 10. Verify finish is protected and maintained during and after installation. Observe that protection against cement, plaster, acids, and other harmful materials is provided.

[ ] [x] [ ] [C] 11. Verify windows are installed to be weathertight. Observe that weeps are provided, if required, and are maintained in a clean condition.

[ ] [x] [ ] [C] 12. Ensure dissimilar metals are isolated.

[ ] [x] [ ] [C] 13. Verify windows are properly adjusted for tolerance, clearance, and operation before glazing.

Legend: Upper Case Letter and **BOLD** = Primary Inspection;   Lower Case = Secondary Inspection
  Main Groups SI: Special Inspections, **AI:** Architectural Inspections, **EI:** Engineering Inspections, **OI:** Other Inspections
  Sub-Groups   **SI:** [T] Testing Laboratory, [H] Hazardous Materials, [Y] Safety   **AI:** No Sub-Groups
     EI: [V] Civil, [S] Structural, [M] Mechanical, [E] Electrical   **OI:** [O] Owner, [C] Contractor, [B] Subcontractor,
       [L] Legal, [G] Government, [F] Fire Protection, [P] Plumbing, [A] Acoustical

Construction Inspection Manual

<u>SI</u> <u>AI</u> <u>EI</u> <u>OI</u>

[ ] [x] [ ] [C] 14. Observe glazing operation. Verify type of sealant is as required and applied in accordance with instructions. See also Section 08800 "Glazing."

[ ] [x] [ ] [C] 15. Verify cleaning of metals and glass is properly performed.

[ ] [x] [ ] [C] 16. Verify screens of type, mesh, and size are provided if required and suit installation.

[ ] [x] [ ] [C] 17. Verify test operable windows for hardware and friction adjustment and ease of operation on completion of installation.

Legend: Upper Case Letter and <u>BOLD</u> = Primary Inspection; Lower Case = Secondary Inspection
Main Groups SI: Special Inspections, **AI:** Architectural Inspections, **EI:** Engineering Inspections, **OI:** Other Inspections
Sub-Groups   **SI:** [T] Testing Laboratory, [H] Hazardous Materials, [Y] Safety   **AI:** No Sub-Groups
  EI: [V] Civil, [S] Structural, [M] Mechanical, [E] Electrical   **OI:** [O] Owner, [C] Contractor, [B] Subcontractor,
   [L] Legal, [G] Government, [F] Fire Protection, [P] Plumbing, [A] Acoustical

DIVISION 8 — DOORS AND WINDOWS — HARDWARE 08700

STANDARDS: AHA (7) AHMA (8) BHMA (21A) UL (93)

SI   AI   EI   OI

[ ] [x] [ ] [C]   1. Recommended order of inspection:

- In hardware storage room before installation
- Door butts and hinges during and after installation
- Locksets, latchsets, and exit devices during and after installation
- Door closers after installation
- Door stops, holders, and push, pull and kickplates after installation.

[ ] [x] [ ] [C]   2. Verify hardware is installed in accordance with manufacturer's templates and instructions.

[ ] [x] [ ] [C]   3. Verify finishes are as required and finishes match as specified.

[ ] [ ] [ ] [C] [B]   4. Verify hardware is removed and/or protected during painting and cleaning operations.

Butts and Hinges

[ ] [ ] [ ] [B] [c]   5. Verify ball bearing, oilite, or nylon type is provided as required.

[ ] [ ] [ ] [C]   6. Verify solid brass, bronze, aluminum, or stainless steel is provided if required.

[X] [x] [ ] [C]   7. Verify fire-door hinges are steel with ball bearings or as otherwise approved for a labeled assembly.

[ ] [x] [ ] [C]   8. Verify mortise-type hinges are mortised flush.

[ ] [x] [ ] [C]   9. Verify sufficient throw is provided to clear trim, and leaf can swing functionally as required.

[ ] [x] [ ] [C]   10. Verify NRP hinges are provided as required and set screws are tightly screwed down.

[ ] [x] [ ] [B] [c]   11. Verify one-half surface hinges are used on composite doors.

[ ] [x] [ ] [C]   12. See that floor hinges are set level and that door is checked for plumbness, with face closed and face and edge open.

Legend: Upper Case Letter and <u>BOLD</u> = Primary Inspection;   Lower Case = Secondary Inspection
  Main Groups SI: Special Inspections, **AI:** Architectural Inspections, **EI:** Engineering Inspections, **OI:** Other Inspections
  Sub-Groups   **SI:** [T] Testing Laboratory, [H] Hazardous Materials, [Y] Safety   **AI:** No Sub-Groups
  EI: [V] Civil, [S] Structural, [M] Mechanical, [E] Electrical   **OI:** [O] Owner, [C] Contractor, [B] Subcontractor, [L] Legal, [G] Government, [F] Fire Protection, [P] Plumbing, [A] Acoustical

Construction Inspection Manual

143

Locksets and Latchsets

<u>SI</u> <u>AI</u> <u>EI</u> <u>OI</u>

[ ] [x] [ ] [C] 13. Predrilled or jig bored provides most accurate installation. After boring, ensure no planing is allowed on lockset edge.

[ ] [x] [ ] [C] 14. Verify mortise for strike provides for full latchbolt projection. Fire assemblies require full throw, and in some instances dead bolts will not latch without full projection.

[ ] [x] [ ] [C] 15. Verify backsets are provided as required and clear stops.

[ ] [x] [ ] [C] 16. Verify cylinder cores are installed with tumblers up.

[ ] [X] [ ] [C] [B] 17. Verify hand-operated hardware operable without requiring grasping and twisting of wrist, such as levers, push-pulls.

[ ] [X] [ ] [C] [B] 18. Verify hand-operated hardware meets accessibility requirements.

[ ] [X] [ ] [C] 19. Confirm access/exit from dead-end spaces such as balconies.

Door Closers

[ ] [x] [ ] [C] 20. Verify closers are attached to wood doors with through bolts and grommet nuts and to metal doors with through bolts unless otherwise specified.

[ ] [x] [ ] [C] 21. Observe operation of closers as soon as possible after installation for proper operation — silent closing and smooth operation at arc opening. Verify panic devices are properly latching.

[ ] [x] [ ] [C] [b] 22. Verify closers are adjusted by hardware supplier representative, if required and after handling system is operational.

[ ] [x] [ ] [C] [b] 23. Verify door operation is within required limits for accessibility, where not in conflict with fire rating requirements.

[ ] [x] [e] [C] 24. Observe that voltage requirements are coordinated with electric power supplied.

Legend: Upper Case Letter and <u>BOLD</u> = Primary Inspection; Lower Case = Secondary Inspection
Main Groups **SI:** Special Inspections, **AI:** Architectural Inspections, **EI:** Engineering Inspections, **OI:** Other Inspections
Sub-Groups **SI:** [T] Testing Laboratory, [H] Hazardous Materials, [Y] Safety **AI:** No Sub-Groups
EI: [V] Civil, [S] Structural, [M] Mechanical, [E] Electrical **OI:** [O] Owner, [C] Contractor, [B] Subcontractor, [L] Legal, [G] Government, [F] Fire Protection, [P] Plumbing, [A] Acoustical

Construction Inspection Manual

Exit Bolts and Flush Bolts

<u>SI</u> <u>AI</u> <u>EI</u> <u>OI</u>

[ ] [x] [ ] [C] 25. Verify exit cross bars are level, with both ends firmly attached to lever arms, and return at same time when depressed or released. Verify top and bottom bolts are fully seated in strikes on vertical rod devices.

[ ] [x] [ ] **[B]** 26. Verify latch bolt enters strike and seats properly on rim or mortise lock
            [c]   devices. If equipped with dead-locking bolt, observe that proper operation is provided.

[ ] [x] [ ] [C] 27. Verify label agrees with door assembly rating and no "dogging" features are allowed.

[ ] [x] [ ] [C] 28. Verify panic bolts have mullion stabilizers at mullions unless otherwise not required, i.e., structural mullions.

Stops, Holders and Plates

[ ] [x] [ ] [C] 29. Verify every door is provided with a door stop as required.

[ ] [x] [ ] [C] 30. Verify stops will suit anticipated conditions if furniture or equipment
            [b]   layout is known.

[ ] [x] [ ] [C] 31. Verify stops or holders to be attached to wallboard, and plaster are screwed to solid blocking.

[ ] [x] [e] **[B]** 32. Verify wiring, and outlet boxes are provided for smoke-sensing or heat-
            [C]   sensing devices and magnetic holders. Verify magnetic holders are installed horizontally in same location as closer to prevent doorwarp unless otherwise required.

Miscellaneous Hardware and Items

[ ] [x] [ ] **[B]** 33. Verify sliding-door hardware capacity generally matches weight of door
            [c]   as required.

[ ] [x] [ ] [C] 34. Verify sliding-door tracks are installed level and door is plumb. If separate tracks are used, verify bracket supports are directly over hangers when door is open or closed (especially required on fire-rated assemblies). Verify spacing and number of brackets are provided as required.

Legend: Upper Case Letter and **BOLD** = Primary Inspection; Lower Case = Secondary Inspection
Main Groups **SI**: Special Inspections, **AI**: Architectural Inspections, **EI**: Engineering Inspections, **OI**: Other Inspections
Sub-Groups   **SI**: [T] Testing Laboratory, [H] Hazardous Materials, [Y] Safety   **AI**: No Sub-Groups
   EI: [V] Civil, [S] Structural, [M] Mechanical, [E] Electrical   **OI**: [O] Owner, [C] Contractor, [B] Subcontractor,
   [L] Legal, [G] Government, [F] Fire Protection, [P] Plumbing, [A] Acoustical

Construction Inspection Manual

<u>SI</u> <u>AI</u> <u>EI</u> <u>OI</u>

[ ] [x] [ ] [C]   35. Verify thresholds are of required size, type, and interlock and are anchored as required.

[ ] [x] [ ] [C]   36. Verify weather stripping and sound stripping allow proper operation of door.

[ ] [x] [ ] [C]   37. Verify all hardware is complete and with required type and number of bolts, screws, and fastening devices installed.

[ ] [x] [ ] [C]   38. Verify keying instructions are followed and keys are delivered to owner as required. Observe that construction locks are removed and permanent cores are provided.

[ ] [**X**] [ ] [**B**]   39. Verify thresholds meet accessibility requirements.
              [C]

Legend: Upper Case Letter and <u>BOLD</u> = Primary Inspection; Lower Case = Secondary Inspection
Main Groups **SI:** Special Inspections, **AI:** Architectural Inspections, **EI:** Engineering Inspections, **OI:** Other Inspections
Sub-Groups **<u>SI</u>:** [T] Testing Laboratory, [H] Hazardous Materials, [Y] Safety **<u>AI</u>:** No Sub-Groups
<u>EI</u>: [V] Civil, [S] Structural, [M] Mechanical, [E] Electrical **<u>OI</u>:** [O] Owner, [C] Contractor, [B] Subcontractor,
[L] Legal, [G] Government, [F] Fire Protection, [P] Plumbing, [A] Acoustical

DIVISION 8 — DOORS AND WINDOWS — GLAZING 08800

STANDARDS: ANSI (16) FGMA (42) GTA (44) UL (93)

SI AI EI OI

[ ] [x] [ ] [C] 1. Types, thickness, quality, pattern, and finish of glass are as required and glass is labeled or otherwise identified.

[ ] [x] [ ] [C] 2. Verify type, materials, and methods of glazing. Verify putty, glazing compound, tape, gasketing, glazier points, screws, shims, separators, beads, and special sections are as required.

[ ] [x] [ ] [B]
              [c] 3. Verify surfaces to receive glass are dry, clean, and properly prepared.

[ ] [x] [ ] [B] 4. Verify wood and steel rabbets and beads are primed before glazing; lacquer and grease are removed from metals; and weathering steel is primed or otherwise prepared.

[ ] [x] [ ] [B]
              [c] 5. Verify required clearance between glass and frames is provided (extremely important for plastic panes).

[ ] [x] [ ] [C] 6. Verify heat-absorbent glass has clean-cut edges. If altered at site, see that this condition is met.

[ ] [x] [ ] [C] 7. Verify no alteration or attempt to alter size or edge of heat-strengthened, tempered, or insulating glass is made on job.

[ ] [x] [ ] [C] 8. Verify glazing blocks and shims are provided for proper positioning and setting as required.

[ ] [x] [ ] [C] 9. Verify embedding requirements, such as puttying and back-puttying, use of points, and use of putty or compound are as required. Observe that corrosion-resistant fasteners are used. Verify glazing compound or sealant is applied in accordance with manufacturer's requirement, including proper rod stock material.

[ ] [x] [ ] [C] 10. Verify plastic panes are protected with covering. Verify covering is removed after installation where exposed to sunlight. Verify plastic panes are protected from paint, tar, plaster, and solvents, and cleaning is performed in strict accordance with manufacturer's recommendations. Look for bubbles or scratches.

[ ] [x] [ ] [C] 11. Verify patterned glass is set in exterior opening with smooth side to exterior. Verify pattern of adjacent panes is consistent.

Legend: Upper Case Letter and **BOLD** = Primary Inspection; Lower Case = Secondary Inspection
Main Groups **SI**: Special Inspections, **AI**: Architectural Inspections, **EI**: Engineering Inspections, **OI**: Other Inspections
Sub-Groups **SI**: [T] Testing Laboratory, [H] Hazardous Materials, [Y] Safety **AI**: No Sub-Groups
EI: [V] Civil, [S] Structural, [M] Mechanical, [E] Electrical **OI**: [O] Owner, [C] Contractor, [B] Subcontractor, [L] Legal, [G] Government, [F] Fire Protection, [P] Plumbing, [A] Acoustical

Construction Inspection Manual

SI AI EI OI

[ ] [x] [ ] [C]   12. Verify gasketing in metal sash is not painted.

[ ] [x] [ ] [C]   13. Verify stop beads are securely fastened and non-removable types are used if required.

[ ] [x] [ ] [C]   14. Verify interior glass is installed using sound-proofing methods required and is otherwise vibration free.

[ ] [x] [ ] [C]   15. Verify mirrors are installed as required — built-in anchorage method, concealed, non-tamperable fastening, centered on fixtures, or otherwise located, frames or rosettes.

[ ] [x] [ ] [C]   16. Verify requirements for maintaining labels and protective identification on glass until final cleaning are met.

[ ] [x] [ ] [C]   17. Verify cleaning of glass is performed properly without scratches, and all surfaces are free of labels, putty, compounds, and paint.

Legend: Upper Case Letter and **BOLD** = Primary Inspection; Lower Case = Secondary Inspection
Main Groups SI: Special Inspections, **AI:** Architectural Inspections, **EI:** Engineering Inspections, **OI:** Other Inspections
Sub-Groups   **SI:** [T] Testing Laboratory, [H] Hazardous Materials, [Y] Safety   **AI:** No Sub-Groups
EI: [V] Civil, [S] Structural, [M] Mechanical, [E] Electrical   **OI:** [O] Owner, [C] Contractor, [B] Subcontractor,
[L] Legal, [G] Government, [F] Fire Protection, [P] Plumbing, [A] Acoustical

DIVISION 8 — DOORS AND WINDOWS — GLAZED CURTAIN WALL 08900

STANDARDS: AA (2) NAAMM (54) PEI (74) SSPC (89) SWI (90)

SI AI EI OI

[ ] [**X**] [ ] [**B**]   1. Refer to Division 8 "Entrances and Storefront 08400," Division 8 "Metal Windows 08500" and Division 8 "Glazing 08800" for applicable items.
        [**C**]

[ ] [x] [ ] [C]   2. Verify components or pre-assembled panels are checked for shipping damage after uncrating; and size, shape, and thickness of metal extrusions or parts match full size details when available. Check that gauges, patterns, and colors are as approved and match samples.

[ ] [x] [ ] [C]   3. Verify protective coating and/or lacquers are provided to proper thickness.

[ ] [x] [ ] [C]   4. Verify joint sealer is provided at shop-assembled joints as required.

[ ] [x] [ ] [C]   5. Verify shop-applied sealant is provided as required.

[ ] [x] [ ] [C]   6. Verify sound deadening material and/or insulation is provided as required.

[ ] [x] [ ] [C]   7. Verify color matches between panels and parts are within specified range.

[ ] [x] [ ] [C]   8. Ensure dissimilar metals and materials are isolated; for example, aluminum, in contact with other metals and cementitious surfaces, may require nylon, polystyrene, or pressure tape, separators or stainless steel bolts.

[ ] [x] [ ] [C]   9. Verify field-applied sealant is of proper type and color and applied where required.

[ ] [x] [ ] [C]   10. Verify expansion joints are provided between units as required.

[ ] [x] [ ] [C]   11. Verify weep holes and drainage systems are provided and are clean before and after erection.

[ ] [x] [ ] [C]   12. Verify erection tolerances are maintained regarding horizontal and vertical alignment and plumbness.

[ ] [x] [ ] [C]   13. Verify reveals and align are of consistent size.

[ ] [x] [s] [C]   14. Verify anchorage to structure is secure for transfer of wind load and is required and permanently tightened after alignment.

Legend: Upper Case Letter and <u>BOLD</u> = Primary Inspection; Lower Case = Secondary Inspection
Main Groups SI: Special Inspections, **AI:** Architectural Inspections, **EI:** Engineering Inspections, **OI:** Other Inspections
Sub-Groups   SI: [T] Testing Laboratory, [H] Hazardous Materials, [Y] Safety   **AI:** No Sub-Groups
     EI: [V] Civil, [S] Structural, [M] Mechanical, [E] Electrical   **OI:** [O] Owner, [C] Contractor, [B] Subcontractor,
     [L] Legal, [G] Government, [F] Fire Protection, [P] Plumbing, [A] Acoustical

Construction Inspection Manual

SI   AI   EI   OI

[ ] [x] [ ] [C]   15. Verify debris, such as spray fireproofing, is removed from within curtain wall sections after erection.

[ ] [x] [ ] [C]   16. Verify exterior is maintained reasonably clean after erection, especially free from cementitious materials.

[ ] [x] [ ] [C]   17. Verify final cleaning is performed as required.

Legend: Upper Case Letter and BOLD = Primary Inspection;   Lower Case = Secondary Inspection
Main Groups SI: Special Inspections, **AI:** Architectural Inspections, **EI:** Engineering Inspections, **OI:** Other Inspections
Sub-Groups   SI: [T] Testing Laboratory, [H] Hazardous Materials, [Y] Safety   **AI:** No Sub-Groups
EI: [V] Civil, [S] Structural, [M] Mechanical, [E] Electrical   **OI:** [O] Owner, [C] Contractor, [B] Subcontractor,
[L] Legal, [G] Government, [F] Fire Protection, [P] Plumbing, [A] Acoustical

DIVISION 9 — FINISHES — LATH AND PLASTER 09200

STANDARDS: ANSI (16) GA (45) LPI (49) MLA (5d) UL (93) PI (72) VI (94).

Framing and Furring

SI AI EI OI

[ ] [x] [ ] [C]   1. Verify materials are non-corrosive where exposed to exterior and damp conditions, if required.

[ ] [x] [ ] [C]   2. Verify stud spacing is as required. Verify studs are doubled-up at jambs. Special reinforcement and heavy gauge studs are provided as required.

[ ] [x] [ ] [C]   3. Verify studs are set to allow for vertical movement such as shrinkage, and slab deflection.

[ ] [x] [ ] [C]   4. Verify studs are friction fit or securely anchored to runner tracks. Soundproofing, such as caulking beads, is provided at floors, and walls as required.

[ ] [x] [ ] [C]   5. Verify locations, layout, and plumbness.

[ ] [x] [ ] [C]   6. Verify channel stiffeners are provided as required.

[ ] [x] [ ] [C]   7. Observe special field conditions of fastening and connection for accuracy.

[ ] [x] [ ] [B] [C]   8. Verify anchorages, blocking, and plates required for other equipment support and fastening are provided and installed.

[ ] [x] [ ] [C] [b]   9. Verify cut-outs and openings are properly framed. Verify that flashing sleeves and saddles are in place.

[ ] [x] [ ] [C]   10. Observe size, gauge, spacing, and fastening of runner and furring channels.

[ ] [x] [ ] [C] [b]   11. Verify hangers of proper type, size, and gauge are provided, and are saddle-tied, bolted, or clipped as required.

[ ] [x] [ ] [C]   12. Verify tie wire material and size for connection of channels to runners is provided and properly tied.

[ ] [x] [ ] [C]   13. Verify ground elevation and layout of furring with controller before installation. Verify installation provides a true plane surface, plumb or level as required.

Legend: Upper Case Letter and **BOLD** = Primary Inspection; Lower Case = Secondary Inspection
Main Groups SI: Special Inspections, **AI:** Architectural Inspections, **EI:** Engineering Inspections, **OI:** Other Inspections
Sub-Groups   SI: [T] Testing Laboratory, [H] Hazardous Materials, [Y] Safety   **AI:** No Sub-Groups
EI: [V] Civil, [S] Structural, [M] Mechanical, [E] Electrical   **OI:** [O] Owner, [C] Contractor, [B] Subcontractor, [L] Legal, [G] Government, [F] Fire Protection, [P] Plumbing, [A] Acoustical

Construction Inspection Manual

SI	AI	EI	OI	
[ ]	[x]	[ ]	[C]	14. Verify grounds and screeds are set for a true, level, and plane surface and to obtain proper depth of plaster. Verify long lengths are used and where required splices are properly provided.
[ ]	[x]	[ ]	[C]	15. Verify wood grounds are provided as required.
[ ]	[x]	[ ]	[C]	16. Verify corner beads, expansion devices, vent screeds, casings, trim, and other accessories are provided and properly installed. Verify long lengths or single lengths are provided. Discuss control and contraction joint type, installation, and method with contractor.
[ ]	[x]	[ ]	[B] [C]	17. Verify frames are provided for access panels as required.
[ ]	[x]	[ ]	[C]	18. Verify connections and provisions are made at corners or adjoining surfaces of different materials.
[ ]	[x]	[ ]	[C]	19. Verify metal lath is of type and gauge required for spacing — galvanized as required, lapped, and tied. Verify ends of galvanized wire ties are bent to prevent rust. On paperbacked lath, see that wire is to wire and paper is to paper. Verify that horizontal and vertical weather lapping is provided.
[ ]	[x]	[ ]	[B] [C]	20. Verify gypsum lath and gypsum wallboard are provided and installed as required: staggered application, staggered back-to-back application, or staggered wall to ceiling. Verify fastenings are provided of type and spacing required. In addition, verify for proper separation between paper backing and lath. Confirm required paper of crimped mesh spacer.
[ ]	[x]	[ ]	[B] [C]	21. Verify over wood-based sheathing, two layers of Grade D paper are installed.
[ ]	[x]	[ ]	[C]	22. Verify termination of paper flashings on order underlayment are not in contact with exterior paving.
[X] **Plaster**	[x]	[ ]	[B] [C]	23. Ensure agency inspection is performed before closing in, if required.
[ ]	[x]	[ ]	[C]	24. Verify plaster mix, proportions, and mixing equipment are as required and adequate.
[ ]	[x]	[ ]	[B] [C]	25. Verify spaces to be plastered have been heated before installation and provision is made for regulated ventilation.

Legend: Upper Case Letter and **BOLD** = Primary Inspection; Lower Case = Secondary Inspection
Main Groups **SI**: Special Inspections, **AI**: Architectural Inspections, **EI**: Engineering Inspections, **OI**: Other Inspections
Sub-Groups   **SI**: [T] Testing Laboratory, [H] Hazardous Materials, [Y] Safety   **AI**: No Sub-Groups
EI: [V] Civil, [S] Structural, [M] Mechanical, [E] Electrical   **OI**: [O] Owner, [C] Contractor, [B] Subcontractor, [L] Legal, [G] Government, [F] Fire Protection, [P] Plumbing, [A] Acoustical

Construction Inspection Manual

SI	AI	EI	OI	
[ ]	[x]	[ ]	**[B]** **[C]**	26. Verify masonry and concrete surfaces that are directly plastered have been roughened, cleaned, and dampened. Bonding materials and requirements are met, including means to avoid telegraphing masonry joints.
[ ]	[x]	[ ]	**[B]** **[C]**	27. Verify setting and curing times are as required. Minimum cement plaster is moist cured for a minimum of 48 hours.
[ ]	[x]	[ ]	**[C]**	28. Verify adjacent spaces and surfaces are protected during plastering operations.
[ ]	[x]	[ ]	**[C]**	29. Verify proper type of plaster required for various types of areas provided.
[ ]	[x]	[ ]	**[C]**	30. Verify adequate lighting is provided for proper workmanship.
[ ]	[x]	[ ]	**[C]**	31. Verify proper basecoat plaster thickness is provided. Confirm scratch coat cover most of the metal or with lath, 1.x. 90 percent or more.
[ ]	[x]	[ ]	**[C]**	32. Verify cement plaster is scratched or scored horizontally only. Verify gypsum plaster is cross scored. Verify brown coats are left flat within ¼ in 5'-0".
[ ]	[x]	[ ]	**[C]**	33. Inspect smoothness and/or texture: Smooth troweled gypsum plaster should reflect a true, even plane when inspected at an oblique angle. Verify texture is provided, using method in accordance with 1 approved sample and is uniform.
[ ]	[x]	[ ]	**[B]** **[C]**	34. Verify color coat is of approved mix and adequately covers undercoat. For integrally colored plaster, closely observe color appearance of sections completed at different times. Confirm need for final color fogging for color consistency.
[ ]	[x]	[ ]	**[B]** **[C]**	35. Cording, if required, is set in expansion screeds, removed at completion of each phase, and completely cleared of all buildup at completion.
[ ]	[x]	[ ]	**[C]**	36. Confirm color fasting of plaster veneers while curing for color match with representative sample.
[ ]	[x]	[ ]	**[C]**	37. Verify clean-up at intervals is performed and complete clean-up and I debris disposal is performed at end of operations. Do not allow excessive debris to accumulate, and see that precautions are taken to prevent tracking to other areas.

Legend: Upper Case Letter and <u>BOLD</u> = Primary Inspection; Lower Case = Secondary Inspection
Main Groups SI: Special Inspections, **AI:** Architectural Inspections, **EI:** Engineering Inspections, **OI:** Other Inspections
Sub-Groups   **SI:** [T] Testing Laboratory, [H] Hazardous Materials, [Y] Safety   **AI:** No Sub-Groups
EI: [V] Civil, [S] Structural, [M] Mechanical, [E] Electrical   **OI:** [O] Owner, [C] Contractor, [B] Subcontractor,
[L] Legal, [G] Government, [F] Fire Protection, [P] Plumbing, [A] Acoustical

SI AI EI OI

[ ] [x] [ ] [C]   38. Verify that alternate curing measures will be used during extremely hot weather.

[ ] [x] [ ] [C]   39. Verify masking is promptly removed at the end of operations.

DIVISION 9 — FINISHES — GYPSUM BOARD 09250

STANDARDS: GA (45) UL (93)

SI AI EI OI

[ ] [x] [ ] [C]   1. Verify material is stored in dry location and does not overload floor systems.

Framing System

[ ] [x] [ ] [C]   2. Verify metal framing materials are galvanized where exposed to exterior and damp conditions, and where otherwise required. Verify wood framing materials are straight and properly cured; chemically treated for fire resistance or other properties, as required.

[ ] [x] [ ] [C]   3. Verify stud spacing is as required; studs are doubled-up at jambs; and special reinforcement and heavy gauge studs are provided as required.

[ ] [x] [ ] [C]   4. Verify studs are set to allow for vertical movement such as shrinkage, and slab deflection.

[ ] [x] [ ] [C]   5. Verify studs are friction fit or fastened as required to securely anchored runner track. Verify soundproofing, such as caulking beads, is provided at floors, and walls if required.

[ ] [x] [ ] [B][C]   6. Verify locations, layout, plumbness and alignment.

[ ] [x] [ ] [C]   7. Verify channel stiffeners are provided as required.

[ ] [x] [ ] [C]   8. Verify special field conditions of fastening and connection are observed for accuracy.

[ ] [x] [ ] [B][C]   9. Verify anchorages, blocking, and plates required for other equipment support and fastening are provided and installed.

[ ] [x] [ ] [C]   10. Verify cut-outs and openings are properly framed.

[ ] [x] [ ] [C]   11. Observe size, gauge, spacing, and fastening of runner and furring channels.

[ ] [x] [ ] [C]   12. Verify hangers of proper type, size, and gauge are provided, and are saddle tied, bolted, or clipped as required.

Legend: Upper Case Letter and **BOLD** = Primary Inspection;   Lower Case = Secondary Inspection
Main Groups SI: Special Inspections, **AI:** Architectural Inspections, **EI:** Engineering Inspections, **OI:** Other Inspections
Sub-Groups   SI: [T] Testing Laboratory, [H] Hazardous Materials, [Y] Safety   **AI:** No Sub-Groups
EI: [V] Civil, [S] Structural, [M] Mechanical, [E] Electrical   **OI:** [O] Owner, [C] Contractor, [B] Subcontractor, [L] Legal, [G] Government, [F] Fire Protection, [P] Plumbing, [A] Acoustical

Construction Inspection Manual

SI　AI　EI　OI

[ ] [x] [ ] [C]　13. Verify tie wire material and size for connection of channels to runners is provided and properly tied.

[ ] [x] [ ] [C]　14. Verify elevation and layout of furring is understood. Verify installation provides a true plane surface, plumb or level as required.

[ ] [x] [ ] [C]　15. Verify corner beads, expansion devices, casings, trim, and other accessories are provided and properly installed. Confirm long lengths or single lengths are provided as specified.

[ ] [x] [ ] [B]
　　　　　　[C]　16. Frames are provided for access panels as required.

[ ] [x] [ ] [C]　17. Verify connections and provisions are made at corners or adjoining surfaces of different materials.

[ ] [x] [ ] [C]　18. Verify holes in metal studs are in alignment.

[ ] [x] [ ] [C]　19. Verify perimeter sealing or treatment is provided as required for sound or thermal isolation.

[ ] [x] [ ] [C]　20. Verify wood studs are in alignment, and out-of-line members are corrected. Verify spacing and construction are as specified.

[ ] [x] [ ] [B]
　　　　　　[C]　21. Verify blocking, bracing, nailers, and back-up to attach gypsum board are provided. Verify whether all edges of gypsum board require continuous blocking. Verify provisions are made for required anchorage and support of other equipment.

[ ] [x] [ ] [B]
　　　　　　[C]　22. Verify wood materials are sufficiently dry to avoid "nail popping" due to shrinkage.

Gypsum Board

[X] [x] [ ] [B]
　　　　　　[C]　23. Verify agency inspection is performed before "closing-in" if required.

[ ] [x] [ ] [C]　24. Verify type, thickness, length. and edges are as required. Verify if horizontal or vertical application is required.

[ ] [x] [ ] [C]　25. Verify type of nail or fastener, gauge, length, and spacing are provided as required. Verify whether special nailing is required.

Legend: Upper Case Letter and <u>BOLD</u> = Primary Inspection; Lower Case = Secondary Inspection
Main Groups SI: Special Inspections, **AI:** Architectural Inspections, **EI:** Engineering Inspections, **OI:** Other Inspections
Sub-Groups　　SI: [T] Testing Laboratory, [H] Hazardous Materials, [Y] Safety　　**AI:** No Sub-Groups
EI: [V] Civil, [S] Structural, [M] Mechanical, [E] Electrical　　**OI:** [O] Owner, [C] Contractor, [B] Subcontractor, [L] Legal, [G] Government, [F] Fire Protection, [P] Plumbing, [A] Acoustical

Construction Inspection Manual

SI   AI   EI   OI

[ ] [x] [ ] [C] 26. Verify installation complies with manufacturer's recommendations or other requirements.

[ ] [x] [ ] **[B]**
             [C] 27. Verify gypsum board is not erected until building is closed-in (depends on weather). Verify ventilation for air circulation is provided, with adequate dry heat.

[ ] [x] [ ] [C] 28. Verify fire-rated gypsum board compound and installation system is provided where required by type of construction, occupancy, or otherwise.

[X] [x] [ ] [C] 29. If fire-rated, verify all penetrations are tight and sealed and otherwise as required by codes.

[ ] [x] [ ] **[B]**
             [C] 30. Verify appropriate gypsum board is used for damp and other special locations if required. Observe that cut edges and cut outs of moisture resistant gypsum board edges are properly sealed. Confirm horizontal joints have factory edges.

[ ] [x] [ ] [C] 31. Verify special lengths are provided if required.

[ ] [x] [ ] [C] 32. Verify special installations required for soundproofing are provided. Observe method and see that isolation is achieved. See that rigid jointing is tight to obstruct passage of sound.

[ ] [x] [ ] **[B]**
             [C] 33. Verify correct sizing for cut-outs, and outlet boxes is performed to avoid patching, sound passage, and thermal loss. Require sawing and do not allow scoring and knock-out.

[ ] [x] [ ] [C] 34. Verify gypsum board is held up from floor 3/8" minimum.

[ ] [x] [ ] [C] 35. Confirm gypsum board is installed with staggered application - back-to-back staggering, wall-to-ceiling staggering, and double-layer staggering. Verify gypsum board on steel stud framing is installed always working towards the open face of studs.

[ ] [x] [ ] [C] 36. Verify vertical joints are aligned with door jambs.

[ ] [x] [ ] [C] 37. Verify fastening is performed as required. Verify fastening is from center outward, paper surfaces not broken, sheets not driven together. Observe that non-metallic cable, plastic, or copper piping is not close to surfaces or damaged.

[ ] [x] [ ] [C] 38. Observe that excessive piecing or jointing is not provided. Ensure damaged sheets are not to be used and are discarded.

Legend: Upper Case Letter and **BOLD** = Primary Inspection;   Lower Case = Secondary Inspection
Main Groups SI: Special Inspections, **AI:** Architectural Inspections, **EI:** Engineering Inspections, **OI:** Other Inspections
Sub-Groups   SI: [T] Testing Laboratory, [H] Hazardous Materials, [Y] Safety   **AI:** No Sub-Groups
EI: [V] Civil, [S] Structural, [M] Mechanical, [E] Electrical   **OI:** [O] Owner, [C] Contractor, [B] Subcontractor, [L] Legal, [G] Government, [F] Fire Protection, [P] Plumbing, [A] Acoustical

Construction Inspection Manual

<u>SI</u> <u>AI</u> <u>EI</u> <u>OI</u>

[ ] [x] [ ] [C]   39. Verify taping system is of type, compound, and method required.

[ ] [x] [ ] [C]   40. Verify number of coatings required is provided, equipment and tools are suitable, sanding between coats is performed, feathering is out 12 inches to 16 inches, and joints will be unnoticeable after finish is applied. Observe that curing time is adequate, and check for bubbles and dimples.

[ ] [x] [ ] **[B]**   41. Verify type of texture or finish specified will be applied.
           [C]

[ ] [x] [ ] [C]   42. Verify types of internal and external metal corners are provided as required. Verify gypsum board accessories are of type required.

[ ] [x] [ ] [C]   43. Verify clean-up at intervals is performed and complete clean-up and debris disposal is performed at end of operations. Do not allow excessive debris to accumulate, and see that precautions are taken to prevent tracking to other areas.

Legend: Upper Case Letter and <u>BOLD</u> = Primary Inspection; Lower Case = Secondary Inspection
  Main Groups SI: Special Inspections, **AI:** Architectural Inspections, **EI:** Engineering Inspections, **OI:** Other Inspections
  Sub-Groups   <u>SI</u>: [T] Testing Laboratory, [H] Hazardous Materials, [Y] Safety   <u>AI</u>: No Sub-Groups
    <u>EI</u>: [V] Civil, [S] Structural, [M] Mechanical, [E] Electrical   <u>OI</u>: [O] Owner, [C] Contractor, [B] Subcontractor,
    [L] Legal, [G] Government, [F] Fire Protection, [P] Plumbing, [A] Acoustical

DIVISION 9 — FINISHES — TILE 09300

STANDARDS: ANSI (16) ASTM (19) CTI (33) FS (41) SPR (83) TCA (91)

General

SI   AI   EI   OI

[ ] [x] [ ] [C]   1. Verify containers are sealed upon delivery with Grade Seals identifying grades of tile as required for glazed interior and ceramic mosaic tiles (quarry, glass mosaic, cement body and marble tiles are not grade sealed).

[ ] [x] [ ] [C]   2. Verify tile color, sizes, patterns, shapes, and type are as approved.

[ ] [x] [ ] [C]   3. Verify trim shapes are appropriate for use and as required; bullnose edges for thinset; radius edges for mortar set.

[ ] [x] [ ] [C]   4. Verify mastergrade certificates are delivered with tile shipment.

[ ] [x] [ ] [C]   5. Verify: grout type and color is approved.

[ ] [x] [ ] [B]
              [C]   6. Tile color is uniform; shading of color within acceptable tolerances.

[ ] [x] [ ] [C]   7. Layout is as approved; generally no cuts smaller than half tile size with cuts balanced and areas centered.

[ ] [x] [ ] [C]   8. Pattern of layout, alignment of trim, juncture with ceiling, and field tile are as specified.

[ ] [x] [ ] [C]   9. Tile joints are straight and true.

[ ] [x] [ ] [C]   10. Tile surfaces are true to plane, level and plumb.

[ ] [x] [ ] [C]   11. Tile corners are flush or level with adjacent tile.

[ ] [x] [ ] [C]   12. Tile edges are on an even plane, smooth to touch.

[ ] [x] [ ] [C]   13. Tile cuts are smooth without jagged or flaked edges.

[ ] [x] [ ] [B]
              [C]   14. Bonding is complete and sound.

[ ] [x] [ ] [B]
              [C]   15. Finished tile is free from pits, chips, cracks or scratches.

[ ] [x] [ ] [C]   16. Grout is uniform in color, tooled uniformly on cushion edged tiles, flush to top of square edged tiles, and smooth, without voids and is hard and durable.

Legend: Upper Case Letter and **BOLD** = Primary Inspection; Lower Case = Secondary Inspection
Main Groups SI: Special Inspections, **AI:** Architectural Inspections, **EI:** Engineering Inspections, **OI:** Other Inspections
Sub-Groups   SI: [T] Testing Laboratory, [H] Hazardous Materials, [Y] Safety   **AI:** No Sub-Groups
EI: [V] Civil, [S] Structural, [M] Mechanical, [E] Electrical   **OI:** [O] Owner, [C] Contractor, [B] Subcontractor, [L] Legal, [G] Government, [F] Fire Protection, [P] Plumbing, [A] Acoustical

Construction Inspection Manual

SI AI EI OI

[ ] [x] [ ] [C] 17. Finished tile surface is cleaned of setting and grouting materials; do not allow muriatic acid on glazed tile.

[ ] [x] [ ] [C] 18. Finished tile is protected from damage.

[ ] [x] [ ] [C] 19. Tile is properly spaced as required.

[ ] [x] [ ] [B][C] 20. Use of glass mesh mortar unit substrate conforms to manufacturer's specifications.

[ ] [x] [ ] [B][C] 21. Use of sound absorbing matting conforms to manufacturer's specifications.

[ ] [x] [ ] [B][C] 22. Surface applied accessories are fastened through tile by drilling tile or grout joints without inducing cracks.

Wall Tile, Mortar Set

[ ] [x] [ ] [B][C] 23. Verify studs and furring are properly anchored and braced.

[ ] [x] [ ] [B][C] 24. Backing for accessories and partitions is in place.

[t] [x] [ ] [B][C] 25. Waterproof membrane is installed if required.

[ ] [x] [ ] [C] 26. Lath is properly located, lapped and tied; galvanized lath and wires in wet areas; galvanized or painted lath in dry areas as specified.

[ ] [x] [ ] [C] 27. Mortar is Portland cement type without gypsum; scratch coat if required fully keyed to lath; mortar bed uniformly floated to cured scratch coat or masonry surface, maximum 3/4" thick, single or double back coats.

[ ] [x] [ ] [C] 28. Bond coat is cement paste on uncured mortar bed or latex-portland-cement or dry set on cured mortar bed.

[ ] [x] [ ] [C] 29. Wall tile is set before floor tile.

[ ] [x] [ ] [B][C] 30. Built-in accessories are set and firmly anchored; aligned with grout joints as required.

[ ] [x] [ ] [C] 31. Tile is protected from movement, 48 hours after setting and 48 hours after grouting; no work on opposite side of wall.

Legend: Upper Case Letter and **BOLD** = Primary Inspection; Lower Case = Secondary Inspection
Main Groups SI: Special Inspections, **AI:** Architectural Inspections, **EI:** Engineering Inspections, **OI:** Other Inspections
Sub-Groups **SI:** [T] Testing Laboratory, [H] Hazardous Materials, [Y] Safety   **AI:** No Sub-Groups
EI: [V] Civil, [S] Structural, [M] Mechanical, [E] Electrical   **OI:** [O] Owner, [C] Contractor, [B] Subcontractor, [L] Legal, [G] Government, [F] Fire Protection, [P] Plumbing, [A] Acoustical

Wall Tile, Thinset

SI AI EI OI

[ ] [x] [ ] [B] [C] 32. Verify wall backing is properly secured and of specified type; water resistant if gypsum board; surface sealed as required; gypsum board penetrations caulked with special sealant.

[ ] [x] [ ] [C] 33. Setting materials are as specified; organic adhesive on gypsum board, dry set or latex-portland-cement on gypsum board cementitious backer unit, masonry or cured mortar bed.

[ ] [x] [ ] [C] 34. Wall tile is set before floor tile.

[ ] [x] [ ] [B] [C] 35. Built-in accessories are set and firmly anchored; aligned with grout joints as required.

[ ] [x] [ ] [C] 36. Tile is protected from movement, 48 hours after setting and 48 hours after grouting; no work on opposite side of wall.

Floor Tile

[ ] [X] [ ] [B] [C] 37. Verify: non-slip tile is provided for required areas.

[ ] [x] [ ] [C] 38. Masonry thresholds or metal edge bars are set to finish grade.

[ ] [x] [ ] [B] [C] 39. Floor drains are set to finish grade at low point.

[ ] [x] [ ] [B] [C] 40. Floor and/or mortar bed are uniformly sloped for drainage and accessibility as required for drainage and accessibility.

[ ] [x] [ ] [C] 41. Expansion joints are located and of type as required.

[ ] [x] [ ] [B] [C] 42. Substrate is clean and dry; without bumps and hollows for thinset application.

[t] [x] [ ] [B] [C] 43. Waterproof membrane or cleavage membrane is properly installed for mortar bed application.

[ ] [x] [ ] [C] 44. Mortar bed is properly reinforced.

[ ] [x] [ ] [C] 45. Mortar bed for plastic bed is installed as work progresses; ensure excessive bed is not placed to avoid initial set.

[ ] [x] [ ] [C] 46. Bond coat is cement paste on plastic mortar bed; dry set or latex-portland-cement on cured mortar bed or concrete; organic adhesive on concrete or plywood; epoxy on cured mortar bed, concrete or plywood.

Legend: Upper Case Letter and **BOLD** = Primary Inspection; Lower Case = Secondary Inspection
Main Groups **SI**: Special Inspections, **AI**: Architectural Inspections, **EI**: Engineering Inspections, **OI**: Other Inspections
Sub-Groups   **SI**: [T] Testing Laboratory, [H] Hazardous Materials, [Y] Safety   **AI**: No Sub-Groups
EI: [V] Civil, [S] Structural, [M] Mechanical, [E] Electrical   **OI**: [O] Owner, [C] Contractor, [B] Subcontractor, [L] Legal, [G] Government, [F] Fire Protection, [P] Plumbing, [A] Acoustical

<u>SI</u>　<u>AI</u>　<u>EI</u>　<u>OI</u>

[ ]　[x]　[ ]　**[B]**　　47. Tile is tamped in to assure good bond; excess bond coat immediately
　　　　　　　　[C]　　　　　removed before set up.

[ ]　[x]　[ ]　**[B]**　　48. Where epoxy grouted, tile is wax coated on face; ensure no wax on back
　　　　　　　　[C]　　　　　or edges.

[ ]　[x]　[ ]　**[C]**　　49. Tile is protected from traffic and dirt, 48 hours after setting and 48 hours
　　　　　　　　　　　　　　after grouting.

Legend: Upper Case Letter and <u>BOLD</u> = Primary Inspection; Lower Case = Secondary Inspection
Main Groups SI: Special Inspections, **AI:** Architectural Inspections, **EI:** Engineering Inspections, **OI:** Other Inspections
Sub-Groups **<u>SI</u>:** [T] Testing Laboratory, [H] Hazardous Materials, [Y] Safety   **<u>AI</u>:** No Sub-Groups
<u>EI</u>: [V] Civil, [S] Structural, [M] Mechanical, [E] Electrical   **<u>OI</u>:** [O] Owner, [C] Contractor, [B] Subcontractor,
[L] Legal, [G] Government, [F] Fire Protection, [P] Plumbing, [A] Acoustical

DIVISION 9 — FINISHES — TERRAZZO 09400

STANDARDS: ASTM (19) NTNA (70)

SI AI EI OI

[ ] [x] [ ] **[B]** 1. Verify substrate to receive terrazzo will provide good bonding and is clean and free of oil, dirt, and other deleterious matter.
           [C]

[ ] [x] [ ] [C] 2. Verify required divider strips, inserts, and reinforcements are provided, and layout and spacing are understood. See that secure anchorage, proper elevation, and tight joints are provided. Verify expansion strips are provided as required.

[ ] [x] [ ] **[B]** 3. Verify precast materials are on hand as approved, and installed as required.
           [C]

[ ] [x] [ ] [C] 4. Verify base beads or temporary screeds are in place and of required height.

[ ] [x] [ ] [C] 5. Verify type and location of nonslip surface specified and indicated is installed.

[ ] [x] [ ] [C] 6. Verify conductive terrazzo is installed as required.

[ ] [x] [ ] [C] 7. Ensure surfaces to receive terrazzo are moistened.

[ ] [x] [ ] [C] 8. Ensure proper curing provisions are observed.

[ ] [x] [ ] [C] 9. Verify grinding operations produce a smooth, true, even surface of good visual appearance.

[ ] [x] [ ] [C] 10. Verify moisture control due to grinding operations is provided in adjacent areas.

[ ] [x] [ ] [C] 11. See that hard-to-get-at spaces are rubbed.

[ ] [x] [ ] [C] 12. Verify surfaces are cleaned of stains, cement, and smears.

[ ] [x] [ ] [C] 13. Verify finish sealer is provided and applied as required.

[ ] [x] [ ] [C] 14. Verify finish floor is protected, properly cured, and not subject to abuse during remainder of construction.

[ ] [x] [ ] [C] 15. Verify color matches aggregate and matrix of color sample.

[ ] [x] [ ] [C] 16. Verify material is the specified thickness.

Legend: Upper Case Letter and **BOLD** = Primary Inspection; Lower Case = Secondary Inspection
Main Groups SI: Special Inspections, AI: Architectural Inspections, EI: Engineering Inspections, OI: Other Inspections
Sub-Groups   **SI**: [T] Testing Laboratory, [H] Hazardous Materials, [Y] Safety   **AI**: No Sub-Groups
EI: [V] Civil, [S] Structural, [M] Mechanical, [E] Electrical   **OI**: [O] Owner, [C] Contractor, [B] Subcontractor, [L] Legal, [G] Government, [F] Fire Protection, [P] Plumbing, [A] Acoustical

Construction Inspection Manual

DIVISION 9 — FINISHES — WOOD FLOORING 09640

STANDARDS: MFMA (50) NOFMA (65) WFI (98)

<u>SI</u> <u>AI</u> <u>EI</u> <u>OI</u>

[ ] [x] [ ] **[B]** 1. Verify areas to receive flooring are closed-in, and adequate temperature of
              [C]   above 50° is provided and maintained. Do not allow overheating.

[ ] [x] [ ] **[B]** 2. Verify interior surfaces such as plaster, concrete, and masonry are dry.
              [C]   Obtain moisture meter test if required.

[ ] [x] [ ] **[C]** 3. Verify delivered material is of grade size, type, and species specified and stored in area at least 72 hours before installation.

Wood Strip Flooring

[ ] [x] [ ] **[C]** 4. Verify floor material temperature has been conditioned.

[ ] [x] [ ] **[B]** 5. Verify subfloor surface is securely nailed, level, even jointed, cleaned,
              [C]   and free of defects that might affect finish flooring.

[ ] [x] [ ] **[B]** 6. Verify felt paper ("slip sheet") or underlayment as required is provided.
              [C]

[ ] [x] [ ] **[C]** 7. Verify expansion space is provided at perimeter of flooring as required.

[ ] [x] [ ] **[C]** 8. Verify joints are driven up tight and tongue undamaged.

[ ] [x] [ ] **[C]** 9. Verify nails of fasteners are of type and size approved and pre-drilling is provided, if required. Nails are diagonally driven, usually 8d, spiral, screw, or cut type as required, and usually spaced at 12"cc. with both ends of each strip nailed.

[ ] [x] [ ] **[C]** 10. Verify end joints are alternated to provide at least two courses between joints.

[ ] [x] [ ] **[C]** 11. Verify specified pattern, border, grain, field direction, feature, and strips before installation.

[ ] [x] [ ] **[B]** 12. Ensure warped and twisted material is not used.
              [C]

[ ] [x] [ ] **[C]** 13. Verify short pieces and varied colors are spread throughout floor, not concentrated or utilized in closets.

[ ] [x] [ ] **[B]** 14. Verify sleepers are of material required, are set and securely anchored
               [C]   over subsurface required, and ventilation method if effective.

Legend: Upper Case Letter and <u>**BOLD**</u> = Primary Inspection; Lower Case = Secondary Inspection
Main Groups SI: Special Inspections, **AI:** Architectural Inspections, **EI:** Engineering Inspections, **OI:** Other Inspections
Sub-Groups   **SI:** [T] Testing Laboratory, [H] Hazardous Materials, [Y] Safety   <u>**AI**</u>: No Sub-Groups
  <u>**EI**</u>: [V] Civil, [S] Structural, [M] Mechanical, [E] Electrical   <u>**OI**</u>: [O] Owner, [C] Contractor, [B] Subcontractor,
  [L] Legal, [G] Government, [F] Fire Protection, [P] Plumbing, [A] Acoustical

Parquet Flooring

SI AI EI OI

[t] [x] [ ] **[B]**   15. Verify surface of substrate, if concrete, is dry, even, level at joints, clean, and otherwise acceptable. Ensure moisture meter test is performed if required.
            [C]

[**X**] [x] [ ] **[B]**   16. Verify vapor barrier, waterproof membrane, or other treatment is provided as required.
            [C]

[ ] [x] [ ] **[C]**   17. Verify adhesives or application materials are non-toxic as approved. Verify installation complies with manufacturer's recommendations or as otherwise required.

[ ] [x] [ ] **[C]**   18. For prefinished material, observe that joints are even and within tolerances required.

Miscellaneous

[ ] [x] [ ] **[C]**   19. Verify sanding is performed using methods and materials required and produces a smooth acceptable surface. Observe sanding in hard-to-get-at spaces. It is recommended that sanding be delayed until other finishing operations are complete.

[ ] [x] [ ] **[C]**   20. Verify fillers, stains, and floor finish materials are provided and applied as recommended by manufacturer or a required.

[ ] [x] [ ] **[C]**   21. Verify protection is provided as required. Observe that floors and stair treads are protected from droppings, paint, and traffic. See that heavy equipment is lifted, not dragged, into place.

[ ] [x] [ ] **[C]**   22. Verify floor waxing and buffing is provided as required.

Legend: Upper Case Letter and <u>BOLD</u> = Primary Inspection; Lower Case = Secondary Inspection
Main Groups SI: Special Inspections, **AI:** Architectural Inspections, **EI:** Engineering Inspections, **OI:** Other Inspections
Sub-Groups   **SI:** [T] Testing Laboratory, [H] Hazardous Materials, [Y] Safety   **AI:** No Sub-Groups
    EI: [V] Civil, [S] Structural, [M] Mechanical, [E] Electrical   **OI:** [O] Owner, [C] Contractor, [B] Subcontractor,
    [L] Legal, [G] Government, [F] Fire Protection, [P] Plumbing, [A] Acoustical

Construction Inspection Manual

DIVISION 9 — FINISHES — RESILIENT FLOORING 09650

STANDARDS: FS (41) NPA (67) RFCA (80) RMA (81)

<u>SI</u> <u>AI</u> <u>EI</u> <u>OI</u>

[ ] [x] [ ] [C] 1. Verify type, size, thickness, pattern, and color of material are as approved. Use single lot for any one area.

[ ] [x] [ ] [C] 2. Verify primer, adhesive, or cement is as required.

[ ] [x] [ ] [C] 3. Verify base complies with approvals concerning size, thickness, cove, color, and type, and molded exterior, interior, and ends.

[ ] [x] [ ] [B] [C] 4. On slabs, verify substrate is free of cracks, holes, trowel marks, and other defects. Surfaces are primed if required.

[ ] [x] [ ] [B] [C] 5. On slabs, verify all patching, grinding, and correction of defects is performed before installation; all areas are dry, hard, and non-powdery, and moisture test is performed as required.

[ ] [x] [ ] [B] [C] 6. Verify on wood subfloors, that nailing or fastening is adequate; non-rising fasteners are used. Test for squeaks. Verify subfloors are filled and sanded before installation, using filler as recommended by manufacturer or as specified, and all defects are corrected.

[ ] [x] [ ] [C] 7. Verify all areas are cleaned before installation.

[ ] [x] [ ] [C] 8. Verify underlayment ("slip sheet") and felt lining are provided as required.

[ ] [x] [ ] [C] 9. Verify floor material is stored at proper temperature before installation as required, and area temperature is maintained during and after installation.

[ ] [x] [ ] [C] 10. Verify rolled material is unrolled at least 24 hours before installation if required.

[ ] [x] [ ] [B] [C] 11. Verify floor material is installed in proper sequence of schedule to minimize damage by other trades.

[ ] [x] [ ] [C] 12. Verify pattern matching is observed in sheet material installation. Observe evenness of color.

[ ] [x] [ ] [C] 13. Verify direction of tile is as specified and layout is as specified. Allow ½ tile minimum at border.

Legend: Upper Case Letter and **BOLD** = Primary Inspection; Lower Case = Secondary Inspection
Main Groups SI: Special Inspections, **AI:** Architectural Inspections, **EI:** Engineering Inspections, **OI:** Other Inspections
Sub-Groups   SI: [T] Testing Laboratory, [H] Hazardous Materials, [Y] Safety   **AI:** No Sub-Groups
EI: [V] Civil, [S] Structural, [M] Mechanical, [E] Electrical   **OI:** [O] Owner, [C] Contractor, [B] Subcontractor, [L] Legal, [G] Government, [F] Fire Protection, [P] Plumbing, [A] Acoustical

Construction Inspection Manual

SI	AI	EI	OI	
[ ]	[x]	[ ]	[C]	14. If more than one lot of tile is used in one area, ensure all material is pre-shuffled to achieve random distribution.
[ ]	[x]	[ ]	[C]	15. Verify cement is applied at proper rate and has proper dryness or tackiness.
[ ]	[x]	[ ]	[C]	16. Observe neatness of cutting and fitting. Verify joints and seams are tight and level.
[ ]	[x]	[ ]	[C]	17. Verify sheet flooring is rolled with 150-pound roller and starting from center, to eliminate air bubbles and wrinkles.
[ ]	[x]	[ ]	[C]	18. On top set base, observe that firm contact is obtained to floor and wall.
[ ]	[x]	[ ]	[C]	19. Verify minimum length of pieces is observed.
[ ]	[x]	[ ]	[C]	20. Verify pre-formed corners and ends stops are provided if required.
[ ]	[x]	[ ]	[C]	21. Ensure gun application of adhesive for base is not allowed.
[ ]	[x]	[ ]	[C]	22. On coved base observe proper trim and use of non-rusting nails as required.
[ ]	[x]	[ ]	[C]	23. Verify that provision for thresholds, breaks, and joining to adjacent materials are understood.
[ ]	[x]	[ ]	[B] [C]	24. Verify level joining at flush floor electrical cover plates, cleanout.
[ ]	[x]	[ ]	[B] [C]	25. Verify excess adhesive, material, and stains are removed immediately after installation.
[ ]	[x]	[ ]	[C]	26. Verify scuffed, broken, or discolored tile is replaced. Recheck at completion of work and observe looseness, bubbles, or substrate defects.
[ ]	[x]	[ ]	[C]	27. Observe application of wax when required. (Do not allow factory finish to be stripped unless specified.)
[ ]	[x]	[ ]	[C]	28. Temporary protective cover is provided if required.

Legend: Upper Case Letter and <u>BOLD</u> = Primary Inspection; Lower Case = Secondary Inspection
Main Groups SI: Special Inspections, **AI:** Architectural Inspections, **EI:** Engineering Inspections, **OI:** Other Inspections
Sub-Groups **SI:** [T] Testing Laboratory, [H] Hazardous Materials, [Y] Safety **AI:** No Sub-Groups
EI: [V] Civil, [S] Structural, [M] Mechanical, [E] Electrical **OI:** [O] Owner, [C] Contractor, [B] Subcontractor, [L] Legal, [G] Government, [F] Fire Protection, [P] Plumbing, [A] Acoustical

Construction Inspection Manual

DIVISION 9 — FINISHES — CARPET 09680

STANDARDS: ASTM (19) CRI (32) FS (41) NPA (67) RFCA (80)

SI   AI   EI   OI

[ ] [x] [ ] [B] [C]   1. Verify substrate is acceptable. Check plywood subfloors for secure nailing to avoid "nail pops" and squeaking. Verify large splits, knots, or unlevel areas are filled, patched with proper material, and sanded flat. Make similar observation of concrete floors, with attention paid to levelness of joints (to be ground smooth).

[ ] [x] [ ] [B] [C]   2. Verify substrate is properly leveled, cleaned and vacuumed.

[ ] [x] [ ] [B] [C]   3. Verify areas are sufficiently finished before carpet is laid, to avoid undue traffic, construction debris, and damage exposure.

[ ] [x] [ ] [C]   4. Verify pad is not exposed to weather. Do not allow wet or damp pad to be installed.

[ ] [x] [ ] [C]   5. Verify supplied materials are as approved — special features such as flame resistance, sound absorptivity, and static. Double check padding requirements.

[ ] [x] [ ] [C]   6. Confirm carpet was unrolled for 24 hours before installation.

[ ] [x] [ ] [B] [C]   7. Verify conditions of thresholds and edges joining other materials before installation. Give special attention to trench duct cover plates and other floor outlets.

[ ] [x] [ ] [C]   8. Verify same color (dye) run is used in one area and matches approved sample.

[ ] [x] [ ] [C]   9. Verify material is laid in same direction and in conformance with seaming diagram.

[ ] [x] [ ] [B] [C]   10. Verify whether seam and pattern layout is understood. If layout is not specified, arrange so that seams are located to avoid high traffic areas or highly visual areas as much as possible. Do not allow cross seams.

[ ] [x] [ ] [C]   11. Verify seam cuts are made along weave line and no dimpling or puckering occurs along finished seam.

[ ] [x] [ ] [C]   12. Confirm normal overage of carpet is provided. Do not allow excessive seams.

Legend: Upper Case Letter and BOLD = Primary Inspection;   Lower Case = Secondary Inspection
Main Groups SI: Special Inspections, **AI:** Architectural Inspections, **EI:** Engineering Inspections, **OI:** Other Inspections
Sub-Groups   SI: [T] Testing Laboratory, [H] Hazardous Materials, [Y] Safety   **AI:** No Sub-Groups
EI: [V] Civil, [S] Structural, [M] Mechanical, [E] Electrical   **OI:** [O] Owner, [C] Contractor, [B] Subcontractor,
[L] Legal, [G] Government, [F] Fire Protection, [P] Plumbing, [A] Acoustical

SI	AI	EI	OI		
[ ]	[x]	[ ]	[C]	13.	Verify conditions of installation at stairways, cut outs, protrusions, access areas.
[ ]	[x]	[ ]	[C]	14.	Confirm use of standard or oversize tape.
[ ]	[x]	[ ]	[C]	15.	Verify carpeted areas are properly cleaned and protected as required. Confirm unauthorized persons are to be kept out.
[ ]	[x]	[ ]	[C]	16.	Scraps over 2 sq. ft. and larger than 8 inches wide may be required for owner maintenance. Verify arrangements made for receipt of excess material.
[ ]	[x]	[ ]	[C]	17.	In large open floor areas, review manufacturer's roll sequence numbers. Give special attention to directional patterns and confirm installation is a indicated on drawings or specifications.

Legend: Upper Case Letter and <u>BOLD</u> = Primary Inspection; Lower Case = Secondary Inspection
Main Groups SI: Special Inspections, **AI:** Architectural Inspections, **EI:** Engineering Inspections, **OI:** Other Inspections
Sub-Groups **SI:** [T] Testing Laboratory, [H] Hazardous Materials, [Y] Safety **AI:** No Sub-Groups
EI: [V] Civil, [S] Structural, [M] Mechanical, [E] Electrical **OI:** [O] Owner, [C] Contractor, [B] Subcontractor, [L] Legal, [G] Government, [F] Fire Protection, [P] Plumbing, [A] Acoustical

Construction Inspection Manual

DIVISION 9 — FINISHES — ACOUSTICAL TREATMENT 09800

STANDARDS: AIMA (1)

Suspension Systems

<u>SI</u> <u>AI</u> <u>EI</u> <u>OI</u>

[ ] [x] [ ] [C]   1. Verify layout of suspended system is understood and coordinated with work of other trades to allow adequate plenum space and avoid interference.

[ ] [x] [ ] [C]   2. Verify hangers of proper material, gauge and spacing are provided as required. Verify intermediate hangers and bridging are provided as required by field conditions.

[ ] [x] [ ] [C]   3. Verify hangers anchored to concrete support systems are installed as required and fastened by twisting around rebar, to devices installed by powder-actuated anchors, to devices with expansion shields drilled into concrete, or other approved method providing adequate support.

[ ] [x] [ ] [C]   4. Verify hangers anchored to wood-support systems are installed as required, fastened by screws, nails, drilled holes, or other approved method that provides adequate support. Observe number of twists. Verify minimum number of twists are provided as required.

[ ] [x] [ ] [C]   5. Verify hangers anchored to steel support systems are installed as required.

[ ] [x] [ ] [C]   6. Verify isolators are provided and installed in accordance with manufacturer's requirements (if required).

[ ] [x] [ ] [C]   7. Verify turnbucklers are provided if required.

[ ] [x] [ ] [C]   8. Verify sway and seismic bracing is provided if required.

[ ] [x] [ ] [C]   9. Verify suspension system, components and accessories are provided and installed as specified.

[ ] [x] [ ] [C]   10. Verify perimeter and edge conditions are provided and installed as required. Ensure joint treatment is consistent.

[ ] [x] [ ] [C]   11. Verify system is installed and adjusted in true alignment, square and level.

[ ] [x] [ ] **[B]**
            [C]   12. Verify system does not receive loading from other equipment, fixtures, or materials not included in its design.

Legend: Upper Case Letter and <u>**BOLD**</u> = Primary Inspection; Lower Case = Secondary Inspection
  Main Groups SI: Special Inspections, **AI**: Architectural Inspections, **EI**: Engineering Inspections, **OI**: Other Inspections
  Sub-Groups   SI: [T] Testing Laboratory, [H] Hazardous Materials, [Y] Safety   **AI**: No Sub-Groups
    EI: [V] Civil, [S] Structural, [M] Mechanical, [E] Electrical   **OI**: [O] Owner, [C] Contractor, [B] Subcontractor,
      [L] Legal, [G] Government, [F] Fire Protection, [P] Plumbing, [A] Acoustical

SI	AI	EI	OI	
[t]	[x]	[ ]	[C]	13. Verify fire-rated systems are installed in accordance with UL requirements. Ensure hold-down clips are required if acoustical material is less than one pound per square foot.
[ ]	[x]	[ ]	[C]	14. Verify access to equipment above ceiling is provided as required. Verify identification of access is provided by use of labels on "T" bars or colored pins, or as required.
[ ]	[x]	[ ]	[C]	15. Verify sound isolation elements above ceiling are provided and installed as required.
[ ]	[x]	[ ]	[C]	16. Verify fire blankets or rated hats over recessed light fixtures and mechanical outlets are provided as required.
[ ]	[x]	[ ]	[C]	17. Verify exposed rivets are painted out.

Acoustical Tile and Board

[ ]	[x]	[ ]	[C]	18. Verify material is of type, thickness, material, pattern and edge condition specified.
[ ]	[x]	[ ]	[C]	19. Verify material is installed in accordance with mounting specified.
[ ]	[x]	[ ]	[C]	20. Verify adhesives and air space are in accordance with manufacturer's instructions.
[ ]	[x]	[ ]	[C]	21. If tile is to be sprayed, verify paint material is approved for application and will not affect acoustical performance of tile.

Acoustical Insulation and Barriers

[ ]	[x]	[ ]	[C]	22. Verify batts are of thickness and density required and are identified.
[ ]	[x]	[ ]	[C]	23. Verify batts are installed tightly to all adjoining surfaces, cutouts, edges and all spaces are completely filled.
[ ]	[x]	[ ]	[C]	24. Verify suspended insulation is well secured, tight fitting, and sealed as required.
[ ]	[x]	[ ]	[C]	25. Verify wall insulation is provided and installed and in locations as required before installation of finish surfaces.
[ ]	[x]	[ ]	[C]	26. Verify accoustical pads are provided at outlet boxes (minimum stocking: 1-2 stud spaces), if required.

Legend: Upper Case Letter and **BOLD** = Primary Inspection; Lower Case = Secondary Inspection
Main Groups SI: Special Inspections, **AI:** Architectural Inspections, **EI:** Engineering Inspections, **OI:** Other Inspections
Sub-Groups **SI:** [T] Testing Laboratory, [H] Hazardous Materials, [Y] Safety **AI:** No Sub-Groups
EI: [V] Civil, [S] Structural, [M] Mechanical, [E] Electrical **OI:** [O] Owner, [C] Contractor, [B] Subcontractor, [L] Legal, [G] Government, [F] Fire Protection, [P] Plumbing, [A] Acoustical

Construction Inspection Manual

DIVISION 9 — FINISHES — PAINTS AND COATINGS 09900

STANDARDS: FS (141) NPVLA (366)

<u>SI</u> <u>AI</u> <u>EI</u> <u>OI</u>

[ ] [x] [ ] **[B]** 1. Verify storage area for painting materials is well ventilated and used rags and debris are removed from storage area.
        [C]

[ ] [x] [ ] **[C]** 2. Verify color schedule is complete and understood. Verify approved color samples are on job, and jobsite "paint-outs" are matched against samples. Before application, confirm stopping points for change of color and finish. Confirm requirements for field mock-ups.

[ ] [x] [ ] **[C]** 3. Verify all materials are new, and materials are products of same manufacturer if required. Verify containers are adequately identified. Disallow containers showing evidence of broken seal.

[ ] [x] [ ] **[B]** 4. Verify surfaces to receive paint are dry. Ensure moisture meter tests on plaster, concrete, or masonry surfaces are made if required. Damp, not wet, surfaces are allowed for water-thinned paints.
        [C]

[ ] [x] [ ] **[B]** 5. Verify surfaces to receive paint are sanded; holes puttied or filled; pitch pocket, knot, and shakes are shellacked or treated and otherwise cleaned of deleterious substances. Verify metal surfaces are treated, primed, or otherwise cleaned as required.
        [C]

[ ] [x] [ ] **[C]** 6. Confirm special requirements for renovation work.

[ ] [x] [ ] **[B]** 7. Verify areas are suitably cleaned and free of conditions affecting drying and finish.
        [C]

[ ] [x] [ ] **[B]** 8. Verify dust control is maintained.
        [C]

[ ] [x] [ ] **[B]** 9. Verify temperature conditions for type of paint are provided, and heating is provided sufficiently in advance in order to have surfaces up to temperature and to avoid condensation.
        [C]

[ ] [x] [ ] **[B]** 10. Verify adequate lighting is provided for proper working conditions.
        [C]

[ ] [x] [ ] **[C]** 11. Verify protection of adjacent areas, surfaces, and items is provided. Hardware, trim, fixtures, and similar items are removed during painting operations or otherwise suitably protected. Clean drop cloths are provided over finished surfaces.

[ ] [x] [ ] **[C]** 12. Observe occasionally the mixing and thinning of paints. Thinning should be controlled and the need demonstrated.

Legend: Upper Case Letter and <u>BOLD</u> = Primary Inspection; Lower Case = Secondary Inspection
  Main Groups SI: Special Inspections, **AI:** Architectural Inspections, **EI:** Engineering Inspections, **OI:** Other Inspections
  Sub-Groups   <u>**SI**</u>: [T] Testing Laboratory, [H] Hazardous Materials, [Y] Safety   <u>**AI**</u>: No Sub-Groups
    <u>**EI**</u>: [V] Civil, [S] Structural, [M] Mechanical, [E] Electrical   <u>**OI**</u>: [O] Owner, [C] Contractor, [B] Subcontractor,
    [L] Legal, [G] Government, [F] Fire Protection, [P] Plumbing, [A] Acoustical

SI	AI	EI	OI	
[ ]	[x]	[ ]	[C]	13. Verify required number of coats is provided. Verify tinting of undercoats is performed if required. Verify opacity is being achieved.
[ ]	[x]	[ ]	[C]	14. Confirm texture and method of application — spray, brush or roller before application.
[ ]	[x]	[ ]	[B] [C]	15. Verify lumps or bumps do not appear on applied coats. These imperfections indicate improper area cleaning and dust control; or paint is drying in cans due to excessive exposure or being on "shelf" too long; or brushes and rollers are dirty. Straining may be done in some instances; otherwise, paint should not be used.
[ ]	[x]	[ ]	[C]	16. Verify workmanship and application are adequate. Do not allow runs, drops, laps, brush marks, "lace curtains," variations in color, texture, and finish.
[ ]	[x]	[ ]	[C]	17. Keep paint records of areas being painted on large jobs.
[ ]	[x]	[ ]	[C]	18. Recommend that an electric trim plate be used for checking outlet boxes to avoid later patching and refinishing.
[ ]	[x]	[ ]	[C]	19. Verify doors receive first coats on both faces of wood at essentially the same time. Observe that tops and bottoms receive treatment.
[ ]	[x]	[ ]	[C]	20. Verify curing time required between coats is provided.
[ ]	[x]	[ ]	[C]	21. Verify sealers, fillers, and stains are applied and treated as required. Verify putty is not applied until after stain or priming and matches stained wood.
[ ]	[x]	[ ]	[C]	22. Verify hard-to-get-at places are painted — bottoms of shelves and back of trim in corners.
[ ]	[x]	[ ]	[C]	23. Verify correction of all unsuitable work is made promptly. Confirm cleanup of area and removal of splatters and smears are made as soon as possible on adjacent surfaces.
[ ]	[X]	[ ]	[C]	24. Verify marking of stairs and platforms for accessibility requirements.

Legend: Upper Case Letter and <u>BOLD</u> = Primary Inspection; Lower Case = Secondary Inspection
Main Groups SI: Special Inspections, AI: Architectural Inspections, EI: Engineering Inspections, OI: Other Inspections
Sub-Groups <u>SI</u>: [T] Testing Laboratory, [H] Hazardous Materials, [Y] Safety   <u>AI</u>: No Sub-Groups
<u>EI</u>: [V] Civil, [S] Structural, [M] Mechanical, [E] Electrical   <u>OI</u>: [O] Owner, [C] Contractor, [B] Subcontractor,
[L] Legal, [G] Government, [F] Fire Protection, [P] Plumbing, [A] Acoustical

Construction Inspection Manual

DIVISION 12 — FURNISHING — MANUFACTURED CASEWORK 12300

STANDARDS: AWI (27) WIC (99)

SI   AI   EI   OI

[ ] [x] [ ] [B] 1. Verify blocking is provided to receive materials as required.
           [C]

[ ] [x] [ ] [C] 2. Verify certificates or grade stamps are provided.

[ ] [x] [ ] [C] 3. Verify materials are not delivered before closing-in building, and are suitably stored.

[ ] [x] [ ] [C] 4. Verify materials have adequate temporary bracing, and skids to prevent wracking, loosened members, or other defects due to handling.

[ ] [x] [ ] [C] 5. Verify species, cut, and finishes are as required. Visually inspect exposed surfaces for evenness. Match with selected flitch.

[ ] [x] [ ] [C] 6. Verify doors are properly fitted with a uniform clearance on all edges.

[ ] [x] [ ] [C] 7. Verify drawers have guides required and operate smoothly.

[ ] [x] [ ] [C] 8. Verify internal features are provided.

[ ] [x] [ ] [C] 9. Verify method of attachment is as required.

[ ] [x] [ ] [C] 10. Verify installation of floor-set cabinets is over a finished floor or mounted directly on subfloor.

[ ] [x] [ ] [C] 11. Verify accessories such as scribe and trim molds are provided.

[ ] [x] [ ] [C] 12. Verify base and toe space will suit adjacent conditions.

[ ] [x] [ ] [C] 13. Verify installation of base cabinets is suitably shimmed to distribute weight uniformly and suit field conditions.

[ ] [x] [ ] [C] 14. Verify installed materials suitably protected against damage.

[ ] [x] [ ] [C] 15. Verify specified hardware is provided. Verify job-installed hardware is as approved.

[ ] [x] [ ] [C] 16. Verify tops are provided as required. Verify cutting of holes for sinks and other appliances is performed as required. Verify cuts are sealed to preclude moisture penetration behind finish.

[ ] [x] [ ] [C] 17. Verify tops to receive other materials, such as tile vinyl and marble have proper provisions.

[ ] [x] [ ] [C] 18. Verify requirements for access for persons with physical disabilities to kitchen counters and lavatory countertops.

Legend: Upper Case Letter and **BOLD** = Primary Inspection; Lower Case = Secondary Inspection
Main Groups SI: Special Inspections, **AI:** Architectural Inspections, **EI:** Engineering Inspections, **OI:** Other Inspections
Sub-Groups   **SI:** [T] Testing Laboratory, [H] Hazardous Materials, [Y] Safety   **AI:** No Sub-Groups
EI: [V] Civil, [S] Structural, [M] Mechanical, [E] Electrical   **OI:** [O] Owner, [C] Contractor, [B] Subcontractor, [L] Legal, [G] Government, [F] Fire Protection, [P] Plumbing, [A] Acoustical

DIVISION 12 — FURNISHING — WINDOW TREATMENTS 12500

STANDARDS: UL (93)

SI AI EI OI

[ ] [x] [ ] **[B]** 1. Verify blocking and backing are provided as required.
              [C]

Drapery

[ ] [x] [ ] [C] 2. Verify stacking and panel overlapping are provided as required.

[ ] [x] [ ] [C] 3. Rods are supported adequately to prevent sagging.

[ ] [x] [ ] [C] 4. Location and height of controls is understood.

[ ] [x] [ ] [C] 5. Material is flame-proofed as required.

[ ] [x] [ ] [C] 6. Non-rusting weights are provided.

[ ] [x] [ ] [C] 7. Carriers are provided as necessary.

[ ] [x] [ ] [C] 8. By-passing arms are provided for by-parting types.

[ ] [x] [ ] [C] 9. Verify fullness required (usually 2½ times).

[ ] [x] [ ] [C] 10. Type and length of batons required are provided.

[ ] [x] [ ] [C] 11. Carriers run freely, especially on curved track.

[ ] [x] [ ] [C] 12. Backing is provided as required.

Legend: Upper Case Letter and **BOLD** = Primary Inspection; Lower Case = Secondary Inspection
Main Groups **SI:** Special Inspections, **AI:** Architectural Inspections, **EI:** Engineering Inspections, **OI:** Other Inspections
Sub-Groups **SI:** [T] Testing Laboratory, [H] Hazardous Materials, [Y] Safety   **AI:** No Sub-Groups
EI: [V] Civil, [S] Structural, [M] Mechanical, [E] Electrical   **OI:** [O] Owner, [C] Contractor, [B] Subcontractor,
[L] Legal, [G] Government, [F] Fire Protection, [P] Plumbing, [A] Acoustical

Construction Inspection Manual

DIVISION 13 — SPECIAL CONSTRUCTION — SUPPRESSION 13900

STANDARDS: CDA (37) NFPA (63) UL (93)

SI   AI   EI   OI

[ ]   [ ]   [m]   [C]   1. Verify temporary protection is provided during construction as required.
            [f]

[X]   [ ]   [M]   [B]   2. Verify approval of system by local/State fire and insurance authorities is received as required before installation, and work is coordinated with other trades to avoid congestion and interference.
           [F]   [C]

[ ]   [ ]   [m]   [B]   3. Verify location of siamese connections, post indicators, hose and threading connections, and alarm system has been approved by local fire department.
           [f]   [C]

Sprinkler System

[ ]   [ ]   [m]   [B]   4. Verify approved layout has been coordinated with other drawings as to head and pipe routing to avoid conflict with structure, lighting fixtures, ductwork, and diffusers. Installation conforms with approved drawings.
           [f]   [C]

[ ]   [ ]   [m]   [B]   5. Verify location of concealed and exposed lines is understood.
           [f]   [C]

[ ]   [ ]   [m]   [B]   6. Verify fire protection system serving other occupied buildings is not interrupted or shut off during construction.
           [f]   [C]

[ ]   [ ]   [m]   [B]   7. Verify main service, distribution, gong or alarm locations, and space provisions.
           [f]   [C]

[ ]   [ ]   [m]   [C]   8. Observe installation for pipe, size, fittings, and valves pipes are reamed, valves are accessible, gravity slope to drain is provided, hangers of proper type and spacing are rigidly installed, branch piping is off top of main, and no unscheduled cutting of structural members occurs.

[t]   [ ]   [m]   [B]   9. Verify inspection test connections and low point plugs. Verify holes through fire walls are plated and sleeved as required by applicable standards.
           [f]   [C]

[t]   [ ]   [m]   [B]   10. Verify heads as required for spaces are provided and installed in accordance with NFPA. Ensure all heads are new, unpainted, properly temperature-rated, and provided with guards where subject to mechanical injury, and have clearances required.
           [f]   [C]

[ ]   [ ]   [m]   [B]   11. Verify drainage valves or plugs allow complete drainage of entire system and are located so as not to cause water damage.
           [f]   [C]

Legend: Upper Case Letter and **BOLD** = Primary Inspection; Lower Case = Secondary Inspection
Main Groups SI: Special Inspections, AI: Architectural Inspections, EI: Engineering Inspections, OI: Other Inspections
Sub-Groups **SI:** [T] Testing Laboratory, [H] Hazardous Materials, [Y] Safety   **AI:** No Sub-Groups
EI: [V] Civil, [S] Structural, [M] Mechanical, [E] Electrical   **OI:** [O] Owner, [C] Contractor, [B] Subcontractor, [L] Legal, [G] Government, [F] Fire Protection, [P] Plumbing, [A] Acoustical

SI　AI　EI　OI

[ ]　[ ]　[m] [B]　　12. For wet pipe systems — check alarm, check valve assembly
　　　　　　[f] [C]　　　　conformance, and water flow indicators for conformance with
　　　　　　　　　　　　　　connection diagram. Observe test of waterflow alarm signal.

[ ]　[ ]　[m] [B]　　13. For dry pipe systems — check valve installation for conformance with
　　　　　　[f] [C]　　　　connection diagram. Verify proper installation of air compressors and
　　　　　　　　　　　　　　tanks is provided. Observe test of alarm signal time, etc.

[ ]　[ ]　[m] [B]　　14. Verify all alarm devices have been provided and are in operable
　　　　　　[f] [C]　　　　condition.

[ ]　[ ]　[m] [B]　　15. Verify electric power, if used, is supplied as required.
　　　　　　[f] [C]

[ ]　[ ]　[m] [B]　　16. Ensure alarm tie-in system is made with fire department if required.
　　　　　　[f] [C]

[ ]　[ ]　[m] [B]　　17. Verify system is completely cleaned, painted, insulated, tested, and
　　　　　　[f] [C]　　　　approved as required. Confirm system is flushed free of debris.

[ ]　[ ]　[m] [B]　　18. Verify spare heads are provided as required.
　　　　　　[f] [C]

[ ]　[ ]　[m] [B]　　19. Verify expansion requirements are provided for at building expansion
　　　　　　[f] [C]　　　　joints and where required.

[ ]　[ ]　[m] [B]　　20. Verify system supports, anchorage and sway bracing is in accordance
　　　　　　[f] [C]　　　　with NFPA.

Other Systems and Equipment

[ ]　[ ]　[m] [B]　　21. Verify wet standpipe cabinets of proper type, size and base length are
　　　　　　[f] [C]　　　　provided, located and installed as required. Confirm valve is not over
　　　　　　　　　　　　　　5'6" above floor.

[ ]　[ ]　[m] [B]　　22. Verify dry standpipes and hose valve shall be located and installed in
　　　　　　[f] [C]　　　　accordance with NFPA.

[ ]　[ ]　[m] [B]　　23. Verify extinguishers and cabinets of proper size, and type provided,
　　　　　　[f] [C]　　　　located and installed as required. Verify that extinguishers are new and
　　　　　　　　　　　　　　are activated so as to be operational.

[ ]　[ ]　[m] [B]　　24. Verify location of alarms.
　　　　　　[f] [C]

[ ]　[X] [M] [B]　　25. Verify that all installations meet accessibility requirements.
　　　　　　[F] [C]

Legend: Upper Case Letter and **BOLD** = Primary Inspection; Lower Case = Secondary Inspection
　Main Groups **SI:** Special Inspections, **AI:** Architectural Inspections, **EI:** Engineering Inspections, **OI:** Other Inspections
　Sub-Groups **SI:** [T] Testing Laboratory, [H] Hazardous Materials, [Y] Safety　**AI:** No Sub-Groups
　EI: [V] Civil, [S] Structural, [M] Mechanical, [E] Electrical　**OI:** [O] Owner, [C] Contractor, [B] Subcontractor,
　[L] Legal, [G] Government, [F] Fire Protection, [P] Plumbing, [A] Acoustical

Construction Inspection Manual

DIVISION 15 — MECHANICAL — PLUMBING FIXTURES AND EQUIPMENT 15400

STANDARDS: AGA (6) AWWA (24) HI (47) PPI (73)

<u>SI</u> <u>AI</u> <u>EI</u> <u>OI</u>

[ ] [ ] [M] [B] 1. Verify grades and location of piping with respect to other features of the building are understood.
         [p] [C]

[ ] [ ] [m] [B] 2. Verify existing lines and conflict with other's work is coordinated to avoid congestion or interference. Verify excavation of stubs or lines to which connection will be made is performed before trenching for new work.
         [p] [C]

[ ] [ ] [m] [B] 3. Verify number, size, and locations of sleeves before foundations, slabs, walls, and floors are placed. Verify adequacy of slopes to receive insulation, caulking, or other requirements.
         [p] [C]

[ ] [ ] [m] [B] 4. Allow no unscheduled cutting of structural members. See that special provisions for pipes passing through or parallel to footings are met as required. Allow no overcutting of holes and weakening of framing. Verify that plates and straps are provided.
         [p] [C]

[ ] [ ] [m] [B] 5. Verify pipe supports, hangers, and anchorages are provided and spaced as required. Verify isolation between pipe and support is provided as required.
         [p] [C]

[ ] [ ] [m] [B] 6. Verify expansion requirements are provided for at building expansion joints and where required.
         [p] [C]

[ ] [ ] [m] [B] 7. Verify protection is provided to keep concrete, trash, and debris out of lines. Verify capping and plugging is as required. Verify lines are cleaned and thoroughly flushed at completion.
         [p] [C]

[ ] [ ] [m] [B] 8. Verify manufacturer's recommendations (in submittals) are adhered to unless otherwise noted.
         [p] [C]

[ ] [ ] [m] [B] 9. Verify pipe supports and earthquake bracing are provided as required.
         [p] [C]

Pipe and Pipe Fittings — General

[ ] [ ] [m] [B] 10. Verify pipes to be threaded are squarely cut, threaded, and reamed properly. Verify equipment used is adequate. Verify joints are wiped clean.
         [p] [C]

Legend: Upper Case Letter and <u>BOLD</u> = Primary Inspection; Lower Case = Secondary Inspection
Main Groups SI: Special Inspections, AI: Architectural Inspections, EI: Engineering Inspections, OI: Other Inspections
Sub-Groups **SI:** [T] Testing Laboratory, [H] Hazardous Materials, [Y] Safety **AI:** No Sub-Groups
EI: [V] Civil, [S] Structural, [M] Mechanical, [E] Electrical **OI:** [O] Owner, [C] Contractor, [B] Subcontractor, [L] Legal, [G] Government, [F] Fire Protection, [P] Plumbing, [A] Acoustical

SI	AI	EI	OI	
[ ]	[ ]	[m] [p]	[B] [C]	11. Verify pipes to be soldered are cut with tool, reamed, brightened, and soldered using flux and solder required. Verify pipes and fittings weakened by overheating are replaced. Verify nonlead solder is used. Verify piping for medical gas service conforms to the requirements of NFPA. Verify piping prepared as required by NFPA including reaming, deburring and brazing without flux. Verify certified welder/brazer requirements.
[ ]	[ ]	[m] [p]	[B] [C]	12. Verify pipes with flanged joints are properly gasketed. Ensure drift pins and spud wrenches are not used.
[ ]	[ ]	[m] [p]	[B] [C]	13. Verify pipes to be welded are properly prepared. Verify welder is certified if required.
[ ]	[ ]	[m] [p]	[B] [C]	14. Verify pipes to be cemented are properly jointed and manufacturer's instructions for pipe, cleaner and cement are followed.
[ ]	[ ]	[m] [p]	[B] [C]	15. Verify installation of valves, unions, and fittings is properly made and access to valves and valve systems is as required.
[ ]	[ ]	[m] [p]	[B] [C]	16. Verify dissimilar metals have dielectric or isolating couplings and no contact of dissimilar metal piping occurs. Verify copper pipes are wrapped with tape against metal studs.
[ ]	[ ]	[m] [p]	[B] [C]	17. Verify pipes and joints are wrapped or coated as required.
[ ]	[ ]	[m] [p]	[B] [C]	18. Verify depth, alignment, and grade of pipes is as required.
[ ]	[ ]	[m] [p]	[B] [C]	19. Verify trench bottom is adequately compacted and pipes are supported or bedded and backfilled as required.
[ ]	[ ]	[m] [p]	[B] [C]	20. Verify future provisions, such as capped lines, and proper location and identification are provided if required.
[ ]	[ ]	[m] [p]	[B] [C]	21. Verify all pipes are inspected for damage, tested, and agency inspected if required before covering up. Observe testing process.

Legend: Upper Case Letter and <u>BOLD</u> = Primary Inspection; Lower Case = Secondary Inspection
Main Groups SI: Special Inspections, **AI:** Architectural Inspections, **EI:** Engineering Inspections, **OI:** Other Inspections
Sub-Groups **SI:** [T] Testing Laboratory, [H] Hazardous Materials, [Y] Safety   **AI:** No Sub-Groups
EI: [V] Civil, [S] Structural, [M] Mechanical, [E] Electrical   **OI:** [O] Owner, [C] Contractor, [B] Subcontractor, [L] Legal, [G] Government, [F] Fire Protection, [P] Plumbing, [A] Acoustical

Construction Inspection Manual

Soil, Waste and Vent Systems

<u>SI</u> <u>AI</u> <u>EI</u> <u>OI</u>

[ ] [ ] [m] **[B]** 22. Verify exterior manholes, lampholes, and cleanouts are located and
        [p] **[C]**   installed as required. Verify cleanouts to grade are provided as required.

[ ] [ ] [m] **[B]** 23. Verify pipes and fittings are of required material, type, size, and weight,
        [p] **[C]**   and are connected and installed as required.

[ ] [ ] [m] **[B]** 24. Verify requirements for dielectric unions. and connectors are met.
        [p] **[C]**

[.] [ ] [m] **[B]** 25. Verify slope of lines and their alignment are as required.
        [p] **[C]**

[ ] [ ] [m] **[B]** 26. Verify no-hub pipe is installed as required — hanger at every other joint,
        [p] **[C]**   except if over 4'0", then at every joint; clamps are provided at base of risers and at every floor penetration, and support is provided at every closet bend, trap, and arm unless otherwise required.

[ ] [ ] [m] **[B]** 27. Verify provisions for settlement and shrinkage are made if required.
        [p] **[C]**   Observe soil stack supports.

[ ] [ ] [m] **[B]** 28. Verify floor drains, areaway drains, and floor sinks are elevated and
        [p] **[C]**   properly located with respect to finish floor and will adequately drain area served. Verify provisions are adequate for connection to membranes, and waterproofness. Square-type floor drains, and cleanouts are aligned with room axis.

[ ] [ ] [m] **[B]** 29. Verify trap primers are provided if required.
        [p] **[C]**

[ ] [ ] [m] **[B]** 30. Verify cleanouts are located to allow access, and locations are as
        [p] **[C]**   understood or required.

[ ] [ ] [m] **[B]** 31. Verify clamping rings are provided as required in floors with
        [p] **[C]**   membranes.

[ ] [ ] [m] **[B]** 32. Verify rough-ins for fixtures and equipment are located and installed as
        [p] **[C]**   required.

[ ] [ ] [m] **[B]** 33. Verify requirements for dielectric unions and connectors are met.
        [p] **[C]**

[ ] [ ] [m] **[B]** 34. Inspect vent piping, combined and concealed in spaces provided, sloped
        [p] **[C]**   on horizontals, and extended through roof; flashed and counterflashed as required.

Legend: Upper Case Letter and <u>BOLD</u> = Primary Inspection; Lower Case = Secondary Inspection
    Main Groups SI: Special Inspections, **AI:** Architectural Inspections, **EI:** Engineering Inspections, **OI:** Other Inspections
    Sub-Groups   **SI:** [T] Testing Laboratory, [H] Hazardous Materials, [Y] Safety   **AI:** No Sub-Groups
    EI: [V] Civil, [S] Structural, [M] Mechanical, [E] Electrical   **OI:** [O] Owner, [C] Contractor, [B] Subcontractor,
       [L] Legal, [G] Government, [F] Fire Protection, [P] Plumbing, [A] Acoustical

Water Supply System

SI AI EI OI

[ ] [ ] [m] **[B]** 35. Verify pipes and fittings are of required material, type, and size; located
        [p] **[C]**   and installed as required. Verify non-lead solder is used.

[ ] [ ] [m] **[B]** 36. Verify exterior lines are installed to depth required; properly bedded and
        [p] **[C]**   backfilled. Verify thrust blocks are provided as required. Confirm coordination is made for meters, shut-offs, hydrants, and boxes.

[ ] [ ] [m] **[B]** 37. Verify exterior lines are installed to depth required, properly bedded and
        [p] **[C]**   backfilled. Verify thrust blocks are provided as required. Confirm coordination is made for meters, shut-offs, hydrants and boxes.

[ ] [ ] [m] **[B]** 38. In large structures, verify shut-off valves are provided if required to
        [p] **[C]**   isolate portions of system.

[ ] [ ] [m] **[B]** 39. Verify rough-ins to fixtures and equipment are located and installed as
        [p] **[C]**   required.

[ ] [ ] [m] **[B]** 40. Verify valves for proper function are used as required, and location and
        [p] **[C]**   accessibility are understood. Verify location and type of access panels. See that water system can be drained at lowest point. Verify all valves are labeled if required.

[ ] [ ] [m] **[B]** 41. Verify air chambers or shock absorbers are provided if required.
        [p] **[C]**

[ ] [ ] [m] **[B]** 42. Verify sound and vibration isolators are provided as required.
        [p] **[C]**

[ ] [ ] [m] **[B]** 43. Verify dielectric fittings are provided for connection of dissimilar
        [p] **[C]**   metals.

[ ] [ ] [m] **[B]** 44. Verify allowance for expansion and contraction is provided.
        [p] **[C]**

[ ] [ ] [m] **[B]** 45. Verify system is tested before concealment or installation of insulation.
        [p] **[C]**   Observe testing process.

[ ] [ ] [m] **[B]** 46. Verify insulation is of required size, weight, thickness, and type and is
        [p] **[C]**   installed as required.

[ ] [ ] [m] **[B]** 47. Verify lines are identified as required.
        [p] **[C]**

[ ] [ ] [m] **[B]** 48. Verify lines are sterilized as required — proper dosage, distribution,
        [p] **[C]**   retention, and final flush-out. Ensure certification is provided.

Legend: Upper Case Letter and <u>BOLD</u> = Primary Inspection; Lower Case = Secondary Inspection
Main Groups SI: Special Inspections, **AI:** Architectural Inspections, **EI:** Engineering Inspections, **OI:** Other Inspections
Sub-Groups   **SI:** [T] Testing Laboratory, [H] Hazardous Materials, [Y] Safety   **AI:** No Sub-Groups
  EI: [V] Civil, [S] Structural, [M] Mechanical, [E] Electrical   **OI:** [O] Owner, [C] Contractor, [B] Subcontractor,
  [L] Legal, [G] Government, [F] Fire Protection, [P] Plumbing, [A] Acoustical

Gas Piping System

SI AI EI OI

[ ] [ ] [m] [B] 49. Verify materials, sizes, and installation are as required.
      [p] [C]

[ ] [ ] [m] [B] 50. Verify location, depth, alignment, and coating of exterior lines is as required. Verify isolation, sleeves, and installation in building are in compliance with codes and safety regulations. See that proper ventilation is provided.
      [p] [C]

[ ] [ ] [m] [B] 51. Verify locations of drip pockets are provided as required.
      [p] [C]

[ ] [ ] [m] [B] 52. Verify plug cocks, gas pressure regulators, earthquake valves and insulating couplings are installed as required. Verify all are labeled if required.
      [p] [C]

Fixtures

[ ] [ ] [m] [B] 53. Verify adequate blocking, backing, and brackets are provided to receive fixtures. Verify chairs and carriers are provided as required. Verify fixtures are rigidly installed.
      [p] [C]

[ ] [x] [M] [B] 54. Verify installed fixtures are undamaged and protected during construction. Ensure use of fixtures is avoided until system is complete and tested.
      [P] [C]

[ ] [x] [M] [B] 55. Verify fixtures are installed with accessories, trim and brass specified. Verify finish is as specified and stops are provided as required. Verify vacuum breakers are provided as required. Verify that code-required laminar-flow type devices are used on all hospital sink fixtures.
      [P] [C]

[ ] [ ] [m] [B] 56. Verify fixtures are installed level. Ensure that the hot water generators are securely anchored. Verify all gauges, valves and strainers, are visible and accessible.
      [p] [C]

[ ] [ ] [m] [B] 57. Check temperature and pressure settings for relief valves. Verify strainers are provided as required; piping is provided to floor drains or exterior from relief valves; and pressure reducing valve is installed and set to pressure as required.
      [p] [C]

[ ] [x] [M] [B] 58. Verify escutcheons, flanges, and cover plates are of type and finish required.
      [P] [C]

Legend: Upper Case Letter and **BOLD** = Primary Inspection; Lower Case = Secondary Inspection
Main Groups SI: Special Inspections, **AI:** Architectural Inspections, **EI:** Engineering Inspections, **OI:** Other Inspections
Sub-Groups   SI: [T] Testing Laboratory, [H] Hazardous Materials, [Y] Safety   **AI:** No Sub-Groups
EI: [V] Civil, [S] Structural, [M] Mechanical, [E] Electrical   **OI:** [O] Owner, [C] Contractor, [B] Subcontractor, [L] Legal, [G] Government, [F] Fire Protection, [P] Plumbing, [A] Acoustical

SI　AI　EI　OI

[ ]　[ ]　[m]　[B]　　59. Observe that floor drains and roof drains with clamping rings are properly installed to membrane and weeps are cleared as provided. Observe that shower drains are similarly installed as required.
　　　　　　[p]　[C]

[ ]　[ ]　[m]　[B]　　60. Verify fixtures are properly cleaned at completion. Verify faucets operate easily and are in proper position.. Flush water-closets for confirming proper operation.
　　　　　　[p]　[C]

[ ]　[X]　[M]　[B]　　61. Verify fixtures allow for use by persons with physical disabilities, including fixture types and space at fixtures.
　　　　　　[P]　[C]

Fuel Oil Piping System

[ ]　[ ]　[m]　[B]　　62. Verify pipe "dope" (sealant) used at screwed joints to be as specified and suitable for use with petroleum product transported.
　　　　　　[p]　[C]

[ ]　[ ]　[m]　[B]　　63. Verify piping for petroleum product (diesel oil, etc.) is material specified and per code requirements.
　　　　　　[p]　[C]

[ ]　[ ]　[m]　[B]　　64. Verify secondary containment material and system is specified and suitable for use for petroleum products.
　　　　　　[p]　[C]

[ ]　[ ]　[m]　[B]　　65. Verify fire protection enclosures or coverings for piping, hangers and bracing support systems is as specified and per code requirements.
　　　　　　[p]　[C]

[ ]　[ ]　[m]　[B]　　66. Verify fuel oil day-tanks and storage tanks to be of type specified, with code required secondary containment, leak-detection, and monitoring features specified.
　　　　　　[p]　[C]

Legend:　Upper Case Letter and BOLD = Primary Inspection;　Lower Case = Secondary Inspection
Main Groups SI: Special Inspections, **AI:** Architectural Inspections, **EI:** Engineering Inspections, **OI:** Other Inspections
Sub-Groups　SI: [T] Testing Laboratory, [H] Hazardous Materials, [Y] Safety　　**AI:** No Sub-Groups
　　EI: [V] Civil, [S] Structural, [M] Mechanical, [E] Electrical　　**OI:** [O] Owner, [C] Contractor, [B] Subcontractor,
　　[L] Legal, [G] Government, [F] Fire Protection, [P] Plumbing, [A] Acoustical

Construction Inspection Manual

DIVISION 15 — MECHANICAL — HEAT-GENERATION EQUIPMENT 15550

LIQUID HEAT TRANSFER 15750

STANDARDS: AGA (6) ASHRAE (21) ASME (22) AWS (23) NFPA (63) U7L (93)

SI AI EI OI

[ ] [ ] [m] [B] 1. Verify location, spacing, and size of all required sleeves, inserts, foundation bolts, foundations, pads, and openings are coordinated with trades involved and are provided.
        [C]

[ ] [ ] [m] [B] 2. Verify space and headroom are provided, and approved equipment is accessible for operation, servicing, cleaning, and repair. See that floor drains are located properly and lighting is suitably located with respect to equipment.
        [C]

[ ] [ ] [m] [B] 3. Verify nameplates, identification, and characteristics of equipment are attached. Confirm they match approved requirements and are not covered by insulation or painted out.
        [C]

[ ] [ ] [m] [B] 4. Verify tube removal, cleaning spaces, or provisions are adequate.
        [C]

[ ] [ ] [M] [B] 5. Verify code clearances to all equipment electric panels are adequate.
        [C]

[ ] [ ] [m] [C] 6. Verify condition of using building equipment for temporary heat is understood and/or approved.

Boilers, Equipment, and Distribution

[ ] [ ] [M] [B] 7. Verify pressure boilers conform with or are identified with ASME code.
        [C]

[ ] [ ] [m] [B] 8. Verify bases or refractory bases are provided as required.
        [C]

[ ] [ ] [m] [B] 9. Verify expansion joints are provided and guided. Check requirement for expansion joint in floor around boiler.
        [C]

[ ] [ ] [m] [B] 10. For oil burning equipment — check size of burner tips, location of electrodes, position of gas or oil pilot, and clearances for removal of burner from furnace.
        [C]

[ ] [ ] [m] [B] 11. For gas burners — approved standard, position of pilot flame and sensing element, regulators and controls provided, regulator installed in a vertical position, and gas vents piped to exterior.
        [C]

Legend: Upper Case Letter and **BOLD** = Primary Inspection; Lower Case = Secondary Inspection
  Main Groups SI: Special Inspections, **AI:** Architectural Inspections, **EI:** Engineering Inspections, **OI:** Other Inspections
  Sub-Groups **SI:** [T] Testing Laboratory, [H] Hazardous Materials, [Y] Safety **AI:** No Sub-Groups
  EI: [V] Civil, [S] Structural, [M] Mechanical, [E] Electrical **OI:** [O] Owner, [C] Contractor, [B] Subcontractor, [L] Legal, [G] Government, [F] Fire Protection, [P] Plumbing, [A] Acoustical

SI	AI	EI	OI	
[ ]	[ ]	[m]	[B] [C]	12. For forced draft fans — check fans for these features: anchorage, alignment, and rotation; accessibility for lubrication; damper operation as required; insulation application; safety control interlocks and air-flow switches.
[ ]	[ ]	[m]	[B] [C]	13. For oil storage tank — approved standard, tank capacity and calibration, required openings, proper anchorage, minimum cover and/or clearance, tank heaters if required, and coatings. Verify buried tanks are provided in double wall configuration with monitoring as required.
[ ]	[ ]	[m]	[B] [C]	14. Verify piping is of material, size, weight and type required; fabrication is performed using proper equipment; lines are reamed, openings are protected, and fittings connections are as approved.
[ ]	[ ]	[m]	[B] [C]	15. Verify piping is hung, guided, or anchored as required; provisions for expansion and contraction are made; pitch of horizontal runs is correct; and provisions for dielectric connection or isolation of dissimilar metals is performed. If hangers are required to be installed over insulation, see that high-density insulation inserts and metal shields are provided as required.
[ ]	[ ]	[m]	[B] [C]	16. Verify valves are installed of type and position required. All fittings such as strainers, checks, gauges, air reliefs, drips, traps, etc., are provided.
[ ]	[ ]	[m]	[B] [C]	17. Verify piping and equipment are insulated with approved materials, and extent and installation are as required. Verify vapor seal is provided as required. Verify thickness and continuity are as required. Band spacing is as required.
[ ]	[ ]	[m]	[B] [C]	18. Verify radiant heating coils are of type, size, and material approved; accurately placed, firmly secured, and positioned; and tested before encasement. Verify test pressure is maintained while concrete is deposited.
[ ]	[ ]	[M]	[B] [C]	19. Verify piping is painted and/or identified for flow and type as required.
[ ]	[ ]	[m]	[B] [C]	20. Verify shut-offs for fuel and water are provided. Verify valves are provided to shut down sections of system if required. Verify valves are labeled if required.

Legend: Upper Case Letter and **BOLD** = Primary Inspection; Lower Case = Secondary Inspection
Main Groups SI: Special Inspections, **AI:** Architectural Inspections, **EI:** Engineering Inspections, **OI:** Other Inspections
Sub-Groups **SI:** [T] Testing Laboratory, [H] Hazardous Materials, [Y] Safety **AI:** No Sub-Groups
EI: [V] Civil, [S] Structural, [M] Mechanical, [E] Electrical **OI:** [O] Owner, [C] Contractor, [B] Subcontractor, [L] Legal, [G] Government, [F] Fire Protection, [P] Plumbing, [A] Acoustical

Construction Inspection Manual

SI	AI	EI	OI		
<u>SI</u>	<u>AI</u>	<u>EI</u>	<u>OI</u>		

[ ] [ ] [m] **[B]**
            [C] 21. Verify safety and relief valves are provided and set to PSIG. Verify discharges are piped to drains.

[ ] [ ] [m] **[B]**
            [C] 22. Verify safety operating controls are provided as required.

[ ] [ ] [m] **[B]**
            [C] 23. Verify combustion air system is provided as required.

[ ] [ ] [m] **[B]**
            [C] 24. Verify breaching and flues are of proper material, construction, and type, and are installed as required.

[ ] [ ] [m] **[B]**
            [C] 25. Verify expansion tanks are located, mounted and anchored as required and provided with accessories and drain.

[ ] [ ] [m] **[B]**
            [C] 26. Verify valves and fittings are insulated as required.

Terminal Units

[ ] [ ] [m] **[B]**
            [C] 27. Heating and ventilating units — verify anchored and provided with vibration isolators as required; access doors are provided and are tight; flexible pipe connectors are provided as required; controls are provided as required; and location and layout is coordinated.

[ ] [ ] [m] **[B]**
            [C] 28. Unit heaters — verify noise level is within approved range; clearances and location are as approved; adequate air distribution is provided; and controls are required are provided.

[ ] [ ] [m] **[B]**
            [C] 29. Base board units — verify location, type, size, mounting, and controls are provided as required. Verify covers, access doors, dampers, and end plates are provided to extent required.

Cleaning, Testing, and Balancing

[ ] [ ] [m] **[B]**
            [C] 30. Verify system is completely clean and flushed of all debris. Operate system in presence of agencies and/or engineers as required.

[ ] [ ] **[M]** **[B]**
            [C] 31. Verify system is completely balanced, and balancing report is available. Coordinate location of all balancing devices.

[ ] [x] **[M]** **[B]**
            [C] 32. Verify system is operating and instruction is given to future operating personnel as required.

Legend: Upper Case Letter and <u>**BOLD**</u> = Primary Inspection; Lower Case = Secondary Inspection
   Main Groups SI: Special Inspections, **AI:** Architectural Inspections, **EI:** Engineering Inspections, **OI:** Other Inspections
    Sub-Groups   **SI:** [T] Testing Laboratory, [H] Hazardous Materials, [Y] Safety   **AI:** No Sub-Groups
      EI: [V] Civil, [S] Structural, [M] Mechanical, [E] Electrical   **OI:** [O] Owner, [C] Contractor, [B] Subcontractor,
        [L] Legal, [G] Government, [F] Fire Protection, [P] Plumbing, [A] Acoustical

Construction Inspection Manual

DIVISION 15 — MECHANICAL — REFRIGERATION EQUIPMENT 15650

STANDARDS: AGA (6) ASHRAE (21) ASTM (19) AWS (23)

SI   AI   EI   OI

[ ] [ ] [M] [B] [C] 1. Verify all materials and equipment are as approved — nameplates, identification, and characteristics are provided and are not covered by insulation or painted out.

[ ] [ ] [m] [B] [C] 2. Verify space is adequate for maintenance, operation, repair, and servicing of equipment. Verify mounting and anchorage methods are provided and located, i.e., pads, hangers.

[ ] [ ] [m] [B] [C] 3. Verify all rotating parts, and belts have guards or provide protection.

[ ] [ ] [m] [B] [C] 4. Verify vibration isolators and flexible connections are provided as approved and required.

[x] [ ] [m] [B] [C] 5. Verify fire separation from fuel-fired equipment is provided if required by code or by specified requirements.

[ ] [ ] [m] [B] [C] 6. Verify freeze protection devices and materials are provided if required.

[ ] [ ] [m] [B] [C] 7. Verify tube removal space and tube cleaning space is adequate.

[x] [ ] [M] [B] [C] 8. Verify code clearances to all equipment electric panels are adequate.

Piping

[ ] [ ] [m] [B] [C] 9. Verify type, size, weight, material, and fittings are as approved and required.

[ ] [ ] [m] [B] [C] 10. Verify installation method of piping is as required — pipes are square cut and reamed; soldering is as required; internal valve parts are protected against heat or removed; joints are thoroughly cleaned and fluxed; and excess flux and acid are removed. Verify solder used is of an approved type.

[ ] [ ] [m] [B] [C] 11. Verify flexible connections are provided as required.

[ ] [ ] [m] [B] [C] 12. Verify unions and flanges installed for maintenance are as required.

[ ] [ ] [m] [B] [C] 13. Verify lines are properly sloped as required.

Legend: Upper Case Letter and **BOLD** = Primary Inspection;   Lower Case = Secondary Inspection
Main Groups **SI**: Special Inspections, **AI**: Architectural Inspections, **EI**: Engineering Inspections, **OI**: Other Inspections
Sub-Groups   **SI**: [T] Testing Laboratory, [H] Hazardous Materials, [Y] Safety   **AI**: No Sub-Groups
EI: [V] Civil, [S] Structural, [M] Mechanical, [E] Electrical   **OI**: [O] Owner, [C] Contractor, [B] Subcontractor, [L] Legal, [G] Government, [F] Fire Protection, [P] Plumbing, [A] Acoustical

Construction Inspection Manual

SI   AI   EI   OI

[ ] [ ] [m] **[B]** 14. Verify air vents are installed at high points and drains installed at low
           [C]   points as required in water lines. Verify proper type of valves are provided, i.e., gate or globe.

[x] [ ] [m] **[B]** 15. Verify balancing cocks are installed as required. Verify pressure gauges,
           [C]   thermal elements, are provided.

[ ] [ ] [m] **[B]** 16. Verify piping is hung, guided or anchored as required; provisions for
           [C]   expansion and contraction are made; pitch of horizontal runs is correct; and provisions are made for dielectric connection or isolation of dissimilar metals. If hangers are required to be installed over insulation, see that high-density insulation inserts and metal shields are provided as required.

[ ] [ ] [m] **[B]** 17. Verify system is checked and tested for leaks.
           [C]

[x] [ ] [m] **[B]** 18. Verify insulation is provided and installed as required. Verify vapor
           [C]   barriers, adhesives, arid sealants are non-combustible, if required. Verify requirements for insulating flanges, fittings, and valves are met.

[ ] [ ] [m] **[B]** 19. Verify piping is properly scaled and flashed as required, when
           [C]   penetrating building elements.

Equipment

[ ] [ ] [m] **[B]** 20. Verify air cooled condenser-air flow is not obstructed, and wind
           [C]   deflectors are provided if required.

[ ] [ ] [m] **[B]** 21. Water-cooled condensers — verify proper flow and no leaks occur.
           [C]

[ ] [ ] [m] **[B]** 22. Evaporative condenser — check for spray coverage, quiet float valve,
           [C]   and water level.

[ ] [ ] [m] **[B]** 23. Reciprocating compressor — check for shaft alignment on direct drive;
           [C]   also, suction and discharge pressures, installation of required gauges, motor amperage under maximum load, and cylinder heat overheating.

[ ] [ ] [m] **[B]** 24. Centrifugal compressor — cheek for alignment of unit, drive and gear
           [C]   box: check for noise and vibration and required gauges.

[ ] [ ] [m] **[B]** 25. Verify receiver location is out of direct sun if installed outside building.
           [C]

Legend: Upper Case Letter and <u>BOLD</u> = Primary Inspection;   Lower Case = Secondary Inspection
  Main Groups SI: Special Inspections, **AI:** Architectural Inspections, **EI:** Engineering Inspections, **OI:** Other Inspections
  Sub-Groups **SI:** [T] Testing Laboratory, [H] Hazardous Materials, [Y] Safety   **AI:** No Sub-Groups
    EI: [V] Civil, [S] Structural, [M] Mechanical, [E] Electrical   **OI:** [O] Owner, [C] Contractor, [B] Subcontractor,
    [L] Legal, [G] Government, [F] Fire Protection, [P] Plumbing, [A] Acoustical

SI	AI	EI	OI	
[ ]	[ ]	[m]	[B] [C]	26. Verify relief valve on receiver: is of size required and discharges to atmosphere.
[ ]	[ ]	[m]	[B] [C]	27. Verify receiver drain, purge valve, liquid level indication, and shutoff valves are provided and/or as required and piped to exterior as required.
[ ]	[ ]	[m]	[B] [C]	28. Cooling tower — verify location and provision for mounting are as approved and required; mist eliminators are provided if required, and overflow and drain piping are provided.
[ ]	[ ]	[m]	[B] [C]	29. Mechanical draft cooling tower — verify unobstructed air intake is provided; fan rotation and speed, belt tension is as required; and weather protection is provided for motor if required.
[ ]	[ ]	[m]	[B] [C]	30. Verify pumps are supported properly, free of excess vibration; piping around is adequately supported; and all gauges and motors are provided.
[ ]	[ ]	[m]	[B] [C]	31. Verify all insulating materials are provided and installed as approved.
[ ]	[ ]	[m]	[B] [C]	32. Observe procedures for testing of systems. Verify future operating personnel are instructed in operation of equipment if required.

Legend: Upper Case Letter and <u>**BOLD**</u> = Primary Inspection; Lower Case = Secondary Inspection
Main Groups SI: Special Inspections, **AI:** Architectural Inspections, **EI:** Engineering Inspections, **OI:** Other Inspections
Sub-Groups **SI:** [T] Testing Laboratory, [H] Hazardous Materials, [Y] Safety   **AI:** No Sub-Groups
EI: [V] Civil, [S] Structural, [M] Mechanical, [E] Electrical   **OI:** [O] Owner, [C] Contractor, [B] Subcontractor, [L] Legal, [G] Government, [F] Fire Protection, [P] Plumbing, [A] Acoustical

Construction Inspection Manual

DIVISION 15 — MECHANICAL — AIR DISTRIBUTION 15800

STANDARDS: AGA (6) ASHRAE (21) ASTM (19) NFMA (62) NFPA (63) SMACNA (82) UL (93)

SI   AI   EI   OI

[ ] [ ] [M] [B]   1. Verify equipment has identification nameplates, and characteristics are as approved.
          [C]

[ ] [ ] [m] [B]   2. Verify approved vibration isolators and flexible connections are furnished and installed as required.
          [C]

[ ] [ ] [m] [B]   3. Verify provision is made for proper mounting and anchorage of equipment including pads and hangers.
          [C]

[ ] [ ] [m] [B]   4. Verify equipment operates without excessive vibration or noise.
          [C]

[ ] [ ] [ ] [B]   5. Verify condition of using building equipment for temporary heat is understood and/or approved.
          [C]

Furnaces

[ ] [ ] [m] [B]   6. Verify: units are located and mounted as required.
          [C]

[ ] [ ] [m] [B]   7. Suitable service access is provided.
          [C]

[x] [ ] [m] [B]   8. Fire-resistive surfaces and spacing are provided as required.
          [C]

[x] [ ] [m] [B]   9. Combustion air provisions are made as required.
          [C]

Air Handling Units and Fans

[ ] [ ] [m] [B]   10. Verify rotation of fan is as required before power is connected.
          [C]

[ ] [ ] [m] [B]   11. Drive method — if belt driven, check means of adjustment. Pulley and belt are aligned. Note bearing and belt numbers before connection of ducts is made to unit.
          [C]

[x] [ ] [m] [B]   12. Verify guards are provided for rotating equipment and belts.
          [C]

[ ] [ ] [m] [B]   13. Verify lubrication of equipment, if required, is accessible and extends to exterior of unit.
          [C]

[ ] [ ] [m] [B]   14. Verify equipment exposed to weather is weatherproof.
        [p] [C]

[x] [x] [ ] [B]   15. Verify roof mounted equipment is properly flashed at curbs. Verify service accessibility is provided.
          [C]

Legend: Upper Case Letter and **BOLD** = Primary Inspection; Lower Case = Secondary Inspection
Main Groups SI: Special Inspections, **AI:** Architectural Inspections, **EI:** Engineering Inspections, **OI:** Other Inspections
Sub-Groups   **SI:** [T] Testing Laboratory, [H] Hazardous Materials, [Y] Safety   **AI:** No Sub-Groups
EI: [V] Civil, [S] Structural, [M] Mechanical, [E] Electrical   **OI:** [O] Owner, [C] Contractor, [B] Subcontractor, [L] Legal, [G] Government, [F] Fire Protection, [P] Plumbing, [A] Acoustical

SI AI EI OI

[ ] [ ] [m] [B] 16. Verify backdraft dampers and/or sound traps are provided as required on
         [C]   exhaust fans. Check for operation, rattle, felt strips, and separate
                  frames as required.

[ ] [ ] [m] [B] 17. Verify exhaust and supplies are oriented to avoid conflict. Verify
         [C]   exhaust air discharges and outside air intakes are separated by code
                  required distances and located to account for prevailing winds.

Filters and Screens

[ ] [ ] [m] [B] 18. Verify required type is furnished and installed. Verify filters are to be
         [C]   clean at completion of final tests.

[ ] [ ] [m] [B] 19. Verify accessibility is provided for removal and replacement of filter as
         [C]   required.

[ ] [ ] [m] [B] 20. Verify method of mounting is as required. Observe that air stream can be
         [C]   distributed over all filter areas.

[ ] [ ] [m] [B] 21. Verify proper amount of adhesive and washing tank for viscous medium
         [C]   type filters are provided.

[ ] [ ] [m] [B] 22. Verify sealing strips are provided as required.
         [C]

[ ] [ ] [m] [B] 23. Verify electrostatic filters have warning lights and interlocks as required
         [C]   — ionizers have free access and do not have loose wires or sparking.

[ ] [ ] [m] [B] 24. Verify automatic sprays provide complete washing and spray coverage.
         [C]

[ ] [ ] [m] [B] 25. Verify traveling screen are observed for oil charge and operation of
         [C]   screen.

[ ] [ ] [m] [B] 26. Renewable roll filters — check for tracking of roll, media runout switch,
         [C]   tinier setting, static pressure control, and tension on media.

[ ] [ ] [m] [B] 27. Verify spare filters are provided as required. Verify equipment filters to
         [C]   be clean and/or replaced as required at date of acceptance.

[ ] [ ] [m] [B] 28. Verify bird and insect screen of proper mesh size and material is
         [C]   provided as required, and isolation is made between dissimilar metal.

Legend: Upper Case Letter and <u>BOLD</u> = Primary Inspection; Lower Case = Secondary Inspection
Main Groups **SI:** Special Inspections, **AI:** Architectural Inspections, **EI:** Engineering Inspections, **OI:** Other Inspections
Sub-Groups **SI:** [T] Testing Laboratory, [H] Hazardous Materials, [Y] Safety **AI:** No Sub-Groups
EI: [V] Civil, [S] Structural, [M] Mechanical, [E] Electrical **OI:** [O] Owner, [C] Contractor, [B] Subcontractor,
[L] Legal, [G] Government, [F] Fire Protection, [P] Plumbing, [A] Acoustical

Ductwork

SI AI EI OI

[ ] [ ] [m] [B] 29. Verify ductwork layout is coordinated with other trades to avoid
           [C]   congestion and interference. A ductwork drawing coordinating plumbing, electrical, and sprinklers, is recommended on complex work.

[ ] [ ] [m] [B] 30. Verify type, material, thickness, and shape are as required. Verify field
           [C]   changes are approved before installation.

[ ] [ ] [m] [B] 31. Verify joint connections are of required type. Check seams and breaks
           [C]   for cracks. Verify joint provides a smooth surface on interior of duct, and laps are in direction of air flow.

[ ] [ ] [m] [B] 32. Verify slope ratio of transitions, radius of curved duct, air turns, and
           [C]   deflectors are provided as required.

[ ] [ ] [m] [B] 33. Verify bracing, reinforcement stiffeners, and hangers are provided and
           [C]   ductwork is installed — all as required.

[ ] [ ] [m] [B] 34. Verify that all volume dampers, branch duct dampers, register or diffuser
           [C]   dampers and splitter dampers are provided as required and operating mechanism is accessible.

[x] [ ] [m] [B] 35. Verify fire dampers and smoke fire dampers of type required are
           [C]   furnished and installed as required by NFPA. Verify that access is provided to dampers.

[ ] [ ] [m] [B] 36. Verify flexible connectors are fabricated and provided where required.
           [C]

[x] [ ] [m] [B] 37. Verify access doors and/or access space is provided at all items requiring
           [C]   servicing, such as fire dampers, automatic dampers, manual dampers, coils, heaters, filters, and thermostats. Verify size is sufficient for access and maintenance.

[ ] [ ] [m] [B] 38. Verify proper sleeves and openings through walls and floors are
           [C]   provided as required and are sealed as required. Allow no unscheduled cutting of structural members without approval.

[ ] [ ] [m] [B] 39. Verify ductwork is properly taped or sealed if required.
           [C]

Legend: Upper Case Letter and **BOLD** = Primary Inspection; Lower Case = Secondary Inspection
Main Groups SI: Special Inspections, AI: Architectural Inspections, EI: Engineering Inspections, OI: Other Inspections
Sub-Groups   **SI:** [T] Testing Laboratory, [H] Hazardous Materials, [Y] Safety   **AI:** No Sub-Groups
EI: [V] Civil, [S] Structural, [M] Mechanical, [E] Electrical   **OI:** [O] Owner, [C] Contractor, [B] Subcontractor, [L] Legal, [G] Government, [F] Fire Protection, [P] Plumbing, [A] Acoustical

SI   AI   EI   OI

Duct Lining and Insulation

[x] [ ] [m] **[B]** 40. Verify ducts are tested for air tightness, if required, before installation of
           [C]   insulation.

[ ] [ ] [m] **[B]** 41. Verify type, thickness, material, extent, and method of fastening and
           [C]   installation are as required.

[ ] [ ] [m] **[B]** 42. Verify sound deadening and vapor barrier are provided as required.
           [C]

[ ] [ ] [m] **[B]** 43. Verify insulation subject to damage is protected as required.
           [C]

[ ] [ ] [m] **[B]** 44. Verify materials are fire retardant or incombustible as required.
           [C]

[ ] [ ] [m] **[B]** 45. Verify vapor barrier integrity is maintained.
           [C]

Outlets, Diffusers, Registers and Grilles

[ ] [ ] [m] **[B]** 46. Verify all ducts, plenums, and equipment are thoroughly cleaned of all
           [C]   debris before supply outlets are installed.

[ ] [x] [m] **[B]** 47. Verify all items are furnished and installed as required, and approved.
           [C]   Verify finishes in areas match as required.

[ ] [ ] [m] **[B]** 48. Verify volume control devices are provided as required and are
           [C]   accessible.

[ ] [ ] [m] **[B]** 49. Verify gaskets are provided and installed as required.
           [C]

[x] [ ] [m] **[B]** 50. Verify items are securely attached and supported as required.
           [C]

Balancing and Testing

[ ] [ ] [m] **[B]** 51. Verify all bearings are lubricated, tension of pulleys and belts is
           [C]   adjusted, guards are in place, and adjustments, and connections are made.

[x] [ ] [m] **[B]** 52. Verify necessary equipment to provide balancing of air flow is available,
           [C]   and each outlet is tested if required. Confirm final air flows, and verify that a report is submitted if required.

[x] [x] [m] **[B]** 53. Objectionable noise due to velocity, distribution, vibration, or inability
           [C]   to achieve required air is to be brought to attention of architect and consultant.

Legend: Upper Case Letter and **BOLD** = Primary Inspection;   Lower Case = Secondary Inspection
Main Groups **SI**: Special Inspections, **AI**: Architectural Inspections, **EI**: Engineering Inspections, **OI**: Other Inspections
Sub-Groups   **SI**: [T] Testing Laboratory, [H] Hazardous Materials, [Y] Safety   **AI**: No Sub-Groups
EI: [V] Civil, [S] Structural, [M] Mechanical, [E] Electrical   **OI**: [O] Owner, [C] Contractor, [B] Subcontractor,
[L] Legal, [G] Government, [F] Fire Protection, [P] Plumbing, [A] Acoustical

Construction Inspection Manual

DIVISION 16 — ELECTRICAL — BASIC ELECTRICAL MATERIALS AND METHODS 16050

STANDARDS: CDA (37) NEC (59) NEMA (60) UL (93)

SI	AI	EI	OI	
[ ]	[ ]	[E]	[B] [C]	1. Verify materials are listed by a nationally recognized testing laboratory.

Raceways

SI	AI	EI	OI	
[ ]	[ ]	[e]	[B] [C]	2. Observe limitations required on use of rigid conduit, thinwall, flexible metal conduit, plastic conduit and other non-metallic conduit: (ex. liquidtight flexible metal conduit.)
[ ]	[ ]	[e]	[B] [C]	3. Conduit size complies with code and is acceptable for the application. Required types of fittings are provided and are compatible with conduit type.
[ ]	[ ]	[e]	[B] [C]	4. Verify conduits are provided and properly installed before slabs on grade are installed. Check thickness of slab to allow for conduit size.
[ ]	[ ]	[e]	[B] [C]	5. Verify slab to allow for conduit size. Stub-ups, and couplings, are installed above finish floor level for free-standing equipment, as required. Exposed conduit should be installed so that bent portion will not extend above floor level.
[ ]	[ ]	[e]	[B] [C]	6. Verify insulating bushings and connector linings are provided as specified.
[ ]	[ ]	[e]	[B] [C]	7. Verify field cutting of conduit — square cut, reamed or filled, and cleaned of oil and filings.
[x]	[ ]	[e]	[B] [C]	8. Verify conduit is secured and fastened as required and in accordance with applicable code. Verify runs in wet areas are spaced at least ¼" off surface. Maintain clearances between dissimilar metals to prevent galvanic action.
[ ]	[ ]	[e]	[B] [C]	9. Verify support of vertical raceways at each floor level is provided in multi-story structures if required.
[ ]	[x]	[e]	[B] [C]	10. Verify areas where exposed conduit is allowed. Exposed conduits should be installed parallel or perpendicular to structure. Verify vertical runs are plumb.
[ ]	[ ]	[e]	[B] [C]	11. Verify raceways are kept plugged during construction.

Legend: Upper Case Letter and **BOLD** = Primary Inspection; Lower Case = Secondary Inspection
Main Groups SI: Special Inspections, **AI:** Architectural Inspections, **EI:** Engineering Inspections, **OI:** Other Inspections
Sub-Groups   SI: [T] Testing Laboratory, [H] Hazardous Materials, [Y] Safety   **AI:** No Sub-Groups
EI: [V] Civil, [S] Structural, [M] Mechanical, [E] Electrical   **OI:** [O] Owner, [C] Contractor, [B] Subcontractor, [L] Legal, [G] Government, [F] Fire Protection, [P] Plumbing, [A] Acoustical

<u>SI</u> <u>AI</u> <u>EI</u> <u>OI</u>

[ ] [ ] [e] **[B]** 12. Ensure no damage or deformation effectively reducing inside area occurs.
             [C]

[ ] [ ] [e] **[B]** 13. Verify coating and surface treatment are provided as required, including connections. No treatment is normally provided if enclosed in concrete, unless otherwise required.
             [C]

[ ] [ ] [e] **[B]** 14. Verify depth of installation in relation to finish grade is minimum per code. Coordinate with other utilities below grade. Verify concrete encasement and sand bed are provided as required. Verify backfill is free of debris and compacted to prevent future settlement.
             [C]

[ ] [ ] [e] **[B]** 15. Verify pull wires are provided for cable installation. Verify use of non-metallic cord or rope for aluminum conduit. Verify pull cords for empty conduits.
             [C]

[ ] [ ] [e] **[B]** 16. Stub-ups for future extensions are provided as required. Location identification is provided and recorded.
             [C]

[ ] [ ] [e] **[B]** 17. Sleeves for future work are provided as required. Verify empty conduits have been labeled for their intended future use.
             [C]

[ ] [ ] [e] **[B]** 18. Means are provided to accommodate contraction and expansion at building expansion joints as required.
             [C]

[x] [ ] [e] **[B]** 19. Seismic bracing provided for all conduit 2½" diameter and larger or in accordance with accepted regulations by the governing authorities.
             [C]

Busways

[ ] [ ] [e] **[B]** 20. Proper support is provided at required intervals. Transverse and longitudinal bracing is provided to prevent lateral movement, in accordance with codes and the governing authorities.
             [C]

[ ] [ ] [e] **[B]** 21. Provision for expansion is in accordance with manufacturer's instructions.
             [C]

[ ] [ ] [e] **[B]** 22. Busway housing is grounded by ground conductor either externally or internally.
             [C]

[ ] [ ] [e] **[B]** 23. Plug-in features and top of devices are as approved and required. Verify handle of overcurrent devices do not exceed height limits above finished floor.
             [C]

Legend: Upper Case Letter and **BOLD** = Primary Inspection; Lower Case = Secondary Inspection
Main Groups SI: Special Inspections, **AI:** Architectural Inspections, **EI:** Engineering Inspections, **OI:** Other Inspections
Sub-Groups   **SI:** [T] Testing Laboratory, [H] Hazardous Materials, [Y] Safety   **AI:** No Sub-Groups
    EI: [V] Civil, [S] Structural, [M] Mechanical, [E] Electrical   **OI:** [O] Owner, [C] Contractor, [B] Subcontractor,
    [L] Legal, [G] Government, [F] Fire Protection, [P] Plumbing, [A] Acoustical

Construction Inspection Manual

SI AI EI OI

[x] [ ] [e] [B]   24. Component sections are legibly identified and marked with voltage, amperage, and name of manufacturer.
         [C]

[ ] [ ] [e] [B]   25. Joints are torqued in accordance with manufacturer's instructions.
         [C]

[ ] [ ] [e] [B]   26. Busway is accessible throughout its installation.
         [C]

[ ] [ ] [e] [B]   27. Trolley busways, trolleys, brushes, contact rollers, and flexible cables have good contact and move freely.
         [C]

[ ] [ ] [e] [B]   28. Busway expansion joints are provided at all building expansion joint locations.
         [C]

Underfloor Raceways

[ ] [ ] [e] [B]   29. Cross-sectional dimensions are as required.
         [C]

[ ] [ ] [e] [B]   30. Sufficient setting depth has been provided at junction boxes.
         [C]

[ ] [ ] [e] [B]   31. Raceways are parallel with floor construction, firmly supported at proper elevation, and in straight alignment.
         [C]

[ ] [ ] [e] [B]   32. No damaged joints are allowed.
         [C]

[ ] [ ] [e] [B]   33. All joints are tight and sealed in accordance with manufacturer's instructions between sections and to junction boxes.
         [C]

[ ] [ ] [e] [B]   34. Inserts, both pre-set and after set, are or will be secure to raceways and set flush with floor.
         [C]

Conductors

[ ] [ ] [e] [B]   35. Type of conductors materials, size, stranding, and type of insulation are as required.
         [C]

[ ] [ ] [e] [B]   36. Pulling of conductors and cables is as required, using suitable equipment and methods. Allow no damage to sheath jackets or insulation.
         [C]

[ ] [ ] [e] [B]   37. Connectors and joints are clean and tight, torqued as necessary.
         [C]

[ ] [ ] [e] [B]   38. Connectors, lugs, clamps, etc., to connect copper and aluminum are approved for specific application; and where subject to moisture are compatible to avoid galvanic corrosion.
         [C]

[ ] [ ] [e] [B]   39. All connections are made in junction or outlet boxes, not in conduits.
         [C]

Legend: Upper Case Letter and <u>BOLD</u> = Primary Inspection; Lower Case = Secondary Inspection
   Main Groups SI: Special Inspections, **AI:** Architectural Inspections, **EI:** Engineering Inspections, **OI:** Other Inspections
   Sub-Groups   **SI:** [T] Testing Laboratory, [H] Hazardous Materials, [Y] Safety   **AI:** No Sub-Groups
     EI: [V] Civil, [S] Structural, [M] Mechanical, [E] Electrical   **OI:** [O] Owner, [C] Contractor, [B] Subcontractor,
        [L] Legal, [G] Government, [F] Fire Protection, [P] Plumbing, [A] Acoustical

SI　AI　EI　OI

[ ] [ ] [E] [B]　40. Verify color coding of wire insulation including neutral and ground as
　　　　　　[C]　　　　specified.

[ ] [ ] [E] [B]　41. Verify ground conductor has been installed in non-metallic conduit.
　　　　　　[C]

[ ] [ ] [E] [B]　42. Grounding conductor is securely attached to equipment, devises, etc, and
　　　　　　[C]　　　　forms complete grounding system.

[ ] [ ] [E] [B]　43. Branch circuit conductors are as specified and suitable for their use.
　　　　　　[C]

[ ] [ ] [E] [B]　44. Vertical cable supports are provided in conduit risers at each floor level.
　　　　　　[C]

Cable Systems

[ ] [ ] [E] [B]　45. Metal-clad cable is installed where specified. Cutting is performed
　　　　　　[C]　　　　without conductor damage, and bushings are installed.

[ ] [ ] [E] [B]　46. Nonmetallic-cable is installed in areas allowed by contract documents
　　　　　　[C]　　　　and permitted by codes. Located so as to prevent driving of nails into
　　　　　　　　　　　　cable. Code gauge protection plates are provided where required.

[ ] [ ] [E] [B]　47. Cable are secured within 12 inches of box or fitting and otherwise at
　　　　　　[C]　　　　internals not exceeding limits for proper support without excessive
　　　　　　　　　　　　sogging. Cable supports should be independent of either support
　　　　　　　　　　　　members.

[ ] [ ] [E] [B]　48. NMC cable is used in wet locations or areas exposed to dampness
　　　　　　[C]　　　　(including exterior masonry walls).

[ ] [ ] [E] [B]　49. Verify cable are installed in spaces allowing accessibility for installation
　　　　　　[C]　　　　and removal.

Outlets

[ ] [ ] [E] [B]　50. Architectural drawings are referred to for comparison of all conditions
　　　　　　[C]　　　　affecting layout of outlets. Coordinate work with other trades.

[ ] [ ] [E] [B]　51. Special equipment outlets have been roughed in per manufacturer's
　　　　　　[C]　　　　rough-in drawings.

[ ] [x] [E] [B]　52. Floor outlets of required type are properly located. Verify dimensions if
　　　　　　[C]　　　　indicated or critical. Request equipment and furniture layout.

[ ] [ ] [E] [B]　53. Ground fault receptacles are provided in locations required by NEC.
　　　　　　[C]

Legend:　Upper Case Letter and <u>**BOLD**</u> = Primary Inspection;　Lower Case = Secondary Inspection
　Main Groups　**SI**: Special Inspections, **AI**: Architectural Inspections, **EI**: Engineering Inspections, **OI**: Other Inspections
　Sub-Groups　　**SI**: [T] Testing Laboratory, [H] Hazardous Materials, [Y] Safety　　**AI**: No Sub-Groups
　　　EI: [V] Civil, [S] Structural, [M] Mechanical, [E] Electrical　　**OI**: [O] Owner, [C] Contractor, [B] Subcontractor,
　　　[L] Legal, [G] Government, [F] Fire Protection, [P] Plumbing, [A] Acoustical

Construction Inspection Manual

SI	AI	EI	OI	
[ ]	[x]	[E]	[B] [C]	54. Wall receptacle, switch outlets, and fixture outlets are mounted at height and location required.
[ ]	[ ]	[E]	[B] [C]	55. Door swings, equipment, and other features are not in conflict for convenience of use.
[ ]	[ ]	[E]	[B] [C]	56. Light outlets in mechanical and equipment rooms are located to suit servicing and maintenance and extend below ducts and ceiling.
[ ]	[ ]	[E]	[B] [C]	57. Verify junction pull, and outlet boxes are of type, size, and location required by code.
[ ]	[ ]	[E]	[B] [C]	58. Verify boxes are secure and independently supported and do not rely on conduits for this support, except as permitted by code.
[ ]	[ ]	[E]	[B] [C]	59. Verify boxes are accessible.
[ ]	[ ]	[E]	[B] [C]	60. Verify cast boxes and special boxes are provided as required in exposed areas. exterior areas, wet locations, and hazardous locations. Verify all boxes exposed to weather are weatherpoof.
[ ]	[ ]	[E]	[B] [C]	61. Verify plaster rings, and extension rings are provided. Verify incombustible surfaces allow ¼ inch space from finish, and combustible shall be flush. Ensure no combustible material is exposed to interior of box.
[ ]	[x]	[E]	[B] [C]	62. Verify architect is notified before closing-in, and agency inspection is provided if required.
[ ]	[ ]	[E]	[B] [C]	63. Verify boxes are sized to allow number of conductors and/or splices in boxes as per code.
[ ]	[ ]	[E]	[B] [C]	64. Verify unused openings are closed.
[ ]	[ ]	[E]	[B] [C]	65. Ensure grounding continuity is maintained, including jumper if required.
[ ]	[ ]	[E]	[B] [C]	66. Verify installed devices are of required type, voltage, amperage, and color.
[ ]	[ ]	[E]	[B] [C]	67. Verify switches are installed in phase conductor of circuit (not neutral) and with "on" position up, except for momentary contact. 3-way and 4-way switches.
[ ]	[ ]	[E]	[B] [C]	68. Verify device plates are of material, type, ganging, and finish required. Identification, and pilot lights are provided as required.

Legend: Upper Case Letter and **BOLD** = Primary Inspection; Lower Case = Secondary Inspection
Main Groups **SI**: Special Inspections, **AI**: Architectural Inspections, **EI**: Engineering Inspections, **OI**: Other Inspections
Sub-Groups **SI**: [T] Testing Laboratory, [H] Hazardous Materials, [Y] Safety   **AI**: No Sub-Groups
EI: [V] Civil, [S] Structural, [M] Mechanical, [E] Electrical   **OI**: [O] Owner, [C] Contractor, [B] Subcontractor, [L] Legal, [G] Government, [F] Fire Protection, [P] Plumbing, [A] Acoustical

Construction Inspection Manual

SI	AI	EI	OI	
[ ]	[x]	[E]	[B] [C]	69. Verify plates completely cover openings and are in contact with finish surface.
[ ]	[x]	[E]	[B] [C]	70. Verify plates are plumb and aligned with wall surfaces.
[ ]	[x]	[E]	[B] [C]	71. Verify surface-mounted boxes are provided with plates for their intended use.
[ ]	[ ]	[E]	[B] [C]	72. Verify neutral of multi-wire circuit will not be interrupted by removal of device or fixture.

Motors

SI	AI	EI	OI	
[ ]	[ ]	[E]	[B] [C]	73. Verify motors have voltage rating, and number of phases to suit supply system.
[ ]	[ ]	[E]	[B] [C]	74. Verify motor rotation is correct for driven machine.
[ ]	[ ]	[E]	[B] [C]	75. Verify motors subject to vibration or mounted on adjustable bases are connected with flexible metal conduit. Verify liquidtight or explosion-proof flexible metal conduit are provided in damp or hazardous locations.
[ ]	[ ]	[E]	[B] [C]	76. Verify flexible metal conduit length is as required and allow flexibility in all possible motor locations.
[ ]	[ ]	[E]	[B] [C]	77. Observe lubrication requirements are met prior to operation.

Motor Control, Disconnects and Starters

SI	AI	EI	OI	
[ ]	[ ]	[E]	[B] [C]	78. Verify manual disconnect switch is provided at each motor and motor starter as required by code.
[ ]	[ ]	[E]	[B] [C]	79. Verify control accessories are furnished as required by operation (start-stop pushbuttons, pilot lights, selector switches, and similar devices).
[ ]	[ ]	[E]	[B] [C]	80. Verify motor full-load rated currents are compared with ratings of motor-running overcurrent protective devices (heater). Verify heaters of proper size are installed in starters.
[ ]	[ ]	[E]	[B] [C]	81. Verify motor controllers are as required for motor being served. Horsepower and voltage rating is to be at least equal to motor controlled.

Legend: Upper Case Letter and **BOLD** = Primary Inspection; Lower Case = Secondary Inspection
Main Groups SI: Special Inspections, **AI:** Architectural Inspections, **EI:** Engineering Inspections, **OI:** Other Inspections
Sub-Groups **SI:** [T] Testing Laboratory, [H] Hazardous Materials, [Y] Safety   **AI:** No Sub-Groups
EI: [V] Civil, [S] Structural, [M] Mechanical, [E] Electrical   **OI:** [O] Owner, [C] Contractor, [B] Subcontractor, [L] Legal, [G] Government, [F] Fire Protection, [P] Plumbing, [A] Acoustical

Construction Inspection Manual

SI	AI	EI	OI	
[ ]	[ ]	**[E]**	**[B]** **[C]**	82. Verify automatic control devices such as thermostats, floats, and pressure switches are adequately rated and as required for operation.
[ ]	[ ]	**[E]**	**[B]** **[C]**	83. Verify magnetic coil voltage is same as control circuit voltage (may be different from motor voltage).
[ ]	[ ]	**[E]**	**[B]** **[C]**	84. Verify motor does not exceed noise limits under operating conditions.

Legend: Upper Case Letter and <u>BOLD</u> = Primary Inspection; Lower Case = Secondary Inspection
Main Groups SI: Special Inspections, **AI:** Architectural Inspections, **EI:** Engineering Inspections, **OI:** Other Inspections
Sub-Groups **SI:** [T] Testing Laboratory, [H] Hazardous Materials, [Y] Safety **AI:** No Sub-Groups
EI: [V] Civil, [S] Structural, [M] Mechanical, [E] Electrical **OI:** [O] Owner, [C] Contractor, [B] Subcontractor, [L] Legal, [G] Government, [F] Fire Protection, [P] Plumbing, [A] Acoustical

DIVISION 16 — ELECTRICAL — LOW VOLTAGE DISTRIBUTION 16400

STANDARDS: CDA (37) NEC (259) NEMA (60) UL (93)

SI   AI   EI   OI

[ ] [ ] [E] [B] 1. Ensure equipment is listed by a nationally recognized testing laboratory, and is tested in accordance with contract documents.
           [C]

Service

[ ] [ ] [E] [B] 2. Verify provisions are made in construction for service entrance system in accordance with utility company requirements and coordinated with drawings. Verify clearances under service drops are provided.
           [C]

[ ] [ ] [E] [B] 3. Verify sleeves and spaces are provided of sizes required.
           [C]

[ ] [ ] [E] [B] 4. Verify meter location and main disconnect location as specified or indicated on contract documents.
           [C]

Transformers

[ ] [ ] [E] [B] 5. Verify pad for exterior transformer is of size required — location and installation method is understood.
           [C]

[ ] [ ] [E] [B] 6. Verify transformer is of type required. Verify noise does not exceed specified limits.
    [a] [C]

[ ] [ ] [E] [B] 7. Verify final connection with flexible conduit.
           [C]

Switchboards and Panelboards

[ ] [ ] [E] [B] 8. Verify space provided is minimum required by code: distance from handle of top switch or breaker to finish floor does not exceed 6 feet 6 inches.
           [C]

[ ] [ ] [E] [B] 9. Verify boards are rigidly and securely anchored to floors and walls with supports of sufficient strength to support weight and resist seismic forces.
    [s] [C]

[ ] [ ] [E] [B] 10. Verify ground fault protection has been provided where required by NEC.
           [C]

[ ] [ ] [E] [B] 11. Verify spaces are provided as required by contract documents for future circuits.
           [C]

Legend: Upper Case Letter and **BOLD** = Primary Inspection; Lower Case = Secondary Inspection
Main Groups **SI**: Special Inspections, **AI**: Architectural Inspections, **EI**: Engineering Inspections, **OI**: Other Inspections
Sub-Groups   **SI**: [T] Testing Laboratory, [H] Hazardous Materials, [Y] Safety   **AI**: No Sub-Groups
EI: [V] Civil, [S] Structural, [M] Mechanical, [E] Electrical   **OI**: [O] Owner, [C] Contractor, [B] Subcontractor,
[L] Legal, [G] Government, [F] Fire Protection, [P] Plumbing, [A] Acoustical

Construction Inspection Manual

SI AI EI OI

[ ] [ ] [E] [B] [C] 12. Verify spare breakers are provided and installed as required by contract documents.

[ ] [ ] [E] [B] [C] 13. Verify location are coordinated with other trades which do not permit other utilities (i.e. ductwork, piping) within code clearances of boards.

Grounding

[ ] [ ] [E] [B] [C] 14. Visually inspect all grounding system conductors, connections, and electrodes as work progresses, and verify continuity of grounding conductors.

[ ] [ ] [E] [B] [C] 15. Verify grounding connectors are accessible for inspection and protected against mechanical injury.

[ ] [ ] [E] [B] [C] 16. If water piping system is used, check that pipe is metallic and no insulating fitting is interposed in pipe between ground wire connection point and interior or exterior pipe system (N.E.C. 250-81 metallic water piping system is always used supplemented if required.)

[ ] [ ] [E] [B] [C] 17. Verify contact surfaces are clean and dry, are metal-to-metal and tight bolt connections are made.

[ ] [ ] [E] [B] [C] 18. Observe size, length, number, material and installation of ground rods.

[ ] [ ] [E] [B] [C] 19. Verify connectors are compatible with metal and pipes use.

[ ] [ ] [E] [B] [C] 20. Verify grounding conductors are connected to both ends of metallic raceway in which it is installed and are both connected to grounding electrode in accordance with code.

[X] [ ] [E] [B] [C] 21. Verify ground fault protection has been provided where required by National Electric Code.

[X] [ ] [E] [B] [C] 22. Test impedance to ground as required by specifications.

Bonding

[x] [ ] [E] [B] [C] 23. Verify bonding raceway systems, cable tray and other enclosures complies with NEC Article #250C.

Legend: Upper Case Letter and **BOLD** = Primary Inspection; Lower Case = Secondary Inspection
Main Groups SI: Special Inspections, **AI:** Architectural Inspections, **EI:** Engineering Inspections, **OI:** Other Inspections
Sub-Groups   **SI:** [T] Testing Laboratory, [H] Hazardous Materials, [Y] Safety   **AI:** No Sub-Groups
   EI: [V] Civil, [S] Structural, [M] Mechanical, [E] Electrical   **OI:** [O] Owner, [C] Contractor, [B] Subcontractor, [L] Legal, [G] Government, [F] Fire Protection, [P] Plumbing, [A] Acoustical

Construction Inspection Manual

DIVISION 16 — ELECTRICAL — LIGHTING 16500

STANDARDS: NEC (59) NEMA (60) UL (93)

SI	AI	EI	OI	
[X]	[ ]	[E]	[B] [C]	1. Verify fixtures are listed by a nationally recognized testing laboratory.
[x]	[X]	[e]	[B] [C]	2. Verify lighting layout is coordinated with architectural drawings, and discrepancies are reported. Verify layout is coordinated with work of other trades.
[ ]	[X]	[E]	[B] [C]	3. Verify fixtures comply with specified fixtures and accessories are provided.
[x]	[ ]	[E]	[B] [C]	4. Verify suspension, supporting, and mounting methods are as required based upon weight, dimensions and seismic restraint of fixture. Observe plumbness and alignment.
[X]	[X]	[E]	[B] [C]	5. Verify mounting height and location are as indicated.
[x]	[x]	[E]	[B] [C]	6. During installation verify defective louvers, cracked glass or plastic, chipped porcelain or finish, distortion, or other defects are corrected before completion. Verify doors are properly aligned, with clearance for operation, verify retaining devices function properly.
[ ]	[ ]	[E]	[B] [C]	7. Verify ballast type is as specified — fluorescent or high-intensity discharge, voltage, power factor, overload protection, proper rating, and low temperature.
[ ]	[ ]	[E] [a]	[B] [C]	8. Verify ballasts noise does not exceed specified limits. Verify required replacements have been made.
[x]	[x]	[E]	[B] [C]	9. Verify lamp type is as required — wattage, energy saving, style, color, characteristics, and long life. Note fluorescent lamp colors at start of installation. Check that lamps are operating properly.
[x]	[ ]	[E]	[B] [C]	10. Verify lamps are new and installed before completion, or reinstalled if required. Verify additional new lamps are provided to compensate for contractor's use of building lighting system if required.
[x]	[x]	[E]	[B] [C]	11. Verify frames and accessories are as required — compatible with adjacent surfaces, no light leaks, weatherproof and corrosion resistant, and finishes match.
[x]	[x]	[E]	[B] [C]	12. Verify fixtures are adjusted or aimed as required for intended illumination.

Legend: Upper Case Letter and **BOLD** = Primary Inspection; Lower Case = Secondary Inspection
Main Groups SI: Special Inspections, AI: Architectural Inspections, EI: Engineering Inspections, OI: Other Inspections
Sub-Groups **SI:** [T] Testing Laboratory, [H] Hazardous Materials, [Y] Safety   **AI:** No Sub-Groups
EI: [V] Civil, [S] Structural, [M] Mechanical, [E] Electrical   **OI:** [O] Owner, [C] Contractor, [B] Subcontractor, [L] Legal, [G] Government, [F] Fire Protection, [P] Plumbing, [A] Acoustical

SI	AI	EI	OI	
[x]	[x]	[E]	[B] [C]	13. Verify fixtures are suitably protected and cleaned at completion of work.
[X]	[ ]	[E]	[B] [C]	14. Verify fixtures are grounded as required by code.
[X]	[x]	[E]	[B] [C]	15. Verify fixtures anchored per contract documents.
[x]	[x]	[E]	[B] [C]	16. Verify fixtures installed in suspended ceilings are secured to the structure above the ceiling. Verify fixtures are independent of ceiling supports as regulated by governing authorities.
[X]	[x]	[E]	[B] [C]	17. Verify earthquake clips and seismic wires per codes and specifications as required by governing authorities.

Legend: Upper Case Letter and **BOLD** = Primary Inspection; Lower Case = Secondary Inspection
Main Groups SI: Special Inspections, **AI:** Architectural Inspections, **EI:** Engineering Inspections, **OI:** Other Inspections
Sub-Groups **SI:** [T] Testing Laboratory, [H] Hazardous Materials, [Y] Safety   **AI:** No Sub-Groups
EI: [V] Civil, [S] Structural, [M] Mechanical, [E] Electrical   **OI:** [O] Owner, [C] Contractor, [B] Subcontractor, [L] Legal, [G] Government, [F] Fire Protection, [P] Plumbing, [A] Acoustical

Construction Inspection Manual

DIVISION 16 — ELECTRICAL — COMMUNICATIONS 16700

STANDARDS: NEC (59) NEMA (60) UL (93)

General

SI   AI   EI   OI

[x] [ ] [E] [B] [C]   1. Verify Boxes, conduits, and fittings conform to Section 16050 requirements.

[x] [ ] [E] [B] [C]   2. Verify barriers are provided between low voltage and line voltage wiring.

[X] [ ] [E] [B] [C]   3. Verify wiring terminations identified.

[X] [ ] [E] [B] [C]   4. Verify all wiring installed in equipment cabinets are installed neatly racked and bundled, and organized with wire ties or in wireways within cabinets.

[x] [X] [E] [B] [C]   5. Verify as-built records kept on all wiring deviations from contract documents.

[x] [ ] [E] [B] [C]   6. Verify backboxes grounded.

[X] [ ] [E] [B] [C]   7. Confirm components are testing agency listed (UL, etc.)

[X] [ ] [E] [B] [C]   8. Devices and systems tested for proper operation.

[ ] [ ] [E] [B] [C]   9. Verify systems are installed by personnel qualified in communication system installation.

[x] [ ] [E] [B] [C]   10. Verify that all equipment is listed for the environment in which it is installed.

[x] [ ] [E] [B] [C]   11. Verify devices mounted so that vibration and jarring will not cause accidental operation or malfunction.

Fire Alarm

[X] [ ] [E] [B] [C]   12. Verify devices are state fire marshal (or governing authority) and testing agency listed.

[x] [ ] [E] [B] [C]   13. Verify system installed by or under the direct supervision of factory trained service technicians, where specified. Check approved shop drawing for installation.

Legend: Upper Case Letter and **BOLD** = Primary Inspection; Lower Case = Secondary Inspection
Main Groups SI: Special Inspections, **AI:** Architectural Inspections, **EI:** Engineering Inspections, **OI:** Other Inspections
Sub-Groups   SI: [T] Testing Laboratory, [H] Hazardous Materials, [Y] Safety   **AI:** No Sub-Groups
EI: [V] Civil, [S] Structural, [M] Mechanical, [E] Electrical   **OI:** [O] Owner, [C] Contractor, [B] Subcontractor, [L] Legal, [G] Government, [F] Fire Protection, [P] Plumbing, [A] Acoustical

Construction Inspection Manual

[X] [ ] [E] [B]   14. Ensure all devices are tested for correct operation.
          [C]

SI **AI** **EI** **OI**

[X] [X] [E] [B]   15. Verify audible alarm devices are heard distinctly in all areas of the
          [C]   building as required by code and governing authorities.

[X] [ ] [E] [B]   16. Refer to NFPA references for testing and maintenance procedures and
          [C]   acceptable power supply sources for normal and emergency power.

[X] [X] [E] [B]   17. Verify visual alarm devices are clearly visible from all points within the
          [C]   visual alarm area and in accordance with code and governing authority
                   requirements.

[X] [ ] [E] [B]   18. Check code requirements and governing authority for installation of
          [C]   wiring in raceways.

[X] [x] [E] [B]   19. Insure that records are kept of all fire alarm system tests and record
          [C]   status of all devices tested. Record all deficiencies observed during the
                   test.

[X] [x] [E] [B]   20. Pre-test check items: a. Verify that all building occupants have been
      [F] [C]   alerted of the test.

                   b. Building ancillary functions to the fire alarm system are bypassed if
                      they are not a part of the test.

                   c. Coordinate test with the local fire department if applicable.

[X] [x] [E] [B]   21. Verify the certification of proper operation of all fire alarm system
      [F] [C]   equipment has been completed.

Legend: Upper Case Letter and BOLD = Primary Inspection; Lower Case = Secondary Inspection
  Main Groups SI: Special Inspections, **AI:** Architectural Inspections, **EI:** Engineering Inspections, **OI:** Other Inspections
  Sub-Groups   **SI:** [T] Testing Laboratory, [H] Hazardous Materials, [Y] Safety   **AI:** No Sub-Groups
     EI: [V] Civil, [S] Structural, [M] Mechanical, [E] Electrical   **OI:** [O] Owner, [C] Contractor, [B] Subcontractor,
      [L] Legal, [G] Government, [F] Fire Protection, [P] Plumbing, [A] Acoustical

Part 4:
Coordination

Coordinating the various disciplines during design is usually the responsibility of either the architect, principal engineer, or construction manager. Coordinating the work of the various suppliers, trades, and subcontractors is the responsibility of the contractor as part of the construction means and methods. However, successful coordination requires all parties of both the design and construction teams to communicate with each other so that the construction can proceed in an orderly manner. Proper form and process needs to be defined and followed, however, to ensure information is properly transmitted and officially received.

While speech is used extensively throughout the construction process, through meetings (scheduled and impromptu) and phone conversations, the nature, scheduling, and legal procedures required during construction require written records with proper distribution to all concerned entities.

Part 4.1 generally outlines the traditional communications procedures that have been recognized by the industry as a type of "protocol" to keep the participants involved as fully informed as possible and to avoid misunderstandings. Naturally, exceptions occur.

Part 4.2 discusses scheduling and the methods typically employed during construction to ensure an orderly process, free of unnecessary delays and costs.

Part 4.3 discusses correspondence and related coordination.

Part 4.4 introduces standards and forms. Appendix D, "Forms," contains examples of recommended forms.

Part 4.1
Communication

The owner communicates directly with the design principal (an architect, engineer or construction manager), and through the design principal with the contractor. The owner does not communicate in writing directly with the construction inspector, consultants, subcontractor, vendors, or suppliers. Verbal communications between the owner and contractor should be avoided except when circumstances require additional clarification, and then only in the presence of the design principal. Written communication should always go directly from the owner to the design principle and from design principal to the contractor.

The design principal communicates directly with the owner, contractor or more specifically the contractor's superintendent), the design consultants, and the construction inspector. The design principal does not communicate directly with the subcontractors, or vendors, or suppliers. When circumstances require additional clarification, verbal communications may occur in the presence of the contractor. Written communications again should only be from the design principal to the contractor and from the contractor to the subcontractors, vendors, and suppliers.

The contractor communicates directly with the design principal and with the contractor's subcontractractors, vendors, and suppliers. The contractor communicates with the owner through the design principal, and with the construction inspector through the contractor's superintendent. The contractor does not directly communicate in writing with the design consultants. Again verbal communication between the contractor and the design consultants should be avoided except in the presence of the design principal and then only when circumstances required additional clarification.

The construction inspector communicates directly with the design principal and the contractor's superintendent. The inspector should not communicate in writing directly with the owner, design consultants, subcontractors, vendors, or suppliers.

Exceptions:

Where the construction inspector is retained by the owner as a permanent or staff employee, the construction inspector may communicate directly with the owner. In all such instances, the design principal should direct the construction inspector's work and be fully informed of communications concerning it.

The design principal may specifically or partially delegate direct communications between design consultants and the construction inspector or the contractor to expedite the work; however, the design principal must be fully informed in writing of all such communications and the design principal must approve in writing any actions from communications directly from the direct contractor to the design consultants

The contractor may specifically or partially delegate direct communications between his subcontractors, vendors, and suppliers and the design principal, design consultants, or the construction inspector. However such direct communications require both the contractor and the design professional be fully informed in writing of any communications and each must approve in writing any resultant actions.

The owner directly communicates with the contractor concerning legal or contractual matters; however, the owner should keep the design principle informed in writing of all direct owner to contractor communications and any resultant actions.

Other direct communications between member of the design and the construction teams may be made where required by laws or codes; however, all responsible parties should be kept informed in writing.

Part 4.2: Scheduling

Either the construction manager or the general contractor should be responsible for preparing the project schedule. Whether a simple bar chart or the most complex computerized critical path format, the project schedule needs to establish the steps necessary to complete the project in a timely manner.

It's not the responsibility of the construction inspector to take an active role in either the creation of the project schedule nor it's enforcement. The construction inspector must, however, maintain a familiarity with the schedule along with its updates in order to ensure personnel and materials required for inspections are available at the proper times.

Where a simple bar chart is used, it is probable that the construction inspector's services will not be a full time effort. In such situations it will be critical for the construction inspector to maintain their own schedule thoroughly coordinated with the bar chart schedule, making sure the timing for each required inspection is established and understood by those responsible for keeping the inspector informed of the work progress. It will also be critical for the construction inspector to be in periodic contact with those responsible for maintaining the project schedule in order to keep their own schedule up-to-date.

More complex projects requiring full time construction inspector services will probably have a critical path schedule, probably on a computer. The construction inspector needs to make sure the inspection dates are kept up-to-date in the project schedule, especially those related to items on the critical path and which would result in project delays where the inspections are not properly coordinated.

Even in the most complicated project with the most advanced critical path project schedule the construction inspector should maintain their own schedule and coordinate their schedule with the project schedule. Full time project involvement does not relieve the construction inspector of verifying that both schedules are kept up-to-date. Paperwork such as scheduling seems the easiest to set aside when seemingly more urgent matters seem to demand attention. Yet failure to maintain the inspection schedule can be costly, especially where construction conceals the work requiring inspection and has to be removed and rebuilt in order to assure proper inspection.

Part 4.3:
Correspondence and Coordination

The orderly continuation of the work requires written correspondence in order to ensure proper coordination. Written correspondence among the construction participants should be used along with the contract documents. Not as official as the agreement, contract documents, and change orders perhaps, but, a necessary part of the process none-the-less, to be prepared with the same consideration as the rest of the documents.

Correspondence can be in the form of letters, memoranda, bulletins, reports, graphs, and even faxes. It is recommended that adequate documentation should be developed as a good practice by all the parties construction participants.

When you write use simple declarative statements and be clear in your intent. Each form of correspondence must include the project name, project number (especially where there might be more than one phase to a project or even the potential of future phases), a date (critical), and the applicable parties to be involved in the correspondence.

The project name should always be consistent throughout correspondence and the contract documents, preferably the same title used on the title sheets of the drawings. Where there is any possibility of confusion the name should include sufficient information to differentiate between multiple projects, such as including a building address, phase number, or similar location identifier. Project numbers can also be helpful in verification, although most projects have different numbers for each of the parties involved.

Date each piece of correspondence. Once a subject becomes important enough to require special correspondence, it is possible that the correspondence will result in various possible decisions. Where multiple correspondences exist on one subject, the dates on the correspondence ensure which is the most recent decision.

Identification of the applicable persons might be as simple as identifying who the correspondence is from and to whom it is directed. It might also include an extensive list of persons who require the information in order to ensure proper coordination. This is frequently done with a "cc" list in a letter or fax. Blind copies might also be appropriate for certain types of correspondence, but the decision to include blind copies should be based on prior mutual agreement between specific persons, such as the owner requesting a blind copy of correspondence between the construction manager and design principal.

Part 4.4:
Forms

Many types of forms have been developed, and it can be said that there is a form for any need. Many organizations individually or through collaboration with other organizations have developed forms in an effort to standardize procedures, but complete unanimity as to type, contents, and arrangements not universally achieved. The American Institute of Architects has developed a complete series of documents, and several of these are reproduced as a part of Appendix D, "Forms."

Although it would be impossible to discuss all possible forms, several widely used forms can be identified:

Application for Payment — A form filled out by the contractor requesting payment due for completed portions of the work. It describes each phase of the work: phase value, previous applications, present application, completion to date, and balance to be completed. Changes are included. The total value of phases, stored materials, previous payments, and retainage is given. It is usually completed in accordance with the contract documents. The contractor certifies that all amounts for items previously requested have been paid. This form is executed by the contractor, signed by the construction inspector, and sent in multiple copies to the architect for his signature.

Certificate for Payment — A form, filled out by the design principal (typically the architect), which is a non-negotiable certificate to the owner confirming that the amount of money requested in the application for payment is due the contractor. It recaps the original contract sum, change orders, and contract sum to date. The listed figures agree with the application as to total completed, total stored, retaining and amounts due. It requires the contractor's signature. This certificate is executed by the architect and distributed with multiple copies to the owner, who issues payment to the contractor and distributes copies to the contractor and other parties.

The design principal receives an executed copy to confirm that payment has been made. The number of copies may vary, since the financing agency, etc., may be involved. The construction inspector should receive a copy of the application for payment and the certificate for payment. This form should include a distribution list (cc).

Change Order — A written order to the contractor, authorizing a change or adjustment in the work. Filled out by the design principal (architect), this form describes the date, the change (sometime by reference to other forms such as construction bulletins or requests for information), the amounts involved, the time increase or decrease, and a recap of the status of the original contract as affected by cost or time. It is typically executed by the design principal (architect), with multiple copies to the contractor, who signs the change order and forwards it to the owner for signature of approval. Executed copies are then distributed to the design principal (architect), contractor, and others involved. It should be remembered that each change order is a contract on its own and requires the same legal consideration as any contract.

Construction Change Directive — Used only in the absence of total agreement on the terms of a change order, this form is a written instruction to the contractor. Filled out by the design principal (architect), this form describes the change prior to agreement on the amounts involved. A formal change order should be prepared to replace the construction change directive order as soon as possible.

Transmittal — An all-purpose form that can be used by all parties. It usually requires minimal additions to transmit items such as shop drawings, samples, and reports. The transmitted item is described and requested actions are checked (√) beside the appropriate action or descriptive item. It indicates distribution to various parties.

Inspector's Daily Report — Filled out by the construction inspector, this form lists the day, weather, site condition, contract time status, work forces on the site, work performed, major equipment material deliveries to the site, phone calls, and visitors, and makes remarks. It is filled out daily and usually sent to the design principal (architect) weekly.

Proper and timely distribution of correspondence is extremely important. The construction inspector should receive copies of all material concerning the contract documents. Although some inspectors may jot notes and items on various media such as scraps of paper, it should be considered critical to develop an organized method to transfer this material to a written record as soon as possible. Remember that information transferred verbally only may result in future confusion.

Here are examples of suggested correspondence:

Owner: Put all matters concerning the project in writing to the design principal (architect). A good policy, if phoning, is to follow up with a letter.

Architect: Promptly produce memoranda of meetings and discussion, including important phone conversations, and distribute copies to the owner, contractor, construction inspector, consultants, and other parties affected. Indicate distribution on correspondence.

Follow up phone conversations in writing for important items as soon as possible.

Keep a log of occurrences and organize an "action" system to act on all items in a timely manner.

Enclose the design consultant's correspondence with the design principal's (architect's) correspondence for all matters affecting the work.

Contractor: Follow up phone conversations in writing for important matters as soon as possible, and make sure distribution list includes all affected parties.

Enclose copies of correspondence received from subcontractors, suppliers, and vendors with correspondence "cover" for all matters affecting the work.

Construction Inspector: Use a field memorandum form consecutively numbered, and list all items to be discussed with the design principal (architect). Use duplicated forms to ensure rapid transfer to all affected parties.

All Parties: React promptly to all correspondence received. Review memoranda and each take action where stipulated. Clarify misunderstandings. Some firms stipulate on their memoranda that all items requiring clarification should be called to the attention of the writer within a specific period of time.

Part 5:
Project Location

When considering the importance of various aspects of construction administration there is a potential for becoming so tied to the process that the obvious is overlooked. Where you are and where the project is are both critical aspects for success.

Construction inspection is not limited to walking around the site and taking notes, regardless of whether you are a full time construction inspector, a design professional, a construction professional (contractor, subcontractor, tradesperson), or any of the multitude of persons who take an active role in the construction inspection process.

As with any project there are tasks which must be done in the company office, things which must be done in the field office, and things which must be done in the field. Failure to recognize the importance of any of these responsibilities can result in problems for the project.

There are also specific concerns which must be considered due to the location of the project itself such as regional considerations, geographic considerations, and climate. Recognition of each of these and how they impact a particular project is critical to successful construction inspection.

Part 5.1:
Office Practice Concerns

Paperwork begins and ends with the main office. Even the largest projects with full time field office staff rely on main office support in accounting, records storage, and administrative assistance.

Each office should establish standard procedures for construction inspection responsibilities and ensure everyone working on a particular project is aware of the standard procedures. For the one person full time field administrator who works out of a field office and a small room in the house with a computer and file cabinet, standard office procedure is based on experience. For the large firm with a regular turnover in personnel, standard office procedure needs to be written down and distributed, with each new employee given the responsibility to read and discuss with the rest of the project team.

The first item of importance is to decide what is being done where. As with any project management, construction administration needs to be organized. Responsibilities need to be established early, with communication channels between those involved clearly defined. Does the construction inspector work directly on the computer, doing their own word processing, scheduling, and record keeping, or is there staff available to free up time for more field work and administrative duties.

Record keeping responsibilities need to be established. Is the construction inspector going to dictate notes into a recorder for secretarial staff to transcribe? Is the transcription going to be done in the field office or the main office? What staff is available where and what is the desired turnaround time?

Where the field inspector is responsible for keeping records, it is beneficial to establish specific times for that responsibility. Just as standard project meeting times are scheduled so should be time to do administrative and record keeping tasks. Where administrative staff is available it becomes just as important to establish the procedure to ensure administrative tasks are being done, files are kept up-to-date, and everyone maintains an appropriate schedule.

Part 5.2:
Field Office Concerns

The field office should be an extension of the main office, even if the main office is a room at home. The question of what is done in the field office versus the main office becomes a project by project issue. What funds are available and what benefits are realized by using the main office versus the field office becomes the primary factor in deciding what is done where.

Regardless of staffing the field office will require the bare essentials: climate control, power, water, sanitary facilities, light, access, and parking. Furniture should include a desk and chair for each staff member, with files and at least one drawing table for laying out the project drawings. Shelving for storage is critical and seldom adequate (although there is a theory that we always tend to fill whatever space is available). Where space is available, a meeting area is helpful, but it is always possible for a few people to meet around a desk.

Field office equipment begins with a telephone and answering machine. The construction inspector may have a cellular (mobile) phone, as may other staff members, but replacing the field office phone with a cellular phone should be done only with consideration of the potential benefits and detriments. Answering services can however, allow the possibility.

A facsimile machine, or fax, is also becoming a standard piece of equipment for the field office, especially where drawings and documents are being transmitted further than across town. Require a designated phone line for the fax.

Then there is the computer and printer, which would seem essential to any field office, even if it is only a laptop computer and battery operated printer. Computers can also serve as the fax machine, the answering machine, and even offer a direct line connection to the main office computer, allowing access to almost limitless information. And, again it is necessary to decide whether or not a to have a designated phone line for the computer. A one person field office with at least three telephone lines may soon be considered essential.

A field office copier can be another consideration, although this too can be eliminated in several circumstances. Storefront type copy services are often readily available. Computers with scanners can provide most of the copier functions. Fax machines are typically capable of some copier capabilities (although few have the ability to accept oversize drawings). There are even some computer printers available which function as printer/fax/copier.

Part 5.3
On-Site Concerns

Most construction inspection is by definition on-site. Horror stories are often traded about projects where inspections were supposed to occur that never happened. One such story was a multiple building high rise apartment complex where it was discovered that the firestopping wasn't installed between the concrete slabs and the curtainwall in more than one building. Another was when a ladder knocked a sprinkler head off which had been glued-on, the piping never having been installed.

Anyone who has ever had any construction inspection/administration responsibilities will probably have their own tales to tell. And, it is critical for everyone to understand that it is impossible for any construction inspection to be all-inclusive. No project can afford to have an inspector in every place at all times.

The key to an inspector sleeping nights is spending the time learning the requirements in the construction documents and keeping a mental record of those items which are most critical to the success of the project. You must stay ahead of the project, know what is supposed to happen next, not just fight what has been done wrong after the fact. This means knowing the contract documents and knowing the project schedule, which also means spending the time necessary in the office before going into the field.

Jobsite Safety:

Is safety everybody's business? If so, what is expected of a construction inspector relative to safety conditions at a job-site? A good deal of confusion exists in the minds of many people on this subject.

As to easily observed hazards, the answer is easy. Any unsafe condition which the construction inspector sees or which is called to the construction inspector's attention should be reported immediately to the contractor's superintendent. Follow up with a written memo and a copy to the owner.

What about hidden hazards or ones not easily detected? Is there a duty on the part of the construction inspector to search them out? The answer is a definite "No!" The construction inspector is not a Safety Engineer or a Safety Inspector. That is not what they are hired for or trained to do.

Jobsite safety measures and procedures are the sole responsibility of the contractor and are normally specified so in the contract documents.

Many safety aspects of a project under construction involve the adequacy of shoring in trenches or scaffolding or falsework which is not easily determined without an engineering analysis of size and placement of support members. This is the complete and sole responsibility of the contractor and the contractor's personnel. Do not try to become the contractor or do the contractor's work.

Find out who the contractor has designated as the site safety engineer or representative-in-charge of site safety. Find out also who the contractor designates as second-in-charge if the contractor's safety representative is absent. These are the ones whose job it is to maintain safe conditions at the jobsite for the workmen, for authorized visitors and for others who have a right to be there including the design principal (architect) design consultants and, of course, for the construction professional too.

Make it clear to the contractor and the contractor's superintendent at the pre-construction conferences that you are not there to review the adequacy of the contractor's safety program.

Explain that as a general rule you are trying to observe and report any discrepancies between the requirements of the contract documents and the contractor's work. Further explain that there is one important exception to that general rule. Although the contract documents do require the contractor to institute and maintain certain safety precautions, that is one contractor activity which you are not hired to review.

Bear in mind that there are two main categories of safety on a construction project, safety during construction and the adequacy of the final result.

The "house-keeping" safety measures and the safety of workmen while assembling the structure are solely the contractor's responsibility. The contractor must conduct construction reviews to determine whether adequacy and compliance with regulated safety requirements are maintained.

The adequacy of the final result, that is, whether it is safe to use after it is finished, presuming the project is constructed precisely as designed, is normally the responsibility of the owner's design professionals (typically the architect) and the design consultant.

Remember that if you volunteer to do something you have not originally agreed to, and you do it negligently, you may be responsible for the damages you cause.

Human safety is a serious matter. Make sure you know at the beginning of the project the identity of those who are specially trained and qualified to handle safety questions. If you have doubts about how the contractor is treating this subject, report what you think there may be deficiencies to the owner and to appropriate governmental safety authorities. But you should not volunteer to become a safety engineer any more than you should volunteer to assume the responsibilities of the design team or contractor.

Once you realize this, it is equally important that others realize it; the contractor's superintendent, the contractor's and subcontractor's personnel, and any others who confusion on this subject might affect the timely completion of the project.

Part 5.4:
Regional Concerns

Regional concerns relate to the politics of the site, both governmental and special interests. One of the advantages to experience is knowing the potential pitfalls and problems that can occur so that you can be prepared. Experience within a particular region is especially beneficial. But, when you don't have regional experience, it is possible to take some precautions.

Few populated areas of the country are so wide open these days that they allow construction to go uncontrolled. It may be still possible to build without regional restrictions in places like Texas, Montana, and Alaska. But, it is just as possible to find yourself awash with regional requirements as can be found in states like New York, Florida, and California.

Controls on hazardous materials, dust abatement, noise abatement, and air quality are common concerns for communities and their authorities regarding construction operations. Work hours can be restricted due to noise or to traffic conditions, specific types of mufflers required on construction equipment for noise control, and engines may be limited to specified idling period for pollution control. Local disposal areas may not be available for construction debris and burning may not be permitted on-site, requiring construction debris to be hauled long distances, possibly in specific types of covered carriers.

Some areas are beginning to require recycling of construction materials rather than allowing them to become landfill. Wood, metal, and paper materials generated by construction operations for blocking, formwork, and packaging may be required to be separated and stored in a manner allowing easy removal for recycling. Other recycling concerns might include restrictions on disposing of paints and plastics, both of which can be delivered to companies capable of recycling these products, but only in areas with the companies having the capability to beneficially reuse the materials.

Special regional concerns might also include concerns for local animal and plant life, including the potential of toxic hazards from construction operations. Other communities might have special concerns for artifacts, requiring local archeologists to be on-site during any excavation to ensure any potential archeological finds are not destroyed, including areas where there might be native American burials.

The construction inspector should be clear regarding responsibilities relating to such regional concerns along with the necessary information (such as phone numbers) on how to comply with the requirements. Part of this should be covered in the construction inspection agreement. Additional information might be gathered from the design team, local authorities, especially local building departments, planning commissions, fire departments, departments of public works, and the local administrative offices (state governor's office, mayor's office, county supervisor's office).

Part 5.5:
Geographic Concerns

Geographic concerns relate to the actual physical site rather than the political aspects of the site. Soil conditions, ground water, flooding, and seismic evens all play an important role.

Known soil conditions should be identified in the contract documents, but, no report can be so thorough as to eliminate the possibility of unexpected conditions. Rock outcroppings, caves, underground water, and variations in the natural topography can all have an impact on a project. The construction inspector should attempt to keep abreast of where unexpected conditions are encountered and make sure the design team is also aware as soon as possible. This is especially critical relating to how changes in the conditions might require changes in the contract documents.

Ground water control is easily handled where information is accurate and conditions don't change appreciably. However, construction inspectors can watch for the unanticipated, such as the impact on adjacent properties.

Where the project is located on a flood plain consideration needs to be given to how the potential of water on-site might impact the construction. On-site storage of materials during the wet season might justify special conditions and observations to ensure materials susceptible to water damage are kept dry.

Construction in seismic areas need to be given some consideration by the construction inspector. In areas where seismic activities are frequent, such as the entire west coast, construction inspectors might want to pay a little extra attention to materials storage. Although safety is not the responsibility of the construction inspector, materials stacked higher than four or five feet might prove a potential hazard in an earthquake. Shoring and bracing should never be "adequate" since additional seismic loads need to be recognized. It isn't the construction inspector's responsibility to design shoring and bracing, however, recognition of potential seismic activity can be justification for indicating a concern regarding systems which might otherwise seem adequate for the normal loads.

Additionally, some areas have special inspections required immediately after seismic events (earthquakes) to establish the extent of potential damage. Even a relatively small earthquake can be a problem for a construction project where the seismic bracing had not be completed.

Part 5.6:
Climate Concerns

Climate is as much a part of construction as the contract documents, codes, regulations, regional concerns and geographic concerns. Few areas have perfect weather all year round. Even those areas where temperature extremes are rare have their own special concerns such as fog, hurricanes, humidity, and even acid rain. One of the construction inspector's most common responsibilities is to record the weather on a daily basis (or at least on every day the construction inspector is on-site).

Cold climates most often bring special concerns relating to snow. Yet any wet conditions can become a problem when the temperature can fall below freezing. Attention should be given to weather reports which tell not only what the immediate conditions are, but also relate to weather predictions. Knowing that temperatures are anticipated to fall below freezing can be just as important to a project as knowing that they are below freezing. Temperatures ranging above and then falling again below freezing should be recorded since the higher temperatures can cause ice and snow to melt and then refreeze as the temperature drops.

Another concern for cold climates is permafrost. That area where the ground is frozen for much of the winter months, requiring insulation to be installed at foundations and even below slabs-on-grade.

Storage of materials in cold climates should also be considered, especially where extended bad weather might result in the project being delayed for an extended period. Materials stored on-site need to be kept elevated above potential standing water and ice, kept covered to protect from rain, snow, and ice, yet allow air circulation to prevent moisture from being trapped. In especially harsh climates it might be beneficial to either store materials off-site indoors, or even to provide temporary storage containers or buildings on-site.

Extremely hot climates can be detrimental to any materials which require curing or drying. Too much heat can remove necessary moisture before curing is complete, concrete might not reach the required strength, plasters have a higher shrinkage and more cracks, and paints may not form proper films. High humidity in a hot climate can also result in problems, especially where the dew point is higher than the night time temperatures which causes moisture to form on surfaces.

Fog is another form of moisture with special concerns. Ocean fog has a high saline content (salts) which is left on surfaces when the moisture evaporates. Salts on metals tend to be corrosive. Salts in cementitious products, including salts in the sand and on aggregates can redesolve later and be carried by water to the surface of the material resulting in efflorescence.

The problems of rain on construction sites it generally understood. Wood should be at least surface dry (S-dry) when used in construction. Many contract documents required kiln dried wood. Either type of wood absorbs water which causes expansion. Framing lumber which gets wet and is not allowed to dry out goes through seasonal changes of expansion and contraction which results in creaks and groans in the structure as the nails move in and out. Another example of rain on construction is water on decks during roofing which can result in future blisters.

Wind can be a problem in itself. Too much wind can be as much of a problem to concrete and plaster as too much sun, causing premature drying, potential loss of strength and cracking. Winds can also become a safety problem during construction where stored materials are not properly secured. A record of extreme winds should be kept along with temperatures.

High winds may provide special concerns, especially relating to jobsite safety. Tropical storm winds, hurricanes, and tornadoes each have their own problems. Storms and hurricanes tend to be predictable. Tornadoes offer little opportunity for special protection beyond standard jobsite concerns for storms, and perhaps good insurance.

Always review manufacturer's instructions, and when there are any questions relating to applicability to specific project conditions, contact the manufacturer's representative.

APPENDIX

APPENDIX A — BIBLIOGRAPHY

The American Institute of Architects, Duties, Responsibilities, and *Limitations of Authority of Full-Time Project Representative*, AIA. Document B352, The American Institute of Architects, 1735 New York Avenue, N.W., Washington, DC 20006.

The American Institute of Architects, *General Conditions of the Contract for Construction*. AIA. Document A201, The American Institute of Architects, 1735 New York Avenue, N.W., Washington, DC 20006.

The American Institute of Architects, *Architect's Handbook of Professional Practice*, Chapter 18, Construction Contract Administration, The American Institute of Architects, 1735 New York Avenue, N.W., Washington, DC 20006.

The American Society for Testing and Materials, *Standards in Building Codes*, The American Society for Testing and Material, 1916 Race Street, Philadelphia, PA 19103.

The American Welding Society, *Inspection Handbook for Manual Metal Arc Welding*: B1.1, American Welding Society, Inc., P.O. Box 351040, Miami, FL 33135. Latest edition.

Birch, Silas B. Jr., *Public Works Inspector's Manual*, Building News, 3055 Overland Avenue, Los Angeles, CA 90034, 1993.

Construction Specifications Institute, MASTERFORMAT, 1995 edition, *List of Titles and Numbers for the Construction Industry*, MP-2-1-88, 601 Madison Street, Alexandria, VA 22314.

International Conference of Building Officials *A Training Manual in Field Inspection of Buildings and Structures*, International Conference of Building Officials, 5360 South Workman Mill Road, Whittier, CA 90601.

Masonry Institute *Reinforced Concrete Masonry Inspector's Manual*, Masonry Institute, 3130 La Selva, Suite 302, San Mateo, CA 94403.

Southern California Chapter, American Public Works Association and Southern California District, Associated General Contractors of California *Standard Specifications for Public Works Construction*, Building News, 1612 Clementine St., Anaheim, CA 92802, 1994.

United States Department of Commerce *Field Inspector Check List for Building Construction*, United Stated Government Printing Office, Superintendent of Documents, Washington, DC 20234, 1942.

APPENDIX B — TERMS AND DEFINITIONS

Addendum: Written or graphic instruments issued prior to the execution of the contract which modify or interpret the bidding documents, including drawings or specifications, by additions, deletions, clarifications, or corrections. Addenda will become part of the contract documents when the construction contract is executed. Plural: *Addenda.*

*Agency:** Administrative subdivision of a public or private organization having jurisdiction over construction of the work.

Application and Certificate for Payment: Contractor's written request for payment of amount due for completed portions of the work, and, if the contract so provides, for materials delivered and suitably stored pending their incorporation into the work.

*Accepted:** Written acknowledgement of review by the architect/engineer or other authority having jurisdiction.

*Architect:** Designation reserved, usually by law, for a person or organization professionally qualified and duly licensed to perform architectural services, including analysis of project requirements, creation and development of project design, preparation of drawings, specifications and bidding requirements, and general administration of the construction contract. As used in this manual, the prime design professional with whom the owner contracts for design services: either the architect or the engineer.

*Architect's Representative:** An individual assigned by the architect to act as his liaison to assist in the administration of the construction contract.

Beneficial Occupancy: Use of a project or portion thereof for the purpose intended.

Building Inspector: A representative of a governmental authority employed to inspect construction for compliance with applicable codes, regulations, and ordinances.

Certificate for Payment: A statement from the architect to the owner confirming the amount of money due the contractor for work accomplished or materials and equipment suitably stored, or both.

Change Order: A written order to the contractor signed by the owner and the architect, issued after the execution of the contract, authorizing a change in the work or an adjustment in the contract sum or the contract time.

---

(1) Many of the terms and definitions have been excerpted from the "Glossary of Construction Industry Terms," copyright 1970 by the American Institute of Architects. Refer to this publication for a comprehensive glossary. However, for the purposes of the manual, certain terms and/or their definitions have been changed or added. An asterisk (*) following the term indicates these exceptions.

Codes:* Regulations, ordinances, or statutory requirements of a governmental unit relating to building construction and occupancy, adopted and administered for the protection of the public health, safety, and welfare.

Consultant:* A individual or organization engaged by the architect/engineer to render professional consulting services complementary to or supplementing his services.

Construction Documents: Working drawings and specifications.

Construction Inspector:* A qualified person engaged to provide full-time inspection of the work. In this manual, may refer to the inspector, specialized inspectors, and staff.

Contract:* The legally enforceable promise or agreement executed by the owner and the contractor for the construction of the work.

Contract Documents: The owner-contractor agreement, the conditions of the contract (general, supplementary, and other conditions), the drawings, the specifications, all addenda issued prior to execution of the contract, all modifications thereto, and any other items specifically stipulated as being included in the contract documents.

Contractor:* The person or organization performing the work and identified as such in the contract.

Date of Substantial Completion: The date certified by the architect when the work or a designated portion thereof is sufficiently complete in substantial accordance with the contract documents so that the owner may occupy the work or designated portion thereof for the use for which it was intended.

Engineer:* Designation. reserved, usually by law, for a person or organization professionally qualified and duly licensed to perform engineering services, including analysis of project requirements, development of project design, preparation of drawings, specifications and bidding requirements, and general administration of the construction contract. See also *Consultant*.

Field Change Order:* A written order used for emergency instruction to the contractor where the time required for the preparation and execution of a formal change order would result in a delay or stoppage of the work. The usage of a field change order may be subject to prior legal approval and limitations. A duly authorized change order replaces a field change order as soon as possible.

Final Acceptance: The owner's acceptance of the project from the contractor upon certification by the architect that it is complete and in substantial accordance with the contract requirements. Final acceptance is confirmed by making of final payment unless otherwise stipulated at the time of making such payment.

Final Inspection: Final review of the project by the architect prior to his issuance of final certificate of payment.

Inspection:* Examination of the work completed or in progress to determine its compliance with contract requirements. The architect ordinarily makes only two inspections of a construction project, one to determine the date of substantial completion, and the other to determine final completion. **These inspections should be distinguished from the more general observations of visually exposed and accessible conditions made by the architect on periodic site visits during the progress of the Work.** A full-time construction inspector makes continuous inspection.

Inspection List: A list of items of work to be completed or corrected by the contractor.

Owner: (1) The architect's client and party to owner-architect agreement; (2) the owner of the project and party to the owner-contractor agreement.

Owner Representative:* A person delegated by the owner to act in his behalf to provide liaison with the architect during the development of the project and construction of the work. The responsibilities delegated should be stipulated to the extent that this person may or can make decisions on the part of the owner.

Progress Payment: Partial payment made during the progress of the work on account of work completed and/or materials received and suitably stored.

Progress Schedule: A diagram, graph, or other pictorial or written schedule showing proposed or actual times of starting and completion of the various elements of the work.

Project: The total construction designed by the architect, of which work performed under the contract documents may be a whole or part.

Punch List: Use *Inspection List*.

Record Drawings:* Construction drawings revised to show significant changes made during the construction process, usually based on marked-up prints, drawings, and other data furnished by the contractor to the architect. Sometimes the term "as-built drawings" has been used in the past. The contractor is responsible for the accuracy of the information provided.

Required:* Need of contract documents, code, agency, normal accepted practice, or other authority unless context implies a different meaning.

Schedule of Values: A statement furnished by the contractor to the architect reflecting the portions of the contract sum allotted for the various parts of the whole and used as a basis for reviewing the contractor's applications for progress payments.

Shop Drawing: Drawings, diagrams, illustrations, schedules, performance charts, brochures, and other data prepared by the Contractor or any subcontractor, manufacturer, supplier, or distributor which illustrate how specific portions of the work shall be fabricated and/or installed.

Standards:* Organizations or public agencies that are recognized or have established by authority, custom general consent, industry standards, or other manner a method, criterion, example, or test for the manufacture, installation, workmanship, or performance of a material or system.

Subcontractor: A person or organization that has a direct contract with a prime contractor to perform a portion of the work at the site.

Substantial Completion: See *Date of Substantial Completion*.

Supplier: A person or organization that supplies materials or equipment for the work, including that fabricated to a special design, but does not perform labor at the site. See also *Vendor*.

Superintendent: Contractor's representative at the site who is responsible for continuous field supervision, coordination, completion of the work, and, unless another person is designated in writing by the contractor to the owner and architect, prevention of accidents.

Vendor: A person or organization that furnishes materials or equipment not fabricated to a special design for the work. See also *Supplier*.

Work: All labor necessary to produce the construction required by the contract documents, and all materials and equipment incorporated or to be incorporated in such construction.

APPENDIX C — CONSTRUCTION INDUSTRY ORGANIZATIONS

In addition to the organizations and agencies listed in Part 2 of this manual, many other professional and trade organizations and agencies listed below represent the members of the construction industry. Many of these groups publish documents and provide other information relating to design, materials, systems, specifications, and construction.

National Organizations

Acoustical Society of America, 120 Wall St., New York, NY 10005, (212) 248-0373.

Adhesive and Sealant Council, 1627 K St. NW, Suite 1000, Washington, DC 20006, (202) 452-1500 - FAX: (202) 452-1501.

Advisory Council on Historic Preservation, The Old Post Office Bldg., 1100 Pennsylvania Ave., N.W., Suite 809, Washington, DC 20004, (202) 606-8503 - FAX: (202) 606-8647.

Air Conditioning and Refrigeration Institute, 4301 N Fairfax Dr., #425, Arlington, VA 22203, (703) 524-8800 - FAX: (703) 524-8800

Air Conditioning and Refrigeration Wholesalers, 1351 South Federal Highway, P.O. Box 640, Deerfield Beach, FL 33441.

Air Conditioning Contractors of America, 1228 17th St., N.W., Washington, DC 20036.

Air Pollution Control Association, P.O. Box 2861, Pittsburgh, PA 15230.

Allied Stone Industries, Carthage Marble Co., Carthage, MO 64836.

American Arbitration Association, 140 W. 51st St., New York, NY 10020, (212) 484-4006, (800) 778-7879.

American Association for Hospital Planning, Century Bldg. Ste. 830, 2341 Jefferson Davis Hwy., Arlington, VA 22202.

American Association of Junior Colleges, National Center for Higher Education, One DuPont Circle, N.W., Washington, DC 20036.

American Association of Museums, 1575 I St. NW Suite 400, Washington, DC 20005, (202) 289-1818 - FAX: (202) 289-6578.

American Association of School Administrators, 1801 N. Moore St., Arlington, VA 22209, (703) 528-0700 - FAX: (703) 841-1543 (800) 771-1162.

American Concrete Institute, P.O. Box 9094, Farmington Hills, IL 48333.

American Concrete Pipe Association, 8300 Boone Blvd., #400, Vienna, VA 22182.

American Construction Inspectors Association, 2275 W. Lincoln Ave., Ste. B, Anaheim, CA 92801.

American Forest Council, 1111 19th St., N.W., #800, Washington, DC 20036, (202) 463-2455.

American Forestry Association, 1319 18th St., N.W., Washington, DC 20036.

American Gas Association, 1515 Wilson Blvd., Arlington, VA 22209, (703) 841-8400.

American Hardware Manufacturers Assoc., 801 N. Plaza Dr., Schaumberg, IL 60173 (312) 885-1025 - FAX: (847) 605-1093.

American Hospital Association, 840 N. Lakeshore Dr., Chicago, IL 60611, (312) 280-6000.

American Institute of Architects, 1735 New York Ave., N.W., Washington. DC 20006, (202) 626-7300 (800) 242-3837.

American Institute of Kitchen Dealers, 124 Main St., Hackettstown, NJ 07840.

American Institute of Landscape Architects, 6810 N. Second Pl., Phoenix, AZ 85012.

American Institute of Planners. 1313 E. 60th St., Chicago, IL 60637.

American Institute of Real Estate Appraisers of the National Assn. of Realtors, 875 N. Michigan Suite 2400, Chicago, IL 60611 (312) 335-4100.

American Insurance Association, 1130 Connecticut Ave., N.W., #1000, Washington, DC 20036, (202) 828-7100 - FAX: (202) 293-1219.

American Iron and Steel Institute, 1101 17th St. N.W., Washington, DC 20005, (202) 452-7100.

American Library Association, 50 E. Huron St., Chicago, IL 60611, (312) 944-6780 (800) 545-2433.

American National Standards Institute, 11 W. 42nd St, 13th Floor, New York, NY 10036 (888) 267-4783 - FAX: (212) 398-0023.

American Plywood Association, 7011 S. 19th St., Tacoma, Washington 98466 (253) 565-6600.

American Public Power Association, 2301 M St., N.W., Washington, DC 20037, (202) 775-8300 FAX: (202) 467-2910.

American Public Works Association, 1313 E. 60th St., Chicago, IL 60637, (773) 667-2200 - FAX: (773) 667-2304.

American Road and Transportation Builders Association, ARBA Bldg., 1010 Massachusetts Ave., Washington, DC 20001, (202) 289-4434.

American Segmental Bridge Institute, 9201 N 25th Ave., Suite 150B, Phoenix, AZ 85021, (602) 997-9964 - FAX: (602) 997-9965.

American Society of Civil Engineers, 1801 Alexander Bell Dr., Reston, VA 20190 (800) 548-2723.

American Society of Golf Course Architects, 221 N. LaSalle St., Chicago, IL 60601, (312) 372-7090 - FAX: (312) 372-6160.

American Society of Heating, Refrigeration and Air Conditioning Engineers, Inc., 1791 Tullie Circle N.E., Atlanta, GA 30329, (404) 636-8400 - FAX: (404) 321-5478 (800) 527-4723.

American Society of Mechanical Engineers, United Engineering Center, 345 E. 47th St., New York, NY 10017 (212) 705-7722 (800) THE-ASME.

American Society of Planning Officials, 1313 E. 60th St., Chicago, IL, 60637, (312) 947-2560.

American Society of Real Estate Counselors, 430 N. Michigan Ave., Chicago, IL 60611, (312) 329-8431 - FAX: (312) 329-8881.

American Society for Testing & Materials, 100 Barr Harbor Dr. W., Conshocken, PA 19428 (610) 832-9585.

American Welding Society, Inc., P.O. Box 351040, 550 NW 42nd Ave., Miami, FL 33135.

American Wood Preservers Association, P.O. Box 286, Woodstock, MD 21663 (410) 465-3169.

Architectural Precast Association, 825 E. 64th St., Indianapolis, IN 46220, (317) 251-1214.

Asphalt Institute, Research Park Dr, Lexington, KY 40512 (606) 288-4961.

Association Builders and Contractors, Inc., 1300 N. 17th St., Roslyn, VA 22209, (703) 812-2000.

Associated Equipment Distributors, 615 W. 22nd St., Oakbrook, IL 60523, (630) 574-0650.

Associated General Contractors of America, 1957 E St., N.W., Washington, DC 20006, (202) 393-2040.

Associated Specialty Contractor, Inc., 3 Bethesda Metro Center #1100, Bethesda, MD 20814, (301) 657-3110.

Association of American Universities, One DuPont Circle, Washington, DC 20036 (202) 387-3760.

Association of University Architects, c/o Forrest M. Kelly, Jr., Physical Planning Officer State University System of Florida Collins Bldg., Tallahassee, FL 32301.

Association of Wall and Ceiling Industries International, 1600 Cameron St., Alexander, VA 22314-2705, (703) 684-2924 FAX: (703) 684-2935.

Association of Women in Architecture, 7440 University Dr., Saint Louis, MO 63130 (314) 621-3484.

Better Heating-Cooling, 35 Russo Pl., Berkeley Heights, NJ, (908) 464-8200.

Builders Hardware Manufacturers Association, 355 Lexington Ave., 17th Fl., New York, NY 10017, (212) 661-4261 - FAX: (212) 370-9047.

Building Congress and Exchange, 2301 N. Charles St., Baltimore, MD 21218.

Building Materials Research Institute, Inc., 15 E. 40th St., New York, NY 10017.

Building Research Institute, 2101 Constitution Ave., N.W., Washington, DC 20418.

Building Stone Institute, 420 Lexington Ave., New York, NY 10017, (212) 490-2530.

Building Systems Research Institute, 2101 Constitution Ave., N.W., Washington, DC 20418.

Building Thermal Envelope Coordinating Council, 101 15th St., N.W., Ste. 700, Washington, DC 20005, (202) 347-5710.

California Association of Realtors, 525 S. Virgil Ave., Los Angeles, CA 90020, (213) 739-8200.

Ceilings and Interior Systems Contractors Association, 1500 Lincoln Hwy, Suite 202, St. Charles, IL 60174 (630) 584-1919.

Cellular Concrete Association, 715 Boylston St., Boston, MD 02116.

Ceramic Tile Distributors Association, 15 Salt Creek Lane, Ste. 422 Hinsdale, IL 60521.

Ceramic Tile Institute, 700 N. Virgil Ave., Los Angeles, CA, 90029.

Committee of Steel Pipe Producers American Iron And Steel Institute, 1000 16th St., N.W., Washington. DC 20036.

Concrete Reinforcing Steel Institute. 933 N. Plluli Grove Road Schaumburg, IL. 60173-4758, (312) 517-1200 - FAX: (312) 517-1206.

Construction Financial Management Assn., 40 Brunswick Ave., Edison, NJ 08818.

Construction Labor Research Council, 1730 M Str NW, Suite 503, Washington, DC 20036.

Construction Specifications Institute, 601 Madison St., Alexandria, VA 22314, (800) 689-2900.

Construction Writers Association, P.O. Box 5586, Buffalo Grove, IL 60089 (847) 398-7756.

Contracting Plasterers Research Institute, 2101 Constitution Ave., N.W., Washington, DC 20418.

Copper Development Association, Inc., 260 Madison Ave - 16th Floor, New York, NY 10016 (212) 251-7200 (800) 232-3282.

Council of Educational Facility Planners, 29 W. Woodruff Ave., Columbus, OH 43210.

Council of Mechanical Specialty Contracting Industries, Inc., 7315 Wisconsin Ave., Washington., DC 20014.

Electrical Association, 140 S. Dearborn St., Chicago, IL 60603.

Electric Power Research Institute, 2000 L Str, Suite 805 NW, Washington, DC 20036 (202) 872-9222 - FAX: (202) 293-2697.

The Energy Bureau, Inc., 41 E. 42nd St., New York, NY 10017.

Engineers Joint Council, 345 E. 47th St., New York, NY 10017.

Federal Housing Administration, Dept. of Housing and Urban Development, 451 7th St.. S.W., Washington, DC 20410 (202) 708-2495 - FAX: (202) 708-2583.

Fine Hardwoods Association, 5603, West Raymond, Ste. 0, Indianapolis, IN 46241, (317) 873-8780.

Flexicore Manufacturers Association, P.O. Box 1807, Dayton, OH 45401, (937) 223-7420.

Food Facilities Consultants Society, 1800 Pickwick Ave., Glenview, IL 60025.

Forest Products Research Society, 2801 Marshall Ct., Madison, WI 53705, (608) 231-1361 (800) 354-7164.

Gardens For All, 180 Flynn Ave., Burlington, VT 05401.

Guild For Religious Architecture, 1913 Architects Bldg., Philadelphia, PA 19103.

Historic American Buildings Survey, 801 19th St., N.W., Washington, DC 20006.

Illuminating Engineers Society, 120 Wall St., New York, 17th Floor, NY 10005 (212) 248-5000.

Information Bureau of Lath/Plaster/Drywall, 3127 Los Feliz Blvd., Los Angeles, CA 90039, (213) 663-2213.

Institute of Electrical and Electronic Engineers, 345 E. 47th St., New York, NY 10017, (212) 705-7900 (800) 678-4333.

Institute of Noise Control Engineering, P.O. Box 3206, Arlington Branch, Poughkeepsie, NY 12603.

Institute of Real Estate Management, 430 N. Michigan Ave., Chicago, IL 60611 (312) 661-1930 (800) 837-0706 - FAX: (800) 338-4736.

International Association of Plumbing and Mechanical Officials, 2001 S. Walnut Dr., Walnut, CA 91789-2825.

International Conference of Building Officials, 14545 Leffingwell, Whittier, CA 90604 (562) 903-1478 - FAX: (561) 903-1480.

International Council of Shopping Centers, 665 Fifth Ave., New York, NY 10022, (212) 421-8181 - FAX: (212) 421-6464.

International Institute of Ammonia Refrigeration, 111 East Wacker Dr., Chicago, IL 60601, (312) 644-6610 - FAX: (312) 565-4658.

International Masonry Institute, 815 15th St., N.W., Washington, DC 20005, (202) 783-3908.

Inter-Society Color Council, Inc., Rensselaer Polytechnic Institute, Troy, NY 12181.

Landscape Architecture Foundation, 636 I Street NE, Washington, DC 20001 (202) 898-2444.

Mason Contractors Association of America, 1550 Spring Rd, Suite 320, Oakbrook, IL 60521 (708) 782-6767.

Metal Buildings Manufacturers Association, c/o Thomas Assoc, 1300 Summer Ave, Cleveland, OH 44115 (216) 241-7333 - FAX: (216) 241-0105.

Model Codes Standardization Council, National Bureau of Standards, Washington, DC 20234.

Mortar Manufacturers Standards Association, 315 S. Hicks Rd., Palatine, IL 60067.

Mortgage Bankers Association of America, 1125 15th St., N.W., Washington, DC 20005, (202) 861-6500.

National Asphalt Pavement Association, 5100 Forbes Blvd. Lanham MD 20706, (301) 731-4748.

National Association of Corrosion Engineers, 1440 South Creek Dr., Houston, TX 77084, (281) 492-0535.

National Association of Decorative Architectural Finishes, 112 N. Alfred St., Alexandria, VA 22314.

National Association of Garage Door Manufacturers, 1300 Summer Ave, Cleveland OH 44115 (216) 241-7333.

National Association of Home Builders National Housing Center, 1201 15th St NW, Washington, DC 20005 (202) 822-0200 (800) 223-2665.

National Association of Home Builders of the J.S., 1201 15th St., N.W., Washington, DC 20005, (202) 822-0200.

National Association of Housing and Redevelopment Officials, 630 Eye St., NW, Washington, DC 20001 (202) 289-3500.

National Association of Realtors, 700 11th Str NW, Washington, DC 20001 (202) 283-1043.

National Association of Store Fixture Manufacturers, 5975 W. Sunrise, Sunrise, FL 33312 (305) 587-9190.

National Board of Boiler and Pressure Vessel Inspectors. 1055 Crupper Ave., Columbus, OH 43229, (614) 888-8320.

National Institute of Standards and Technology, Fire and Building Research Labs, Gaithersburg, MD 20899.

National Concrete Masonry Association, 2302 Horse Pen Rd., Herndon, VA 22071 (703) 713-1900.

National Construction Association, 1730 M St., N.W., Suite 503, Washington, DC 20036, (202) 466-8880.

National Corporation for Housing Partnership, 1133 15th St., N.W., Washington, DC 20005, (202) 216-2900.

National Crushed Stone Association, 1415 Elliott Pl., N.W., Washington, DC 20007.

National Decorating Products Assn., 415 Ax Minister, St. Louis, MO 63026.

National Electrical Contractors Association, Inc., 3 Betheseda, MD 20814.

National Fire Protection Association, 1 Batterymarch Park, Quincy, MA 02269 (800) 344-3555.

National Housing Conference, 815 15th St NW Suite 711, Washington, DC 20005 (202) 393-5772.

National Institute of Building Sciences, 1201 L St., N.W., #400, Washington, DC 20005, (202) 289-7800.

National Petroleum Council, 1625 K St., N.W., Ste. 600, Washington, DC 20006, (202) 393-6100.

National Precast Concrete Association, 10333 N. Meridian St. Suite 272, Indianapolis, IN 46290 (317) 571-9500.

National Ready Mixed Concrete Association, 900 Spring St., Silver Springs, MD 20910 (301) 587-1400 - FAX: (301) 585-4219.

National Roofing Contractors Association, 10255 W. Higgins Rd, Suite 600, Rosemont, IL 60018 (708) 299-9070.

National Science Foundation, 4201 Wilson Blvd, Arlington, VA 22230 (703) 306-1070.

National Slag Association, 900 Spring St., Silver Springs, MD 20910 (301) 587-1400.

National Wood Flooring Association, 233 Old Meramec Stations Rd, Manchester, MO 63021 (314) 391-5161.

North American Wholesale Lumber Association, 3601 Algonquin Rd, Suite 400, Rolling Meadows, IL 60008 (708) 870-7470.

Painting and Decorating Contractors of America, 3913 Old Lee Highway, Ste. 33-B, Fairfax, VA 22030, (703) 359-0826 - FAX: (703) 359-2576 (800) 332-7322.

Plastering Information Bureau, 21243 Ventura Blvd, Suite 115, Woodland Hills, CA 91364.

Plastic in Construction Council, 355 Lexington, New York, NY 10001.

Plumbing and Drainage Institute, P.O. Box 93, Indianapolis, IN 46206, (317) 251-5298.

Portland Cement Association, 5420 Old Orchard Rd., Skokie, IL 60077. (800) 868-6733.

Prestressed Concrete Institute, 175 W. Jackson Blvd., Chicago, IL 60604, (312) 786-0300.

Red Cedar Shingle and Handsplit Shake Bureau, 515 116th Ave. N E., Ste. 275, Bellevue, WA 98004, (425) 453-1323.

Scaffold Industry Association, Inc., 14039 Sherman Way, Van Nuys, CA 91405, (818) 782-2012 - FAX: (818) 786-3027.

Scaffolding and Shoring Institute, c/o Thomas Associates, Inc., 1300 Summer, OH 44115, (216) 241-7333.

Screen Manufacturers Association, 2850 S. Ocean Blvd., No. 311, Palm Beach, FL 33480 (407) 533-0991.

Sealed Insulating Glass Manufacturers Association, 401 N. Michigan Ave, Chicago, IL 60611, (312) 644-6610.

Sheet Metal and Air Conditioning Contractors National Assn., Inc., 4201 Lafayette Center Dr., Chantilly, VA 20151 (703) 803-2989.

Sheet Metal and Air Conditioning Contractors' National Association, Inc., 4201 Lafayette Center Dr., Chantilly, VA 20151 (703) 803-2980.

Society of the Plastic Industry, 1275 K St., N.W., Washington, DC 20005, (202) 371-5200.

Solar Energy Industries Association, 122 C St. NW, 4th Floor, Washington, DC 20001, (202) 383-2600.

Southern Cypress Manufacturers Association, 400 Penn Center Blvd, Pittsburgh, PA 15235 (412) 829-0770.

Stained Glass Association of America, P.O. Box 22642, Kansas City, MO 64113 (816) 333-6690.

Steel Door Institute, 30200 Detroit Rd., Cleveland, OH 44145, (216) 226-0010.

Stucco Manufacturers Association, 507 Evergreen, Pacific Grove, CA 93950 (408) 649-3466.

Truss Plate Institute, 583 D'Onofrio Dr. Suite 200, Madison, WI 53719 (608) 833-5900.

United Brotherhood of Carpenters and Joiners of America, 101 Constitution Avenue, N.W., Washington, DC 20001, (202) 546-6206.

United States Conference of Mayors, 1620 Eye St., N.W., Washington, DC 20006, (202) 293-7330.

United States League of Savings Institutions, 111 E. Wacker Dr., Chicago, IL 60601, (312) 644-3100.

Urban Institute, 2100 M St., N.W., Washington, DC 20037, (202) 624-7062.

Vermiculite Association, 11 S. La Salle St., Suite 1400, Chicago, IL 60603 (312) 201-0101.

Waferboard Assn., P.O. Box 724533, Atlanta, GA 30339.

Wallcovering Information Bureau, 66 Morris Ave., Springfield, NJ 07081.

Wallcovering Wholesalers Association, 401 N. Michigan Ave, Chicago, IL 60611 (312) 245-1083.

Western Red Cedar Lumber Association, 1500 Yeon Bldg., Portland, OR 97204.

Western Wood Products Association, 522 SW 5th Ave Suite 500, Portland, OR 97204 (503) 224-3930.

Wood and Synthetic Flooring Institute. 1800 Pickwick Avenue, Glenview, IL 60025.

Wood Truss Council of America, 1 WTCA Center, 6425 Normandy Ln, Madison, WI 53719 (608) 274-3329.

CALIFORNIA ORGANIZATIONS

Air Conditioning and Refrigeration Contractors Association of Southern California, 401 Shatto Pl., Los Angeles, CA 90020, (213) 738-7238 (213) 738-5260.

American Public Works Association, Northern California Chapter

American Subcontractors Association, Los Angeles/Orange County Chapter, c/o Philip B. Greer, Atkinson, Andelson, et al, 13304 E. Alondra Blvd., Suite 200, Cerritos, CA 90701.

Associated Builders and Contractors, Golden Gate Chapter, 11875 Dublin Blvd, Suite 258 Dublin, CA 94568, (510) 829-9230

Associated General Contractors of California, East Bay District, 1390 Willow Pass Road, Suite 1030, Concord, CA 94520 (510) 827-2422

Associated General Contractors of California, State Office, 3095 Beacon Blvd., West Sacramento, CA 95691, (916) 371-2422.

Associated Plumbing and Mechanical Contractors of Sacramento, 50 Fullerton Ct. #100, Sacramento, CA 95825, (916) 452-4917 FAX: (916) 452-0532.

Associated Roofing Contractors of the Bay Area Counties, 8301 Edgewater Dr., Oakland, CA 94621, (510) 635-8800.

Associated Tile Contractors of Southern California, 2736 S. La Cienega Blvd., Los Angeles, CA 90034.

Builders Exchange of Alameda, 3055 Alvarado St., San Leandro, CA 94577, (510) 483-8880.

Builders Exchange of Contra Costa, 1900 Bates Ave., Suites E & F, Concord, CA 94520 (510) 685-8630.

Builders Exchange of Modesto, P.O. Box 4307, Modesto, CA 95352, (209) 522-9031.

Builders Exchange of Monterey Peninsula, 343 Ocean Avenue, Monterey, CA 93940, (408) 373-3033 - FAX: (408) 373-8682.

Builders Exchange of Napa/Solano, 135 Camino Dorado, Napa, CA 94558, (707) 255-2515.

Builders Exchange of the Peninsula, 735 Industrial Road, San Carlos, CA 94070, (650) 591-4486 FAX: (650) 591-8108.

Builders Exchange of Salinas Valley, 590-A Brunched Avenue, Suite A, Salinas, CA 93901, (408) 758-1624 - FAX: (408) 758-6203.

Builders Exchange of San Francisco, 850 S. Van Ness Avenue, San Francisco, CA 94110, (415) 282-8220.

Builders Exchange of Santa Clara, 400 Reed St., Santa Clara, CA 95050, (408) 727-4000.

Builders Exchange of Santa Cruz, 2555 So. Cal Dr., Santa Cruz, CA 95065 (408) 476-6349.

Builders Exchange of Stockton, 7500 N. West Lane (plans only), Stockton, CA 95210, P.O. Box 8040 (letters only), Stockton CA 95208, (209) 478-1005 - FAX: (209) 478-2132.

California Association of Realtors, 525 So. Virgil Avenue, Los Angeles, CA 90020, (213) 365-9256 - FAX: (213) 365-9256.

California Building Industry Association, 1107 9th Street, Suite 1060, Sacramento, CA 95814, (916) 443-7933 - FAX: (916) 443-1960.

California Conference of Masonry Contractor Associations, 7844 Madison Ave., Ste. 140, Fair Oaks, CA 95628.

California Contractors Association, 6055 E. Washington Blvd., Suite 200, Los Angeles, CA 90040, (213) 726-3511 - FAX: (213) 726-2366.

California, Division of State Architecture.

California Landscape Contractors Association, 2021 N St. #300, Sacramento, CA 95814, (916) 448-2522.

California OSHPD.

California Wall and Ceiling Contractors Association, 1111 Town and Country Rd. #45, Orange, CA 92668.

Ceramic Tile Institute of Northern California, 10408 Fair Oaks Blvd., Fair Oaks, CA 95628, (916) 965-8453 - FAX: (916) 965-8454.

Concrete Masonry Association of California and Nevada, 6060, Sunrise Vista Dr., Citrus Heights, CA 95610, (916) 722-1700.

Concrete Pumpers Association of Southern California, 1567 Colorado Blvd., Los Angeles, CA 90041, (213) 257-5266.

Construction Industry Research Board, 2511 Empire Avenue, Burbank, CA 91504, (818) 8341-8210.

Contractors Bonding Association, 529 W. Imperial Way, Suite 5, Los Angeles, CA 90044.

El Dorado Builders Exchange, 2808 Mallard Ln. #B, Placerville, CA 95667, (530) 622-8642.

Electric Power Research Institute, 3412 Hillview Avenue, Palo Alto, CA 94304, (415) 855-2000.

Electric Contractors of California and Nevada, 7700 Edgewater Dr. #640, Oakland, CA. 94621.

Engineering Contractors Association, 8310 Florence Avenue, Downey, CA 90240, (562) 861-0929.

Floor Covering Institute, 400 Reed St., Santa Clara, CA 95050, (408) 727-4320.

Fresno Builders Exchange, P.O. Box 111, Fresno, CA 3707, CA 95667, (209) 237-1831.

Independent Roofing Contractors of California, 3478 Buskirk Avenue #1040, Pleasant Hill, CA 94523, (510) 939-3715.

Kern County Builders Exchange, 1121 Baker St., Bakersfield, CA 93305, (805) 324-5364.

Los Angeles County Painting and Decorating Contractors Association, Inc., 1106 Colorado Blvd., Los Angeles, CA 90041, (213) 258-8136 - FAX: (213) 258-2279.

Marin Builders Exchange, 110 Belvedere St., San Rafael, CA 94901, (415) 456-3222.

Masonry Contractors Association of Sacramento, 7844 Madison Avenue, Suite 140, Fair Oaks, CA 95628, (916) 966-7666.

Mechanical Contractors Legislative Council of California, 7 Crow Canyon Court, Ste. 200, San Ramon, CA 94583.

Merced-Mariposa Builders Exchange, P.O. Box 761, Merced, CA 95341, (209) 722-3612.

Minority Contractors Association of Los Angeles, 3707 W. Jefferson, Los Angeles, CA 90016, (213) 737-7952.

National Association of Women in Construction of Los Angeles, P.O. Box 90935, Pasadena, CA 91109.

National Association of Women in Construction, 4865 Pasadena Avenue, Sacramento, CA 95841, (916) 483-2724.

National Electrical Contractors Association, Los Angeles Chapter, 401 Shatto Pl., Los Angeles, CA 90020, (213) 487-7313 - FAX: (213) 388-5230.

North Coast Builders Exchange, 216 W. Perkins St., Ukiah, CA 954:82, (707) 462-9019.

North Coast Builders Exchange, 987 Airway Ct., Santa Rosa, CA 95403, (707) 542-9502.

Northern California Drywall Contractors Association, 12241 Saratoga-Sunnyvale Road, Saratoga, CA 95070, (408) 255-1544.

Northern California Engineering Contractors Association, 3354 Regional Prkwy, Santa Rosa, CA 95403, (707) 525-1910.

Pacific Coast Builders Conference, 605 Market Street, San Francisco, CA 94105, (415) 821-3307.

Painting and Decorating Contractors of California, 3504 Walnut Avenue, Suite A, Carmichael, CA 95608, (916) 972-1055 - FAX: (916) 972-9831.

Painting and Decorating Contractors of Central Coast Counties, 4050 Ben Lomond Dr., Palo Alto, CA 94306, (650) 493-6200.

Painting and Decorating Contractors of Sacramento, 3913 Old Lee Highway, Suite 33B, Fairfax, VA 22030 (800) 332-7322.

Peninsula Builders Exchange, 735 Industrial Rd., San Carlos, CA 94070, (650) 591-4486.

Roofing Contractors Association of Southern California, 6280 Manchester Blvd, Suite 104, Buena Park, CA 90621 (714) 522-4694.

Roofing Industry Council, 400 Reed St., Suite D, Santa Clara, CA 95050.

Sacramento Builders Exchange, 1331 T Street, Sacramento, CA 95814, (916) 442-8991.

San Francisco Builders Exchange, 850 S. Van Ness Avenue, San Francisco, CA 94110, (415) 282-8220.

San Luis Obispo County Building Contractors Association, 3563 Sueldo St., Suite G, San Luis, CA 93401 (805) 543-7016.

Santa Barbara Contractors Association, P.O. Box 41622 Santa Barbara, CA 93410, (805) 964-9175.

Santa Maria Valley Contractors Association, 2003 N. Preisker Ln. Santa Maria, CA 93454, (805) 925-1191.

Shasta Building Exchange, 2990 Innsbruck, Redding, CA 96003 (530) 221-2140.

Society of American Military Engineers, Orange County Post, c/o Tim Kashuba, Moffatt and Nichol, 250 Wardlow Rd., Long Beach, CA 90807, (213) 426-9551 - FAX: (213) 424-7489.

Southern California Builders Association, 4552 Lincoln, Suite 207, Cypress, CA 90630 (714) 995-5841.

Southern California Drywall Contractors Association, 111 Town and Country Rd., Suite 45, Orange, CA 92668, (714) 998-8125.

Southern California Environmental Balancing Bureau, P.O. Box 605, Santa Ynez, CA 93460.

Ventura County Contractors Association, P.O. Box 7365, Oxnard, CA 93031, (805) 981-8088.

Western Electrical Contractors Association, Sacramento Valley Chapter, 7500 14th Avenue #25, Sacramento, CA 95820, (916) 453-0112.

Western States Ceramic Tile Contractors Association, 5004 E. 59th Pl., Maywood, CA 90270, (213) 560-1673.

APPENDIX D — FORMS

The following data and forms are included:

1. Application and Certificate for Payment
   AIA Document G 702, 1992 Edition, ©1992
   The American Institute of Architects.

2. Change Order
   AIA Document G 701, 1987 Edition
   ©1987 The American Institute of Architects.

3. Construction Change Directive
   AIA Document G 714, 1987 Edition
   ©1979 The American Institute of Architects.

4. Shop Drawing and Sample Record
   AIA Document G 712, 1972 Edition
   ©1972 The American Institute of Architects.

5. Continuation Sheet
   AIA Document G 703, 1992 Edition
   ©1992 The American Institute of Architects.

6. Certificate of Substantial Completion
   AIA Document G 704, 1992 Edition
   ©1992 The American Institute of Architects.

7. Project Representative
   AIA Document B 352, Duties, Responsibilities and Limitations of Authority of the Architect's Project Representative, 1993 Edition
   ©1993 The American Institute of Architects.

8. Field Change Order
   Developed by the Editorial Committee.

9. Inspector's Daily Report:
   Developed by the Editorial Committee.

10. **CSI Masterformat,** MP-2-88, 1988 Edition
    ©1988 Construction Specifications Institute.

Construction Inspection Manual

Reproduced with permission of The American Institute of Architects under license number #98126. This license expires October 31, 1999. FURTHER REPRODUCTION IS PROHIBITED. Because AIA Documents are revised from time to time, users should ascertain from the AIA the current edition of this document. Copies of the current edition of this AIA document may be purchased from The American Institute of Architects or its local distributors. The text of this document is not "model language" and is not intended for use in other documents without permission of the AIA.

APPLICATION AND CERTIFICATE FOR PAYMENT
AIA DOCUMENT G702 (Instructions on reverse side)

PAGE ONE OF ___ PAGES

TO OWNER: **PROJECT:**   **APPLICATION NO.:**   Distribution to:
☐ OWNER
☐ ARCHITECT
☐ CONTRACTOR

PERIOD TO:
PROJECT NOS.:

FROM CONTRACTOR: **VIA ARCHITECT:**   **CONTRACT DATE:**

CONTRACT FOR:

CONTRACTOR'S APPLICATION FOR PAYMENT

Application is made for payment, as shown below, in connection with the Contract.
Continuation Sheet, AIA Document G703, is attached.

1. ORIGINAL CONTRACT SUM .... $
2. Net change by Change Orders .... $
3. CONTRACT SUM TO DATE (Line 1 ± 2) .... $
4. TOTAL COMPLETED & STORED TO DATE .... $
   (Column G on G703)
5. RETAINAGE:
   a. ____% of Completed Work $_____
      (Columns D + E on G703)
   b. ____% of Stored Material $_____
      (Column F on G703)
   Total Retainage (Line 5a + 5b or
   Total in Column I of G703) .... $
6. TOTAL EARNED LESS RETAINAGE .... $
   (Line 4 less Line 5 Total)
7. LESS PREVIOUS CERTIFICATES FOR PAYMENT .... $
   (Line 6 from prior Certificate)
8. CURRENT PAYMENT DUE .... $
9. BALANCE TO FINISH, INCLUDING RETAINAGE
   (Line 3 less Line 6) $

CHANGE ORDER SUMMARY	ADDITIONS	DEDUCTIONS
Total changes approved in previous months by Owner		
Total approved this Month		
TOTALS		
NET CHANGES by Change Order		

The undersigned Contractor certifies that to the best of the Contractor's knowledge, information and belief the Work covered by this Application for Payment has been completed in accordance with the Contract Documents, that all amounts have been paid by the Contractor for Work for which previous Certificates for Payment were issued and payments received from the Owner, and that current payment shown herein is now due.

CONTRACTOR:

By: _____ Date: _____

State of:
County of:
Subscribed and sworn to before
me this ___ day of ___
Notary Public:
My Commission expires:

ARCHITECT'S CERTIFICATE FOR PAYMENT

In accordance with the Contract Documents, based on on-site observations and the data comprising this application, the Architect certifies to the Owner that to the best of the Architect's knowledge, information and belief the Work has progressed as indicated, the quality of the Work is in accordance with the Contract Documents, and the Contractor is entitled to payment of the AMOUNT CERTIFIED.

AMOUNT CERTIFIED .... $ _____
(Attach explanation if amount certified differs from the amount applied for. Initial all figures on this Application and on the Continuation Sheet that are changed to conform to the amount certified.)

ARCHITECT:
By: _____ Date: _____

This Certificate is not negotiable. The AMOUNT CERTIFIED is payable only to the Contractor named herein. Issuance, payment and acceptance of payment are without prejudice to any rights of the Owner or Contractor under this Contract.

G702-1992

AIA DOCUMENT G702 • APPLICATION AND CERTIFICATE FOR PAYMENT • 1992 EDITION • AIA® • ©1992 • THE AMERICAN INSTITUTE OF ARCHITECTS, 1735 NEW YORK AVENUE, N.W., WASHINGTON, DC 20006-5292 • WARNING: Unlicensed photocopying violates U.S. copyright laws and will subject the violator to legal prosecution.

CAUTION: You should use an original AIA document which has this caution printed in red. An original assures that changes will not be obscured as may occur when documents are reproduced.

Construction Inspection Manual

249

Reproduced with permission of The American Institute of Architects under license number #98126. This license expires October 31, 1999. FURTHER REPRODUCTION IS PROHIBITED. Because AIA Documents are revised from time to time, users should ascertain from the AIA the current edition of this document. Copies of the current edition of this AIA document may be purchased from The American Institute of Architects or its local distributors. The text of this document is not "model language" and is not intended for use in other documents without permission of the AIA.

CHANGE ORDER
AIA DOCUMENT G701

OWNER	☐
ARCHITECT	☐
CONTRACTOR	☐
FIELD	☐
OTHER	☐

PROJECT:
(name, address)

TO CONTRACTOR:
(name, address)

CHANGE ORDER NUMBER:

DATE:

ARCHITECT'S PROJECT NO:

CONTRACT DATE:

CONTRACT FOR:

The Contract is changed as follows:

SAMPLE

Not valid until signed by the Owner, Architect and Contractor.

The original (Contract Sum) (Guaranteed Maximum Price) was . . . . $
Net change by previously authorized Change Orders . . . . $
The (Contract Sum) (Guaranteed Maximum Price) prior to this Change Order was . . . . $
The (Contract Sum) (Guaranteed Maximum Price) will be (increased) (decreased)
  (unchanged) by this Change Order in the amount of . . . . $
The new (Contract Sum) (Guaranteed Maximum Price) including this Change Order will be . . $
The Contract Time will be (increased) (decreased) (unchanged) by (   ) days.
The date of Substantial Completion as of the date of this Change Order therefore is

NOTE: This summary does not reflect changes in the Contract Sum, Contract Time or Guaranteed Maximum Price which have been authorized by Construction Change Directive.

ARCHITECT	CONTRACTOR	OWNER
Address	Address	Address
BY	BY	BY
DATE	DATE	DATE

AIA DOCUMENT G701 • CHANGE ORDER • 1987 EDITION • AIA® • ©1987 • THE
AMERICAN INSTITUTE OF ARCHITECTS, 1735 NEW YORK AVE., N.W., WASHINGTON, D.C. 20006

G701—1987

Reproduced with permission of The American Institute of Architects under license number #98126. This license expires October 31, 1999. FURTHER REPRODUCTION IS PROHIBITED. Because AIA Documents are revised from time to time, users should ascertain from the AIA the current edition of this document. Copies of the current edition of this AIA document may be purchased from The American Institute of Architects or its local distributors. The text of this document is not "model language" and is not intended for use in other documents without permission of the AIA.

CONSTRUCTION CHANGE DIRECTIVE

AIA DOCUMENT G714

OWNER	☐
ARCHITECT	☐
CONTRACTOR	☐
FIELD	☐
OTHER	☐

(Instructions on reverse side. This document replaces AIA Document G713, Construction Change Authorization.)

PROJECT:
(name, address)

TO CONTRACTOR:
(name, address)

DIRECTIVE NO:
DATE:
ARCHITECT'S PROJECT NO:
CONTRACT DATE:
CONTRACT FOR:

You are hereby directed to make the following change(s) in this Contract:

SAMPLE

PROPOSED ADJUSTMENTS

1. The proposed basis of adjustment to the Contract Sum or Guaranteed Maximum Price is:

   ☐ Lump Sum (increase) (decrease) of $_____.

   ☐ Unit Price of $_____ per _____.

   ☐ as provided in Subparagraph 7.3.6 of AIA Document A201, 1987 edition.

   ☐ as follows:

2. The Contract Time is proposed to (be adjusted) (remain unchanged). The proposed adjustment, if any, is (an increase of _____ days) (a decrease of _____ days).

When signed by the Owner and Architect and received by the Contractor, this document becomes effective IMMEDIATELY as a Construction Change Directive (CCD), and the Contractor shall proceed with the change(s) described above.

Signature by the Contractor indicates the Contractor's agreement with the proposed adjustments in Contract Sum and Contract Time set forth in this Construction Change Directive.

ARCHITECT
Address

BY _____
DATE _____

OWNER
Address

BY _____
DATE _____

CONTRACTOR
Address

BY _____
DATE _____

AIA **CAUTION: You should sign an original AIA document which has this caution printed in red. An original assures that changes will not be obscured as may occur when documents are reproduced.**

AIA DOCUMENT G714 • CONSTRUCTION CHANGE DIRECTIVE • 1987 EDITION • AIA® • ©1987 • THE AMERICAN INSTITUTE OF ARCHITECTS, 1735 NEW YORK AVENUE, N.W., WASHINGTON, D.C. 20006   G714-1987

WARNING: Unlicensed photocopying violates U.S. copyright laws and is subject to legal prosecution.

Construction Inspection Manual

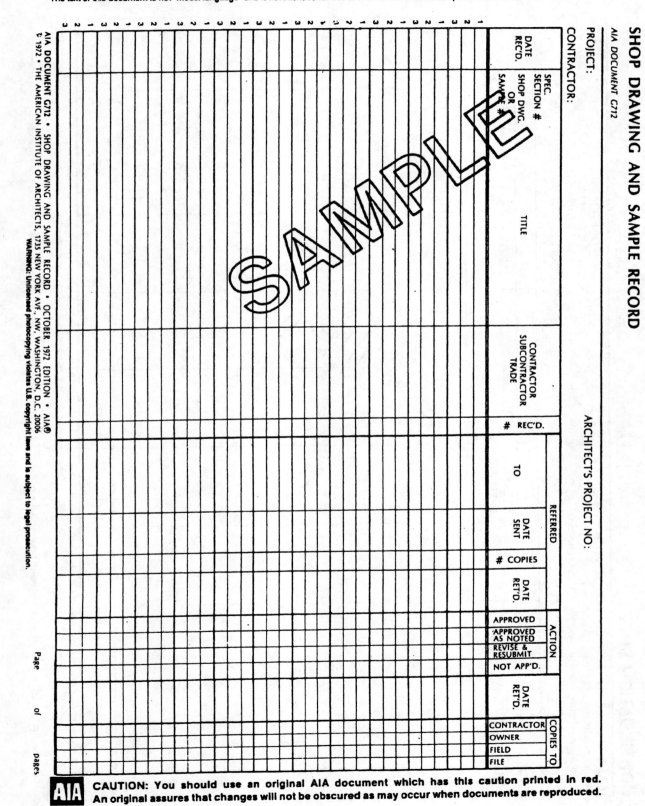

Construction Inspection Manual

Reproduced with permission of The American Institute of Architects under license number #96126. This license expires October 31, 1999. FURTHER REPRODUCTION IS PROHIBITED. Because AIA Documents are revised from time to time, users should ascertain from the AIA the current edition of this document. Copies of the current edition of this AIA document may be purchased from The American Institute of Architects or its local distributors. The text of this document is not "model language" and is not intended for use in other documents without permission of the AIA.

CONTINUATION SHEET

AIA DOCUMENT G703 (Instructions on reverse side)

AIA Document G702, APPLICATION AND CERTIFICATE FOR PAYMENT, containing Contractor's signed Certification, is attached.
In tabulations below, amounts are stated to the nearest dollar.
Use Column I on Contracts where variable retainage for line items may apply.

APPLICATION NO.:
APPLICATION DATE:
PERIOD TO:
ARCHITECT'S PROJECT NO.:

PAGE   OF   PAGES

A	B	C	D	E	F	G		H	I
			WORK COMPLETED		MATERIALS PRESENTLY STORED (NOT IN D OR E)	TOTAL COMPLETED AND STORED TO DATE (D+E+F)	% (G ÷ C)	BALANCE TO FINISH (C − G)	RETAINAGE (IF VARIABLE RATE)
ITEM NO.	DESCRIPTION OF WORK	SCHEDULED VALUE	FROM PREVIOUS APPLICATION (D + E)	THIS PERIOD					

SAMPLE

AIA DOCUMENT G703 • CONTINUATION SHEET FOR G702 • 1992 EDITION • AIA® • ©1992 • THE AMERICAN INSTITUTE OF ARCHITECTS, 1735 NEW YORK AVENUE, N.W., WASHINGTON, D.C. 20006-5292 • WARNING: Unlicensed photocopying violates U.S. copyright laws and will subject the violator to legal prosecution.

G703-1992

CAUTION: You should use an original AIA document which has this caution printed in red. An original assures that changes will not be obscured as may occur when documents are reproduced.

Reproduced with permission of The American Institute of Architects under license number #98126. This license expires October 31, 1999. FURTHER REPRODUCTION IS PROHIBITED. Because AIA Documents are revised from time to time, users should ascertain from the AIA the current edition of this document. Copies of the current edition of this AIA document may be purchased from The American Institute of Architects or its local distributors. The text of this document is not "model language" and is not intended for use in other documents without permission of the AIA.

CERTIFICATE OF SUBSTANTIAL COMPLETION

AIA DOCUMENT G704

(Instructions on reverse side)

OWNER ☐
ARCHITECT ☐
CONTRACTOR ☐
FIELD ☐
OTHER ☐

PROJECT:
(Name and address)

PROJECT NO.:
CONTRACT FOR:
CONTRACT DATE:

TO OWNER:
(Name and address)

TO CONTRACTOR:
(Name and address)

DATE OF ISSUANCE:
PROJECT OR DESIGNATED PORTION SHALL INCLUDE:

The Work performed under this Contract has been reviewed and found, to the Architect's best knowledge, information and belief, to be substantially complete. Substantial Completion is the stage in the progress of the Work when the Work or designated portion thereof is sufficiently complete in accordance with the Contract Documents so the Owner can occupy or utilize the Work for its intended use. The date of Substantial Completion of the Project or portion thereof designated above is hereby established as

which is also the date of commencement of applicable warranties required by the Contract Documents, except as stated below:

_____

A list of items to be completed or corrected is attached hereto. The failure to include any items on such list does not alter the responsibility of the Contractor to complete all Work in accordance with the Contract Documents.

ARCHITECT   BY   DATE

The Contractor will complete or correct the Work on the list of items attached hereto within _____ days from the above date of Substantial Completion.

CONTRACTOR   BY   DATE

The Owner accepts the Work or designated portion thereof as substantially complete and will assume full possession thereof at _____ (time) on _____ (date).

OWNER   BY   DATE

The responsibilities of the Owner and the Contractor for security, maintenance, heat, utilities, damage to the Work and insurance shall be as follows:
(Note—Owner's and Contractor's legal and insurance counsel should determine and review insurance requirements and coverage.)

 CAUTION: You should sign an original AIA document which has this caution printed in red. An original assures that changes will not be obscured as may occur when documents are reproduced.

DUTIES, RESPONSIBILITIES AND LIMITATIONS OF AUTHORITY OF THE ARCHITECT'S PROJECT REPRESENTATIVE

AIA DOCUMENT B352

Recommended as an Exhibit When an Architect's Project Representative is Employed

1. GENERAL

1.1 The Architect's Project Representative shall be stationed at the site and shall be responsible for assisting the Architect in the administration of the Contract. Through the observations of the Project Representative, the Architect shall endeavor to provide further protection for the Owner against defects and deficiencies in the Work. Apart from such further protection, the rights, responsibilities and obligations of the Architect as described in the Agreement Between the Owner and Architect shall not be modified by the furnishing of such Project Representative.

1.2 Communications by the Architect's Project Representative relating to administration of the Contract shall in general be restricted to the Architect and Contractor. The Project Representative shall communicate with the Owner and Contractor under the direction of the Architect and with the Architect's full knowledge. The Project Representative shall not communicate with Subcontractors or material suppliers except with the full knowledge and approval of the Contractor and Architect.

2. DUTIES AND RESPONSIBILITIES

The Project Representative shall:

2.1 Perform on-site observations of the progress and quality of the Work as may be reasonably necessary to determine in general if the Work is being performed in a manner indicating that the Work when completed will be in conformance with the Contract Documents. Notify the Architect immediately if, in the Project Representative's opinion, Work does not conform to the Contract Documents or requires special inspection or testing.

2.2 Monitor the Contractor's construction schedules on an ongoing basis and alert the Architect to conditions that may lead to delays in completion of the Work.

2.3 Receive and respond to requests from the Contractor for information and, when authorized by the Architect, provide interpretations of Contract Documents.

2.4 Receive and review requests for changes by the Contractor, and submit them, together with recommendations, to the Architect. If they are accepted, prepare Architect's Supplemental Instructions, incorporating the Architect's Modifications to the Contract Documents.

2.5 Attend meetings as directed by the Architect and report to the Architect on the proceedings.

2.6 Observe tests required by the Contract Documents. Record and report to the Architect on test procedures and test results; verify testing invoices to be paid by the Owner.

2.7 Maintain records at the construction site in an orderly manner. Include correspondence, Contract Documents, Change Orders, Construction Change Directives, reports of site meetings, Shop Drawings, Product Data, and similar submittals; supplementary drawings, color schedules, requests for payment; and names, addresses and telephone numbers of the Contractors, Subcontractors and principal material suppliers.

2.8 Maintain a log book of activities at the site, including weather conditions, nature and location of Work being performed, verbal instructions and interpretations given to the Contractor, and specific observations. Record any occurrence or Work that might result in a claim for a change in Contract Sum or Contract Time. Maintain a list of visitors, their titles, and time and purpose of their visit.

2.9 Assist the Architect in reviewing Shop Drawings, Product Data and Samples. Notify the Architect if any portion of the Work requiring Shop Drawings, Product Data or Samples is commenced before such submittals have been approved by the Architect. Receive and log Samples required at the site, notify the Architect when they are ready for examination, and record the Architect's approval or other action; maintain custody of approved Samples.

2.10 Observe the Contractor's record copy of the Drawings, Specifications, addenda, Change Orders and other Modifications at intervals appropriate to the stage of construction and notify the Architect of any apparent failure by the Contractor to maintain up-to-date records.

2.11 Review Applications for Payment and forward to the Architect with recommendations for disposition.

2.12 Review the list of items to be completed or corrected which is submitted by the Contractor with a request for issuance of a Certificate of Substantial Completion. Review the Work. If the list is accurate, forward it to the Architect for final disposition; if not, so advise the Architect and return the list to the Contractor for correction.

2.13 Assist the Architect in conducting inspections to determine the date or dates of Substantial Completion and the date of final completion.

2.14 Assist the Architect in receipt and transmittal to the Owner of documentation required of the Contractor at completion of the Work.

3. LIMITATIONS OF AUTHORITY

The Architect's Project Representative, in acting on behalf of the Owner, shall not exceed the authority of the Architect under the Agreement Between the Owner and Architect. The Project Representative shall NOT:

3.1 Authorize deviations from the Contract Documents.

3.2 Approve substitute materials or equipment except as authorized in writing by the Architect.

3.3 Personally conduct or participate in tests or third party inspections except as authorized in writing by the Architect.

3.4 Assume any of the responsibilities of the Contractor's superintendent or of Subcontractors.

3.5 Expedite the Work for the Contractor.

3.6 Have control over or charge of or be responsible for construction means, methods, techniques, sequences or procedures, or for safety precautions and programs in connection with the Work.

3.7 Authorize or suggest that the Owner occupy the Project in whole or in part.

3.8 Issue a Certificate for Payment or Certificate of Substantial Completion.

3.9 Prepare or certify to the preparation of a record copy of the Drawings, Specifications, addenda, Change Orders and other Modifications.

3.10 Reject Work or require special inspection or testing except as authorized in writing by the Architect.

3.11 Accept, distribute, or transmit submittals made by the Contractor that are not required by the Contract Documents.

3.12 Order the Contractor to stop the Work or any portion thereof.

AIA DOCUMENT B352 • ARCHITECT'S PROJECT REPRESENTATIVE • AIA® • ©1993 • THE AMERICAN INSTITUTE OF ARCHITECTS, 1735 NEW YORK AVENUE, N.W., WASHINGTON, D.C. 20006-5292 • **WARNING: Unlicensed photocopying violates U.S. copyright laws and will subject the violator to legal prosecution.**

B352—1993

Construction Inspection Manual

FIELD CHANGE ORDER

Field Change Order No: _____

Project No: _____

Date: _____

This form to be used only for emergency instructions to a contractor where the time required for preparation and execution of a formal Change Order would result in delay or stoppage of the work. A duly authorized Change Order shall replace this Field Order as soon as possible and shall bear appropriate reference to the Field Order.

PROJECT _____
_____
_____

TO CONTRACTOR _____ Contract Date: _____
_____

You are hereby authorized to make the following changes in your contract for the above project.

_____
_____
_____
_____
_____
_____
_____
_____
_____
_____
_____
_____
_____
_____

Agreed Cost/Agreed Credit _____
  or
Maximum Cost/Minimum Credit _____

(To be used where agreed cost or credit cannot be immediately determined. The final agreed amount shall not be more than the maximum cost nor less than the minimum credit.)

_____   _____   _____
FOR THE ARCHITECT   FOR THE OWNER   FOR THE CONTRACTOR

DISTRIBUTION: ☐ Owner
               ☐ Architect
               ☐ Contractor
               ☐ Inspector
               ☐ Other

INSPECTOR'S DAILY REPORT

PROJECT _____ PROJECT NO. _____

REPORT NO. _____

Contract Time _____ Date _____

Days Elapsed _____ Day: M   T   W   Th   F   S

Extensions _____ Weather: Clear____ Overcast____ Rain____ Mist____ Hot____ Cold____ Warm____ Foggy____

Days Left _____ Site Condition: Clear____ Muddy____ Dusty____ Other_____

Temperature Range: _____

CONTRACTOR'S LABOR FORCE	Supervision	Carpenters	Laborers	Other	TOTAL

SUBCONTRACTORS

WORK PERFORMED: _____

_____

_____

_____

_____

_____

_____

_____

_____

_____

_____

_____

_____

REMARKS or ACTION: _____

_____

_____

MAJOR DELIVERIES: _____

_____

PHONE CALLS: _____

VISITORS: _____

INSPECTOR _____

MASTERFORMAT

LEVEL TWO NUMBERS AND TITLES

INTRODUCTORY INFORMATION
00001 PROJECT TITLE PAGE
00005 CERTIFICATIONS PAGE
00007 SEALS PAGE
00010 TABLE OF CONTENTS
00015 LIST OF DRAWINGS
00020 LIST OF SCHEDULES

BIDDING REQUIREMENTS
00100 BID SOLICITATION
00200 INSTRUCTIONS TO BIDDERS
00300 INFORMATION AVAILABLE TO BIDDERS
00400 BID FORMS AND SUPPLEMENTS
00490 BIDDING ADDENDA

CONTRACTING REQUIREMENTS
00500 AGREEMENT
00600 BONDS AND CERTIFICATES
00700 GENERAL CONDITIONS
00800 SUPPLEMENTARY CONDITIONS
00900 ADDENDA AND MODIFICATIONS

FACILITIES AND SPACES
FACILITIES AND SPACES

SYSTEMS AND ASSEMBLIES
SYSTEMS AND ASSEMBLIES

CONSTRUCTION PRODUCTS AND ACTIVITIES

DIVISION 1 GENERAL REQUIREMENTS
01100 SUMMARY
01200 PRICE AND PAYMENT PROCEDURES
01300 ADMINISTRATIVE REQUIREMENTS
01400 QUALITY REQUIREMENTS
01500 TEMPORARY FACILITIES AND CONTROLS
01600 PRODUCT REQUIREMENTS
01700 EXECUTION REQUIREMENTS
01800 FACILITY OPERATION
01900 FACILITY DECOMMISSIONING

DIVISION 2 SITE CONSTRUCTION
02050 BASIC SITE MATERIALS AND METHODS
02100 SITE REMEDIATION
02200 SITE PREPARATION
02300 EARTHWORK
02400 TUNNELING, BORING, AND JACKING
02450 FOUNDATION AND LOAD-BEARING ELEMENTS
02500 UTILITY SERVICES
02600 DRAINAGE AND CONTAINMENT
02700 BASES, BALLASTS, PAVEMENTS, AND APPURTENANCES
02800 SITE IMPROVEMENTS AND AMENITIES
02900 PLANTING
02950 SITE RESTORATION AND REHABILITATION

DIVISION 3 CONCRETE
03050 BASIC CONCRETE MATERIALS AND METHODS
03100 CONCRETE FORMS AND ACCESSORIES
03200 CONCRETE REINFORCEMENT
03300 CAST-IN-PLACE CONCRETE
03400 PRECAST CONCRETE
03500 CEMENTITIOUS DECKS AND UNDERLAYMENT
03600 GROUTS
03700 MASS CONCRETE
03900 CONCRETE RESTORATION AND CLEANING

DIVISION 4 MASONRY
04050 BASIC MASONRY MATERIALS AND METHODS
04200 MASONRY UNITS
04400 STONE
04500 REFRACTORIES
04600 CORROSION-RESISTANT MASONRY
04700 SIMULATED MASONRY
04800 MASONRY ASSEMBLIES
04900 MASONRY RESTORATION AND CLEANING

DIVISION 5 METALS
05050 BASIC METAL MATERIALS AND METHODS
05100 STRUCTURAL METAL FRAMING
05200 METAL JOISTS
05300 METAL DECK
05400 COLD-FORMED METAL FRAMING
05500 METAL FABRICATIONS
05600 HYDRAULIC FABRICATIONS
05650 RAILROAD TRACK AND ACCESSORIES
05700 ORNAMENTAL METAL
05800 EXPANSION CONTROL
05900 METAL RESTORATION AND CLEANING

DIVISION 6 WOOD AND PLASTICS
- 06050 BASIC WOOD AND PLASTIC MATERIALS AND METHODS
- 06100 ROUGH CARPENTRY
- 06200 FINISH CARPENTRY
- 06400 ARCHITECTURAL WOODWORK
- 06500 STRUCTURAL PLASTICS
- 06600 PLASTIC FABRICATIONS
- 06900 WOOD AND PLASTIC RESTORATION AND CLEANING

DIVISION 7 THERMAL AND MOISTURE PROTECTION
- 07050 BASIC THERMAL AND MOISTURE PROTECTION MATERIALS AND METHODS
- 07100 DAMPPROOFING AND WATERPROOFING
- 07200 THERMAL PROTECTION
- 07300 SHINGLES, ROOF TILES, AND ROOF COVERINGS
- 07400 ROOFING AND SIDING PANELS
- 07500 MEMBRANE ROOFING
- 07600 FLASHING AND SHEET METAL
- 07700 ROOF SPECIALTIES AND ACCESSORIES
- 07800 FIRE AND SMOKE PROTECTION
- 07900 JOINT SEALERS

DIVISION 8 DOORS AND WINDOWS
- 08050 BASIC DOOR AND WINDOW MATERIALS AND METHODS
- 08100 METAL DOORS AND FRAMES
- 08200 WOOD AND PLASTIC DOORS
- 08300 SPECIALTY DOORS
- 08400 ENTRANCES AND STOREFRONTS
- 08500 WINDOWS
- 08600 SKYLIGHTS
- 08700 HARDWARE
- 08800 GLAZING
- 08900 GLAZED CURTAIN WALL

DIVISION 9 FINISHES
- 09050 BASIC FINISH MATERIALS AND METHODS
- 09100 METAL SUPPORT ASSEMBLIES
- 09200 PLASTER AND GYPSUM BOARD
- 09300 TILE
- 09400 TERRAZZO
- 09500 CEILINGS
- 09600 FLOORING
- 09700 WALL FINISHES
- 09800 ACOUSTICAL TREATMENT
- 09900 PAINTS AND COATINGS

DIVISION 10 SPECIALTIES
- 10100 VISUAL DISPLAY BOARDS
- 10150 COMPARTMENTS AND CUBICLES
- 10200 LOUVERS AND VENTS
- 10240 GRILLES AND SCREENS
- 10250 SERVICE WALLS
- 10260 WALL AND CORNER GUARDS
- 10270 ACCESS FLOORING
- 10290 PEST CONTROL
- 10300 FIREPLACES AND STOVES
- 10340 MANUFACTURED EXTERIOR SPECIALTIES
- 10350 FLAGPOLES
- 10400 IDENTIFICATION DEVICES
- 10450 PEDESTRIAN CONTROL DEVICES
- 10500 LOCKERS
- 10520 FIRE PROTECTION SPECIALTIES
- 10530 PROTECTIVE COVERS
- 10550 POSTAL SPECIALTIES
- 10600 PARTITIONS
- 10670 STORAGE SHELVING
- 10700 EXTERIOR PROTECTION
- 10750 TELEPHONE SPECIALTIES
- 10800 TOILET, BATH, AND LAUNDRY ACCESSORIES
- 10880 SCALES
- 10900 WARDROBE AND CLOSET SPECIALTIES

DIVISION 11 EQUIPMENT
- 11010 MAINTENANCE EQUIPMENT
- 11020 SECURITY AND VAULT EQUIPMENT
- 11030 TELLER AND SERVICE EQUIPMENT
- 11040 ECCLESIASTICAL EQUIPMENT
- 11050 LIBRARY EQUIPMENT
- 11060 THEATER AND STAGE EQUIPMENT
- 11070 INSTRUMENTAL EQUIPMENT
- 11080 REGISTRATION EQUIPMENT
- 11090 CHECKROOM EQUIPMENT
- 11100 MERCANTILE EQUIPMENT
- 11110 COMMERCIAL LAUNDRY AND DRY CLEANING EQUIPMENT
- 11120 VENDING EQUIPMENT
- 11130 AUDIO-VISUAL EQUIPMENT
- 11140 VEHICLE SERVICE EQUIPMENT
- 11150 PARKING CONTROL EQUIPMENT
- 11160 LOADING DOCK EQUIPMENT
- 11170 SOLID WASTE HANDLING EQUIPMENT
- 11190 DETENTION EQUIPMENT
- 11200 WATER SUPPLY AND TREATMENT EQUIPMENT
- 11280 HYDRAULIC GATES AND VALVES
- 11300 FLUID WASTE TREATMENT AND DISPOSAL EQUIPMENT
- 11400 FOOD SERVICE EQUIPMENT
- 11450 RESIDENTIAL EQUIPMENT
- 11460 UNIT KITCHENS
- 11470 DARKROOM EQUIPMENT

11480 ATHLETIC, RECREATIONAL, AND THERAPEUTIC EQUIPMENT
11500 INDUSTRIAL AND PROCESS EQUIPMENT
11600 LABORATORY EQUIPMENT
11650 PLANETARIUM EQUIPMENT
11660 OBSERVATORY EQUIPMENT
11680 OFFICE EQUIPMENT
11700 MEDICAL EQUIPMENT
11780 MORTUARY EQUIPMENT
11850 NAVIGATION EQUIPMENT
11870 AGRICULTURAL EQUIPMENT
11900 EXHIBIT EQUIPMENT

DIVISION 12 FURNISHINGS
12050 FABRICS
12100 ART
12300 MANUFACTURED CASEWORK
12400 FURNISHINGS AND ACCESSORIES
12500 FURNITURE
12600 MULTIPLE SEATING
12700 SYSTEMS FURNITURE
12800 INTERIOR PLANTS AND PLANTERS
12900 FURNISHINGS RESTORATION AND REPAIR

DIVISION 13 SPECIAL CONSTRUCTION
13010 AIR-SUPPORTED STRUCTURES
13020 BUILDING MODULES
13030 SPECIAL PURPOSE ROOMS
13080 SOUND, VIBRATION, AND SEISMIC CONTROL
13090 RADIATION PROTECTION
13100 LIGHTING PROTECTION
13110 CATHODIC PROTECTION
13120 PRE-ENGINEERED STRUCTURES
13150 SWIMMING POOLS
13160 AQUARIUMS
13165 AQUATIC PARK FACILITIES
13170 TUBS AND POOLS
13175 ICE RINKS
13185 KENNELS AND ANIMAL SHELTERS
13190 SITE-CONSTRUCTED INCINERATORS
13200 STORAGE TANKS
13220 FILTER UNDERDRAINS AND MEDIA
13230 DIGESTER COVERS AND APPURTENANCES
13240 OXYGENATION SYSTEMS
13260 SLUDGE CONDITIONING SYSTEMS
13280 HAZARDOUS MATERIAL REMEDIATION
13400 MEASUREMENT AND CONTROL INSTRUMENTATION
13500 RECORDING INSTRUMENTATION
13550 TRANSPORTATION CONTROL INSTRUMENTATION
13600 SOLAR AND WIND ENERGY EQUIPMENT
13700 SECURITY ACCESS AND SURVEILLANCE
13800 BUILDING AUTOMATION AND CONTROL
13850 DETECTION AND ALARM
13900 FIRE SUPPRESSION

DIVISION 14 CONVEYING SYSTEMS
14100 DUMBWAITERS
14200 ELEVATORS
14300 ESCALATORS AND MOVING WALKS
14400 LIFTS
14500 MATERIAL HANDLING
14600 HOISTS AND CRANES
14700 TURNTABLES
14800 SCAFFOLDING
14900 TRANSPORTATION

DIVISION 15 MECHANICAL
15050 BASIC MECHANICAL MATERIALS AND METHODS
15100 BUILDING SERVICES PIPING
15200 PROCESS PIPING
15300 FIRE PROTECTION PIPING
15400 PLUMBING FIXTURES AND EQUIPMENT
15500 HEAT-GENERATION EQUIPMENT
15600 REFRIGERATION EQUIPMENT
15700 HEATING, VENTILATING, AND AIR CONDITIONING EQUIPMENT
15800 AIR DISTRIBUTION
15900 HVAC INSTRUMENTATION AND CONTROLS
15950 TESTING, ADJUSTING, AND BALANCING

DIVISION 16 ELECTRICAL
16050 BASIC ELECTRICAL MATERIALS AND METHODS
16100 WIRING METHODS
16200 ELECTRICAL POWER
16300 TRANSMISSION AND DISTRIBUTION
16400 LOW-VOLTAGE DISTRIBUTION
16500 LIGHTING
16700 COMMUNICATIONS
16800 SOUND AND VIDEO

APPENDIX E – INSPECTOR'S BASIC BOOKSHELF

Following is a descriptive listing of books and reference documents which the well-equipped construction inspector should have in his library. Many of these books are updated and republished on a regular frequency. It is important to have the latest edition available in one's library. Where appropriate, the date of the latest edition available at the time of the publication of this manual is called out in the descriptive listing.

Each book listed appears under a Divisional heading from the Technical Items Checklist for which it is a particularly important reference. There may be more than one book listed under a particular Division; in which case it is at the discretion of the inspector as to whether he wishes to possess one or all of these particular books.

All of the books which are listed are stocked and sold by the publisher of this manual, **Building News**. They may be purchased over-the-counter at the company's professional store located at 1612 S. Clementine St., Anaheim, CA 92802.

Division 2 – Sitework

Standard Specifications, 1992 Edition. Published by State of California, Department of Transportation (Caltrans), Sacramento, California.

Standard Specifications for Public Works Construction (The Greenbook), 1997 Edition. Published by the Building News for Joint Co-operative Committee of the Southern California Chapter of American Public Works Association and the Southern California Districts, Associated General Contractors of California, every three years.

Public Works Inspector's Manual, Fifth Edition, 1993. Silas B. Birch, Jr., author. Published by Building News.

Manual of Traffic Controls for Construction and Maintenance, 1990 Edition (pocket-size booklet). Regulations of State of California, Department of Transportation, Sacramento, California. Published by Building News.

Division 3 – Concrete

ACI's Manual of Concrete Practice, 1998 Edition. Published by American Concrete Institute, Detroit, Michigan.
   Part 1: *Materials and General Properties of Concrete*
   Part 2: *Construction Practice and Inspection*
   Part 3: *Use of Concrete in Buildings – Design, Specifications and Related Topics*
   Part 4: *Bridges, Substructures, Sanitary and Other Special Structures – Structural Properties*
   Part 5: *Masonry – Precast Concrete – Special Processes*
ACI Manual of Concrete Inspection.

Placing Reinforcing Bars. Conforms to ACI Building Code. Published by Concrete Reinforcing Steel Institute, Schaumburg, Illinois 60173.

Division 4 – Masonry

Reinforced Concrete Masonry Construction Inspector's Handbook. Second Edition. Published by Masonry Institute of America, Los Angeles, California 90057 and International Conference of Building Officials, Whittier, California 90601.

Masonry Design Manual, Fourth Edition. Published by Masonry Industry Advancement Committee, Los Angeles, California.

Residential Masonry Fireplace and Chimney Handbook, 1994 Edition. Published by Masonry Institute of America, Los Angeles, California 90057.

Division 5 – Metals

Manual of Steel Construction – Allowable Stress Design, Ninth Edition. Published by American Institute of Steel Construction, Chicago, Illinois 60680.

Division 6 – Carpentry

Manual of Millwork. Standard of the Industry. Published by Woodwork. Institute of California, Fresno, California 93733.

Architectural Woodwork – Quality Standards, Guide Specifications and Quality Certification Program, Seventh Edition. Published by the Architectural Woodwork Institute, Arlington, Virginia 22206.

Timber Design and Construction Sourcebook. A Comprehensive Guide to Methods and Practice. By Karl-Heinz Goetz, Dieter Hoor. Karl Moehler, Julius Natterer, with Peter F. Martecchini. Published 1989 by McGraw-Hill Publishing Co., New York, N.Y.

Design of Wood Structures, Third Edition. By Donald E. Breyer. Published 1998 by McGraw-Hill Publishing Company.

Timber Construction Manual, Fourth Edition, 1994. Published for American Institute of Timber Construction, Englewood, Colorado, by John Wiley & Sons, New York.

Western Woods Use Book, Structural Data and Design Tables. Third Edition. Published by Western Wood Products Association, Portland, Oregon 97204.

West Coast Lumber Standard Grading Rules #16. Published by West Coast Lumber Inspection Bureau, Portland, Oregon 97223.

Western Lumber Grading Rules 88. Published by Western Wood Products Association, Portland, Oregon 97203.

Construction Inspection Manual

Division 7 – Thermal and Moisture Protection

Architectural Sheet Metal Manual. Published by Sheet Metal and Air Conditioning Contractors Association, Vienna, Virginia 22182.

Building Energy Efficiency Standards. Published by California Energy Commission, Sacramento, California 95814.

Energy Conservation Manual for New Residential Buildings. Published by California Energy Commission, Sacramento, California 95814.

Division 8 – Doors and Windows

NFPA 80: Fire Doors and Windows. Published by National Fore Protection Association, Quincy, Massachusetts 02269.

Division 9 – Finishes

Gypsum Construction Handbook. Published by U.S. Gypsum Company, Chicago, Illinois.

Plaster and Drywall Systems Manual. Third Edition. Written by J.R. Gorman, Walter Pruter and James J. Rose. Published by Building News, Los Angeles, California 90064.

Plastering Skills. By F. Van Den Branden and Thomas Hartsell. Published by American Technical Publishers, Homewood, Illinois 60430.

Ceiling Systems Handbook. Revised 10th Printing. Published by Ceilings & Interior Systems, Contractors Association, Deerfield, Illinois 60015.

Ceramic Tile Manual, Third Edition. Published by Ceramic Tile Institute, Los Angeles, California.

Paint Handbook. By Guy E. Weismantel. Published by McGraw-Hill Book Company, New York, N.Y. 10020.

Division 15 – Mechanical

ASHRAE Handbook. Published by American Society of Heating, Refrigerating and Air Conditioning Engineers, Atlanta, Georgia 30329.

ASHRAE Fundamentals Handbook

ASHRAE Handbook of Applications

ASHRAE Handbook of Refrigeration Systems & Applications

ASHRAE Equipment Handbook

ASHRAE Systems Handbook

SMACNA HVAC Duct Construction Standards, Metal and Flexible. Published by Sheet Metal and Air Conditioning Contractors National Association, Vienna Virginia 22182.

Model ("Uniform") Codes

Uniform Plumbing Code. Published by International Association of Plumbing and Mechanical Officials, Walnut, California.

Uniform Mechanical Code. Published by International Association of Plumbing and Mechanical Officials, Walnut California.

SBCCI Standard Mechanical Code. Published by Southern Building Code Congress International, Birmingham, Ala. 35213.

SBCCI Standard Plumbing Code. Published by Southern Building Code Congress International, Birmingham, Ala. 35213.

BOCA Basic National Plumbing Code. Published by Building Officials and Code Administrators International.

BOCA Basic National Mechanical Code. Published by Building Officials and Code Administrators International.

International Plumbing Code. Published by International Code Council, Whittier, California 90601-2298.

International Mechanical Code. Published by International Code Council, Whittier, California 90601-2298.

NFPA 13 – Installation of Sprinkler Systems. Published by National Fire Protection Assn., Quincy, Massachusetts 02269.

NFPA Automatic Sprinkler Systems Handbook, Fourth Edition. Published by National Fire Protection Assn., Quincy, Massachusetts.

Division 16 – Electrical

National Electrical Code (NFPA 70). Published by National Fire Protection Association, Quincy, Massachusetts 02269.

National Electrical Code Handbook. Published by National Fire Protection Association, Quincy, Massachusetts 02269.

California Electrical Code. Title 24, Part 3 of California Code of Regulations. Published for California Building Standards Commission.

Ferm's Fast Finder Index to the National Electrical Code.

Model ("Uniform") Building Codes

Uniform Building Code. Published by International Conference of Building Officials, Whittier, California.

BOCA National Building Code. Published by Building Officials & Code Administrators International, Inc., Country Club Hills, Illinois 60477.

SBCCI Standard Building Code. Published by Southern Building Code Congress International, Birmingham, Alabama 35213.

ASTM Standards in Building Codes, 4-Volume Set. Published by American Society for Testing and Materials, Philadelphia, Pennsylvania.

CABO One- and Two-Family Dwelling Code. Published by Council of American Building Officials, Falls Church, Va. 22041.

Dictionaries – Reference

BNi Building News Construction Dictionary Illustrated. Published by BNi Publications, Inc. 1612 S. Clementine St., Anaheim, California 92802.

BNi Building News Construction Dictionary Pocket Edition. Published by BNi Publications, Inc. 1612 S. Clementine St., Anaheim, California 92802.

Means Illustrated Construction Dictionary. Published by R.S. Means Co., Kingston, Massachusetts 02364.

Dictionary of Architecture and Construction, by Cyril M. Harris. Published by McGraw-Hill Publishing Company, New York, N.Y. 10020.

Compilation of ASTM Standard Definitions. Published by American Society for Testing and Materials, Philadelphia, Pennsylvania 19103.

Construction Dictionary. Published by National Association of Women in Construction, Arizona.

Inspection

Construction Inspection Manual, Seventh Edition. Published for the California Construction Advancement Program, by Building News, Los Angeles, California 90064.

Field Inspection Manual. Published by International Conference of Building Officials, Whittier, California 90601.

Construction Inspection Handbook, by James J. O'Brien. Published by Van Nostrand Reinhold, New York, N.Y. 10003.

Field Inspection Handbook, Brock & Sutcliffe. Published by McGraw-Hill Brook Company, New York, N.Y.

Construction Inspection, Second Edition, by James E. Clyde. Published by John Wiley & Sons, New York, N.Y.

APPENDIX F — INSPECTOR'S BASIC TOOLS

Basic Tools, usable in most divisions:

- 12-16 ft. steel tape
- 50-100 ft. steel tape
- 2 or 2½ ft. carpenter's level
- 10 ft. straight edge
- Machinist's mirror with handle
- Pocket scale (architect's)
- Pocket scale (engineer's)
- Calipers and dividers (machinist's type)
- Pocket mirror, for flashing in wall forms, etc.
- 3-cell flashlight (for days with no sun)
- Magnifying glass
- 50 ft. string line
- Pocket knife
- Magnet
- Camera/Camcorder
- Pocket calculator
- Electronic measuring device

Special Tools, required in certain divisions only:

Division	Tools
2	Hand level (locke type)
2	Folding rule (reading in tenths/foot)
2	Asphalt thermometer (to 500°F.)
3	Slump cone and rod
3	Cylinder carriers and two 5-gallon buckets
3	Pocket thermometer (32° to 125° ±)
5	Weld gauges (set)
5	Welding hood or shield
5	Tempil stix
7	Asphalt thermometer (to 500°F.)
8	Glass thickness gauge (use with flashing)
9	Adjustable depth gauge (for fireproofing)
15	Micrometer
15	2 thermometers (32° to 220°F.)
15	Anemometer or draft gauge
15	Sling psychrometer
15	Chlorine solution test papers
16	Receptacle tester
16	"Wiggy"
16	Clamp-on ammeter
16	Phase rotation meter

APPENDIX G — INSPECTOR'S BASIC PROJECT FILE

I. General Information:
Clarification
Field Orders
Proposed Changes
Quotations and Acceptances (of proposed changes)
Change Orders
Cost Breakdown (Schedule of Values)
Certificate of Payment
Request for Payment
Testing Lab Invoices
T & M (Time and Material)
20-Day Preliminary Notice To Protect Lien Rights
Equipment Furnished by Owner
Segregated Contracts
Color and Material Selections

II. Correspondence:
Between Architect and Owner
Between Architect and Inspector
Between Architect and Contractor
 (General Correspondence)
Testing Laboratory
Soils Laboratory
Architects/Consultants
 (Mechanical, Electrical, Structural, etc.)

III. Government Agencies and Programs:
Fire Marshal
USPHS (United States Public Health Service)
HEW (Health Education Welfare)
HUD (United States Department of Housing and Urban Development)
Department of Public Health
City or County Building Official
Local Affirmative Action Program
Certified Payrolls (including Davis-Bacon, Landrum-Griffin, Copeland Acts, etc.)

IV. Field Information:
Transmittals Incoming
Transmittals Outgoing
Field Sketches
Field Memoranda
Job Meetings
Schedule
Deficiency List
Job Problems (written questions to architect)
Daily Comments (to architect or owner)
Job Security and Safety

V. Project Closeout:
Special Tools
Valve Schedule
Electrical On-line Diagram
Schematic of Mechanical System
Instruction of Owner's Personnel
Certificates of Compliance
Notice of Cessation
Notice of Completion
Guarantees
Spare Parts or Materials (received from contractors)
Instruction Manuals
Receipts from Owner (for spare parts, etc.)
Record Drawings

VI. Technical Information:
Use CSI format to develop files
Make a folder marked "Testing Lab Reports" for any division where appropriate.

APPENDIX H — MISCELLANEOUS CONSTRUCTION AIDS

Cylinder Casting
(Reproduced through courtesy of Master Builders, Inc.)

Note: For complete and related procedures see ASTM Designations: C 470 Single-Use Molds for Forming 6 by 12-in. Concrete Compression Test Cylinders; C 31 Standard Method of Making and Curing Concrete Compressive and Flexural Strength Test Specimens in the Field; C 94 Standard Specifications for Ready Mixed Concrete; and C 172 Standard Method of Sampling Fresh Concrete.

Use Only Non-Absorbent Waterproof Molds

For casting concrete cylinders in the field, use only approved non-absorbent waterproof molds, 6" (15 cm) in diameter by 12" (30 cm) high, with base plates or bottoms. They should be placed on a smooth, firm, level surface for filling and cast in the area where they are to be stored during the first 24 hours and where they will be protected from vibration, jarring, striking, etc.

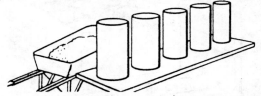

Take 3-Part Sample; Combine And Remix

Three samples of the concrete should be obtained, at regularly spaced intervals, directly from the mixer discharge. Combine the samples in a wheelbarrow, buggy or metal pan and remix with a shovel to ensure uniformity of the 3-part sample.

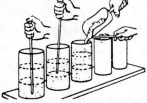

Fill Molds In Three Equal Layers And Rod Each Layer 25 Times

Fill molds in three equal layers and uniformly rod each layer 25 times with a ⅝" bullet-nosed rod. When rodding the second and third layers, the rod should just break through into the layer beneath. Fill all molds uniformly — that is, place and rod the bottom layer in all cylinders, then place and rod the second layer, etc. The third layer should contain an excess amount of concrete which is struck off smooth and level after rodding.

01400 QUALITY CONTROL

Cylinder Casting — (Cont.)

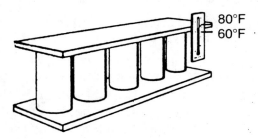

Protect Cylinders From Moisture Loss, Movement And Temperature Extremes

Cover the tops of the cylinders to prevent loss of moisture by evaporation. Do not disturb or move cylinders for 24 hours after casting. Protect them against temperatures that fall below 60°F (16C) or exceed 80°F (27C). Cylinders left on the job for several days and exposed to high or low temperatures will give substandard results. Additional cylinders used for determining when forms may be stripped or when concrete may be put into service should be removed from the molds after 24 hours and then job-cured adjacent to and under the same conditions as the concrete they represent.

Cure And Handle Cylinders With Care

After 24 hours, cylinders for acceptance tests should be placed in moist curing at 73.4°F ± 3°F (23 ± 1.7C) or sent to a laboratory for similar standard curing. Careful handling during moving is necessary since cylinders which are allowed to rattle around in a box, at the back of a car, or pick-up truck, can suffer considerable damage.

Important:

Alway Use Accepted Standards — Standard test procedures were developed to establish lines of uniformity and reproducibility. Only specimens tested according to accepted, reliable standards, such as those established by the American Society for Testing and Materials, give valuable indications of the uniformity and potential quality of the concrete in a structure.

01400   QUALITY CONTROL

Concrete Slump Test
(Reproduced through courtesy of Master Builders, Inc.)

Purpose of Test: To determine the consistency of fresh concrete and to check its uniformity from batch to batch. This test is based on ASTM C 143: Standard Method of Test for Slump of Portland Cement Concrete.

Take two or more representative samples — at regularly spaced intervals — from the middle of the mixer discharge; do not take samples from beginning or end of discharge. Obtain samples within 15 minutes or less. **Important:** Slump test must be made within 5 minutes after taking samples.

Combine samples in a wheelbarrow or appropriate container and remix before making test.

Dampen slump cone with water and place it on a flat, level, smooth, moist, non-absorbent, firm surface.

1. Stand on two foot pieces of cone to hold it firmly in place during Steps 1 through 4. Fill cone mold ⅓ full by volume [2½" (63.5mm) high] with the concrete sample and rod it with 25 strokes using a round, bullet-nosed steel rod of ⅝" (16mm) diameter x 24" (61mm) long. Distribute rodding strokes evenly over entire cross section of the concrete by using approximately half the strokes near the perimeter (outer edge) and then progressing spirally toward the center.

2. Fill cone ⅔ full by volume [6" (23mm) or half the height] and again rod 25 times with rod just penetrating into, but not through, the first layer. Distribute strokes evenly as described in Step 1.

01400 **QUALITY CONTROL**

Concrete Slump Test — (Cont.)

3. Fill cone to overflowing and again rod 25 times with rod just penetrating into but not through the second layer. Again distribute strokes evenly.

4. Strike off excess concrete from top of cone with the steel rod, so that the cone is exactly level full. Clean the overflow away from the base of the cone mold.

5. Immediately after completion of Step 4, the operation of raising the mold shall be performed in 5 to 10 seconds by a steady upward lift with no lateral or torsional notion being imparted to the concrete. The entire operation from the start of the filling through removal of the mold shall be carried out without interruption and shall be completed within an elapsed time of 2½ minutes.

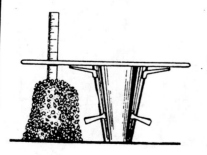

6. Place the steel rod horizontally across the inverted mold, so the rod extends over the slumped concrete. Immediately measure the distance from bottom of the steel rod to the original center of the top of the specimen. This distance, to the nearest ¼ inch (6mm), is the slump of the concrete.

01400 QUALITY CONTROL

Construction Inspection Manual

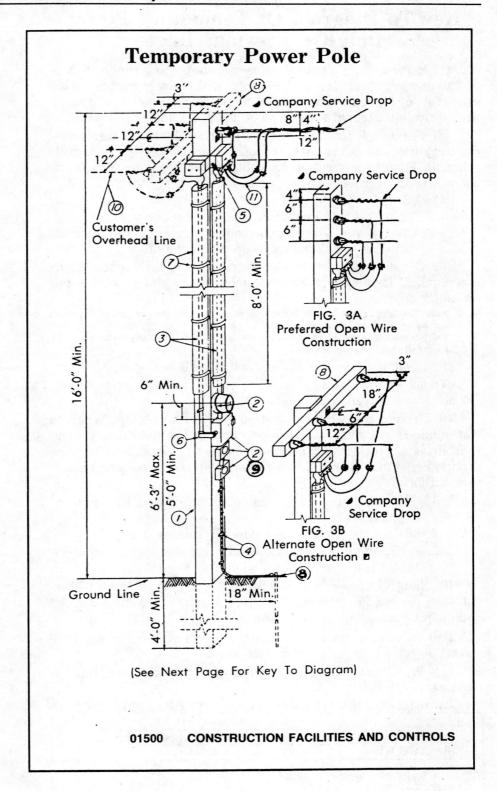

Temporary Power Pole

FIG. 3A
Preferred Open Wire Construction

FIG. 3B
Alternate Open Wire Construction

(See Next Page For Key To Diagram)

01500   **CONSTRUCTION FACILITIES AND CONTROLS**

Key To Diagram Of Temporary Power Pole On Previous Page

(1) **Service Pole.** (1) One piece self-supporting timber 6" x 6" x 20' minimum, or (2) 25-foot pole with 5-inch minimum top diameter, or (3) self-supporting 20-foot minimum metal pole meeting equivalent strength requirements for (1) and (2) or equal.

(2) **Service Entrance Raceway.** (1) Galvanized rigid steel conduit, or (2) electrical metallic conduit, or (3) polynivyl chloride schedule 80 plastic conduit without protective covering. Metal conduit shall be covered with wood moulding or fiber conduit.

(4) **Ground:**
    A. No. 8 AWG minimum armored copper wire.
    B. No. 8 AWG minimum copper wire covered with ½" minimum galvanized rigid iron conduit.

(5) 4" x 4" x 6" wood block, bolted to pole in ½" gain. Block not required for polyvinyl chloride schedule 80 plastic conduit nor on metal pole.

> Note: Covering of metallic conduit and wood blocks over the tops of the risers will not be required on metal poles, provided the metal pole is effectively grounded, and provided all metallic conduits are adequately bonded to the metal pole with approved clamps or connectors.

(6) Conduit fitting, threaded, with cover and gasket.

(7) Extend protective covering to bottom of service heads. Do not leave conduits exposed.

(8) Ground to metallic water line where available. Where a water line is not available, one or more of the following ground rods driven 8 feet into the ground shall be installed to provide ground resistance within limits required by the governing inspection authority:
1. ¾" minimum inside diameter galvanized iron pipe or conduit.
2. ¾" minimum outside diameter solid iron rod.
3. ½" minimum outside diameter rod or copper clad steel, solid brass or copper.

(9) Raintight boxes and receptacles of approved type. Different voltages require receptacles that are not interchangeable for equipment grounding purposed only.

(10) A continuous conductor shall be provided for any secondary service pole or distribution point.

(11) Wire, insulated, size as required (24" minimum extension from service head).

ALLOWABLE CURRENT-CARRYING CAPACITY OF FLEXIBLE CORD

Size AWG Wire	Rubber Types S, SO, SR, SJ, SJO / Thermoplastic Types ST, SRT, SJT — Amperes
16	10
14	15
12	20
10	25
8	35
6	45
4	60
2	80

Approved Configurations For Plugs And Receptacles

			15 AMPERE		20 AMPERE		30 AMPERE		50 AMPERE		60 AMPERE	
			RECEPTACLE	PLUG	RECEPTACLE	PLUG	RECEPTACLE	PLUG	RECEPTACLE	PLUG	RECEPTACLE	PLUG
2-POLE 2-WIRE	125 V	1	1-15R	1-15P		1-20P		1-30P				
	250 V	2		2-15P	2-20R	2-20P	2-30R	2-30P				
	277 V AC	3				(RESERVED FOR FUTURE CONFIGURATIONS)						
	600 V	4				(RESERVED FOR FUTURE CONFIGURATIONS)						
2-POLE 3-WIRE GROUNDING	125 V	5	5-15R	5-15P	5-20R	5-20P	5-30R	5-30P	5-50R	5-50P		
	250 V	6	6-15R	6-15P	6-20R	6-20P	6-30R	6-30P	6-50R	6-50P		
	277 V AC	7	7-15R	7-15P	7-20R	7-20P	7-30R	7-30P	7-50R	7-50P		
	347 V AC	24	24-15R	24-15P	24-20R	24-20P	24-30R	24-30P	24-50R	24-50P		
	480 V AC	8				(RESERVED FOR FUTURE CONFIGURATIONS)						
	600 V AC	9				(RESERVED FOR FUTURE CONFIGURATIONS)						
3-POLE 3-WIRE	125/250 V	10			10-20R	10-20P	10-30R	10-30P	10-50R	10-50P		
	3ø 250 V	11	11-15R	11-15P	11-20R	11-20P	11-30R	11-30P	11-50R	11-50P		
	3ø 480 V	12				(RESERVED FOR FUTURE CONFIGURATIONS)						
	3ø 600 V	13				(RESERVED FOR FUTURE CONFIGURATIONS)						
3-POLE 4-WIRE GROUNDING	125/250 V	14	14-15R	14-15P	14-20R	14-20P	14-30R	14-30P	14-50R	14-50P	14-60R	14-60P
	3ø 250 V	15	15-15R	15-15P	15-20R	15-20P	15-30R	15-30P	15-50R	15-50P	15-60R	15-60P
	3ø 480 V	16				(RESERVED FOR FUTURE CONFIGURATIONS)						
	3ø 600 V	17				(RESERVED FOR FUTURE CONFIGURATIONS)						
4-POLE 4-WIRE	3ø 208Y/120 V	18	18-15R	18-15P	18-20R	18-20P	18-30R	18-30P	18-50R	18-50P	18-60R	18-60P
	3ø 480Y/277 V	19				(RESERVED FOR FUTURE CONFIGURATIONS)						
	3ø 600Y/347 V	20				(RESERVED FOR FUTURE CONFIGURATIONS)						
4-POLE 5-WIRE GROUNDING	3ø 208Y/120 V	21				(RESERVED FOR FUTURE CONFIGURATIONS)						
	3ø 480Y/277 V	22				(RESERVED FOR FUTURE CONFIGURATIONS)						
	3ø 600Y/347 V	23				(RESERVED FOR FUTURE CONFIGURATIONS)						

SEE NEMA PUBLICATION NO. WD6 FOR DIMENSIONS.
©1989 BY NATIONAL ELECTRICAL MANUFACTURERS ASSOCIATION.

16050   ELECTRICAL MATERIALS AND METHODS

Approved Configurations For Plugs And Receptacles

		15 AMPERE		20 AMPERE		30 AMPERE		50 AMPERE		60 AMPERE	
		RECEPTACLE	PLUG	RECEPTACLE	PLUG	RECEPTACLE	PLUG	RECEPTACLE	PLUG	RECEPTACLE	PLUG
2-POLE 2-WIRE	125 V	L1-15R	L1-15P								
	250 V			L2-20R	L2-20P						
	277 V AC			(RESERVED FOR FUTURE CONFIGURATIONS)							
	600 V			(RESERVED FOR FUTURE CONFIGURATIONS)							
2-POLE 3-WIRE GROUNDING	125 V	L5-15R	L5-15P	L5-20R	L5-20P	L5-30R	L5-30P	L5-50R	L5-50P	L5-60R	L5-60P
	250 V	L6-15R	L6-15P	L6-20R	L6-20P	L6-30R	L6-30P	L6-50R	L6-50P	L6-60R	L6-60P
	277 V AC	L7-15R	L7-15P	L7-20R	L7-20P	L7-30R	L7-30P	L7-50R	L7-50P	L7-60R	L7-60P
	347 V AC			L24-20R	L24-20P						
	480 V AC			L8-20R	L8-20P	L8-30R	L8-30P	L8-50R	L8-50P	L8-60R	L8-60P
	600 V AC			L9-20R	L9-20P	L9-30R	L9-30P	L9-50R	L9-50P	L9-60R	L9-60P
3-POLE 3-WIRE	125/250 V			L10-20R	L10-20P	L10-30R	L10-30P				
	3 ø 250 V	L11-15R	L11-15P	L11-20R	L11-20P	L11-30R	L11-30P				
	3 ø 480 V			L12-20R	L12-20P	L12-30R	L12-30P				
	3 ø 600 V					L13-30R	L13-30P				
3-POLE 4-WIRE GROUNDING	125/250 V			L14-20R	L14-20P	L14-30R	L14-30P	L14-50R	L14-50P	L14-60R	L14-60P
	3 ø 250 V			L15-20R	L15-20P	L15-30R	L15-30P	L15-50R	L15-50P	L15-60R	L15-60P
	3 ø 480 V			L16-20R	L16-20P	L16-30R	L16-30P	L16-50R	L16-50P	L16-60R	L16-60P
	3 ø 600 V					L17-30R	L17-30P	L17-50R	L17-50P	L17-60R	L17-60P
4-POLE 4-WIRE	3 ø 208Y/120 V			L18-20R	L18-20P	L18-30R	L18-30P				
	3 ø 480Y/277 V			L19-20R	L19-20P	L19-30R	L19-30P				
	3 ø 600Y/347 V			L20-20R	L20-20P	L20-30R	L20-30P				
4-POLE 5-WIRE GROUNDING	3 ø 208Y/120 V			L21-20R	L21-20P	L21-30R	L21-30P	L21-50R	L21-50P	L21-60R	L21-60P
	3 ø 480Y/277 V			L22-20R	L22-20P	L22-30R	L22-30P	L22-50R	L22-50P	L22-60R	L22-60P
	3 ø 600Y/347 V			L23-20R	L23-20P	L23-30R	L23-30P	L23-50R	L23-50P	L23-60R	L23-60P

SEE NEMA PUBLICATION NO. WD6 FOR DIMENSIONS.
©1989 BY NATIONAL ELECTRICAL MANUFACTURERS ASSOCIATION.

16050   ELECTRICAL MATERIALS AND METHODS

Criteria For Stairs, Ladders And Ramps Or Inclines
Table Of Risers And Runs For Stairs

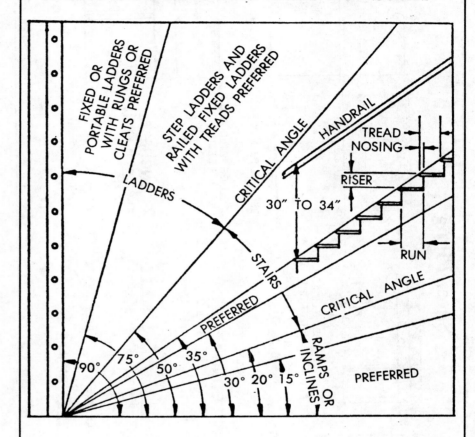

TABLE OF RISERS AND RUNS FOR STAIRS
(Run + Riser = 17½")

Angle with Horizontal	Riser in Inches	Run in Inches	Angle with Horizontal	Riser in Inches	Run in Inches
22°—0'	5	12½	36°—52'	7½	10
23°—14'	5¼	12¼	38°—29'	7¾	9¾
24°—38'	5½	12	40°—08'	8	9½
26°—00'	5¾	11¾	41°—44'	8¼	9¼
27°—33'	6	11½	43°—22'	8½	9
29°—03'	6¼	11¼	45°—00'	8¾	8¾
30°—35'	6½	11	46°—38'	9	8½
32°—08'	6¾	10¾	48°—16'	9¼	8¼
33°—41'	7	10½	49°—54'	9½	8
35°—16'	7½	10¼			

06100   **ROUGH CARPENTRY**

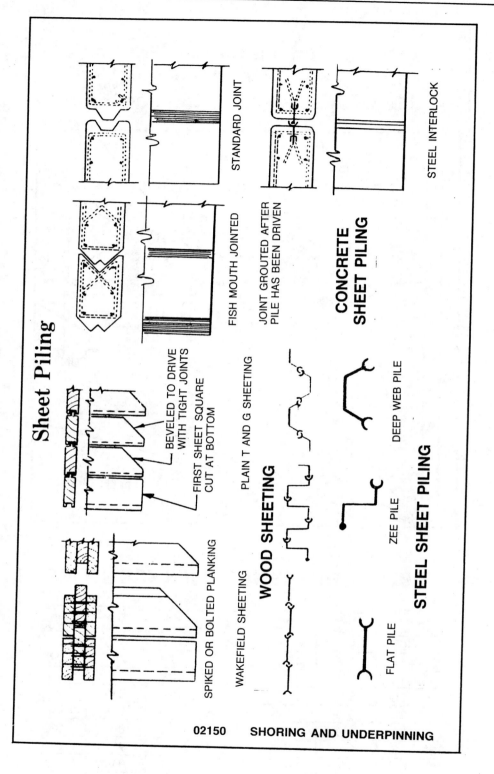

02150 SHORING AND UNDERPINNING

Construction Inspection Manual

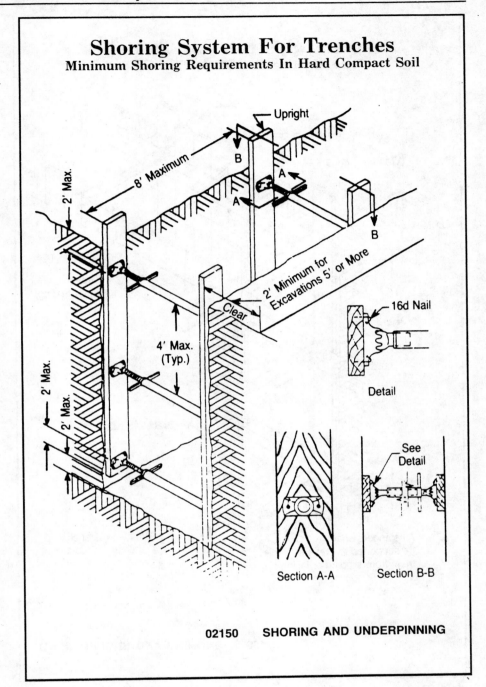

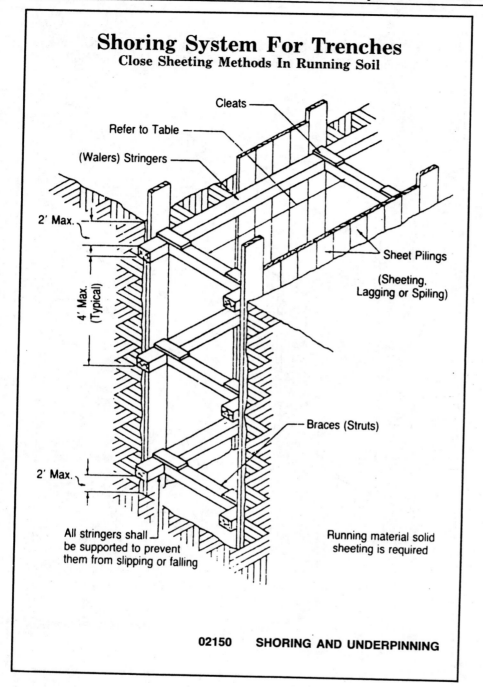

Construction Inspection Manual

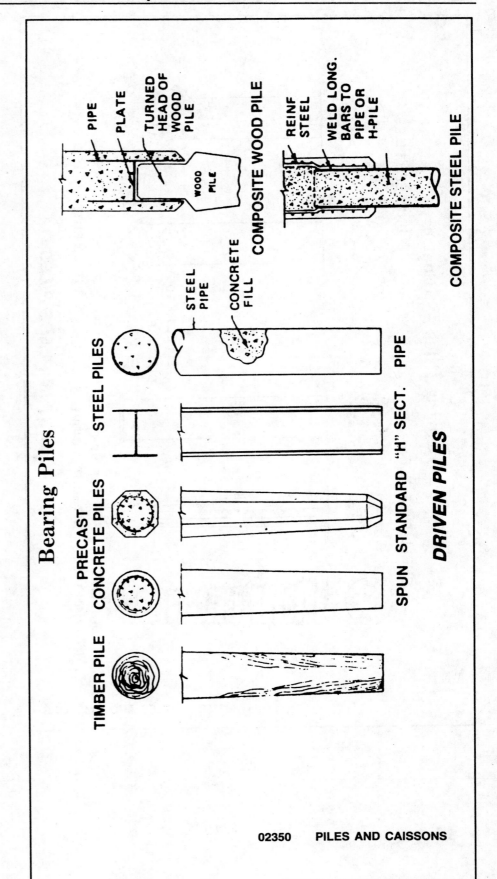

02350   PILES AND CAISSONS

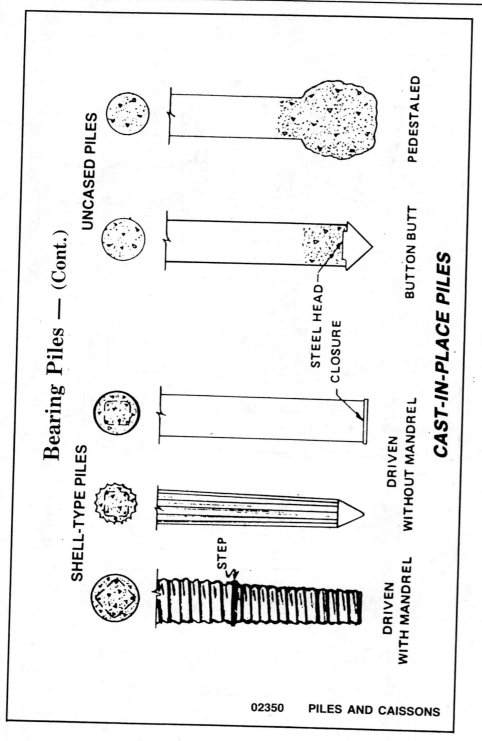

Construction Inspection Manual

Form Nonmenclature

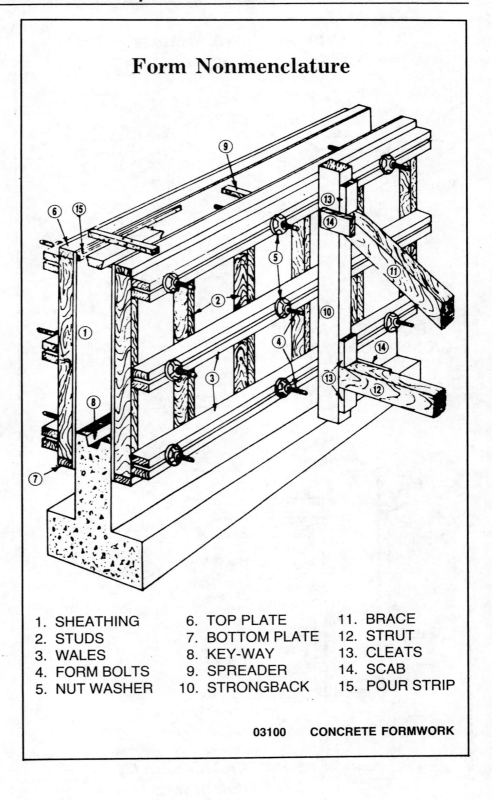

1. SHEATHING
2. STUDS
3. WALES
4. FORM BOLTS
5. NUT WASHER
6. TOP PLATE
7. BOTTOM PLATE
8. KEY-WAY
9. SPREADER
10. STRONGBACK
11. BRACE
12. STRUT
13. CLEATS
14. SCAB
15. POUR STRIP

03100   **CONCRETE FORMWORK**

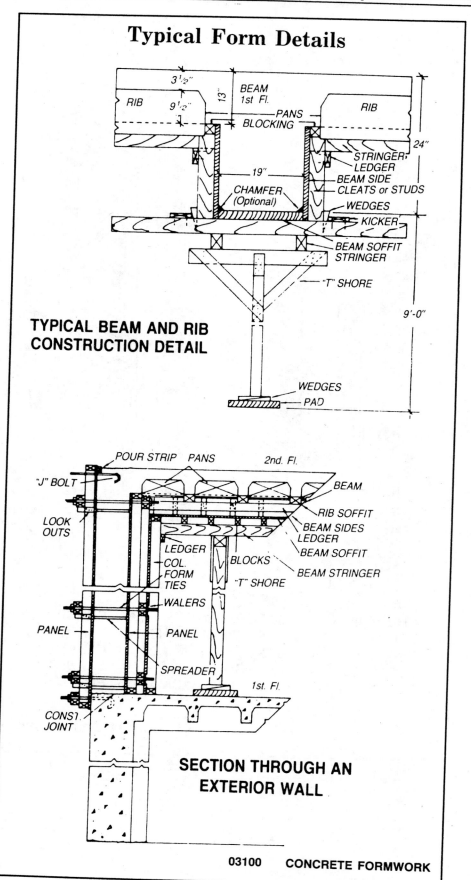

Falsework Nomenclature

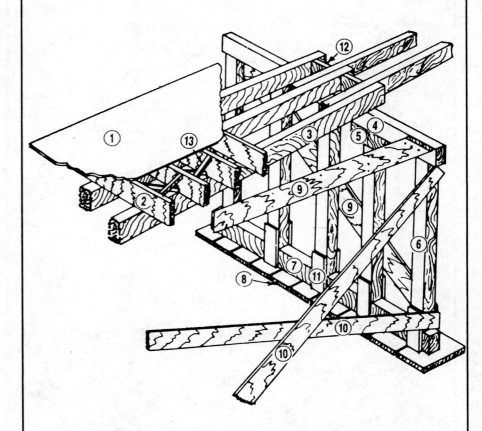

1. SHEATHING
2. JOIST
3. STRINGER
4. CAP
5. CORBEL
6. POST
7. SILL
8. FOOTING
9. SWAY BRACE
10. LONGITUDINAL BRACE
11. SCAB
12. BLOCKING
13. BRIDGING

03100   CONCRETE FORMWORK

Typical Pan-Joist Form Construction

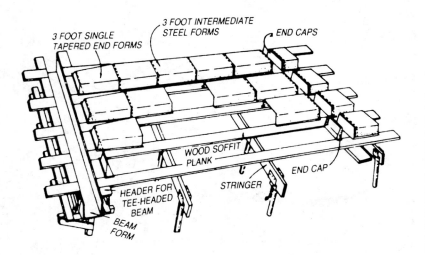

Typical Waffle Slab Form Construction

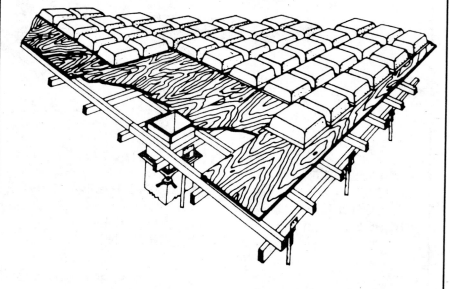

03100   CONCRETE FORMWORK

Basic Principles Of Concrete Reinforcement

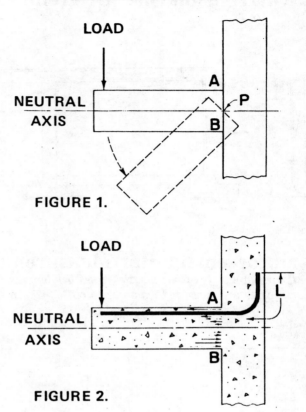

FIGURE 1.

FIGURE 2.

In Figure 1, the horizontal concrete member, when loaded vertically near its free end, would tend to rotate about point P, a point on an imaginary line drawn through the geometric center of the horizontal member called the NEUTRAL AXIS. It can be readily seen that the movement of the member to the position represented by the dotted lines will place the contact surface represented by PA in tension while simultaneously placing PB in compression. This is typical for all structural members under a bending load. The cross-sectional area on one side of the neutral axis will be in tension while the area on the other side of the neutral axis is in compression. The values of tension and compression increase in a straight line ratio from zero at the neutral axis to a maximum at the outer surface of the member.

By placing a steel bar as shown in Figure 2 in the area of maximum tension, the tension forces can be effectively resisted by the steel bar which, by preventing rotation of the member, results in the maximum utilization of the compression resistance of the concrete in the lower portion (below the nautral axis) of the member. In many cases, steel is placed in compression areas where stresses are too high for the concrete to resist the compression forces alone.

03200   CONCRETE REINFORCEMENT

Location Of Main Steel Reinforcement On Tension Side Of Member

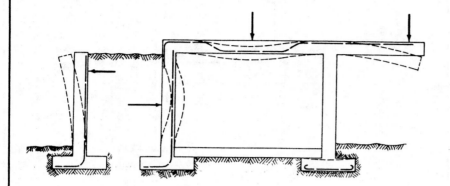

Reinforcement Bar Measurement

All measurements of bar lengths, truss bar lengths and column cores are measured from out to out as shown below:

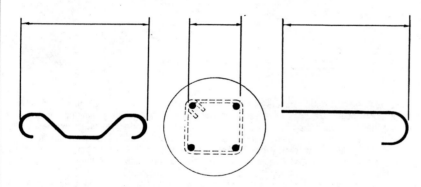

TRUSS BAR   COLUMN CORE   HOOK BAR

03200   CONCRETE REINFORCEMENT

Principles Of Reinforcement Concrete Beams

In the above figure, a plain beam has broken as a result of a load applied at the center. The break first occurs at the bottom, concrete being weak in tension.

The addition of rebar, strong in tension, resists such a break. When such a simply reinforced concrete beam is loaded until it begins to break, cracks appear due to a combination of tension and vertical shear. This stress is known as diagonal tension.

Cracking is best resisted by rebars at right angles to the cracks. This is impractical since it would require very complicated placing of rebar. The compromise then is to let some of the longitudinal bars be straight and these will be at right angles to the cracks near the center. Other bars are bent (called double bent or truss bars) to resist cracks towards the ends and approximate a right angle to the direction of cracks.

Finally, to complete the reinforcement, stirrups (vertical bars in the shape of a "U" hooked at their upper ends) are added to resist diagonal tension and to firmly anchor the longitudinal steel to the compressed part of the beam. These stirrups are called web reinforcement. In most cases it is not necessary to use web reinforcement for the entire length of the beam, shear being maximum at the supports and decreasing toward the center.

03200 CONCRETE REINFORCEMENT

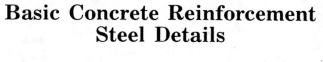

Basic Concrete Reinforcement Steel Details

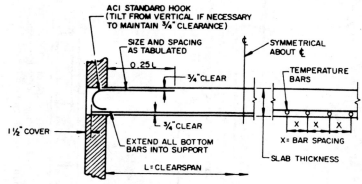

SINGLE SPAN, SIMPLY SUPPORTED

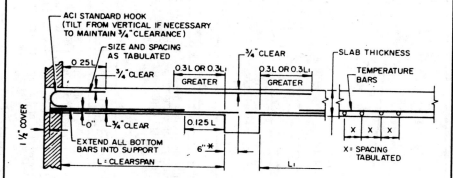

END SPAN, SIMPLY SUPPORTED

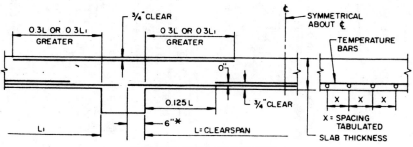

INTERIOR SPAN, CONTINUOUS

* MIN. 6", UNLESS OTHERWISE SPECIFIED BY THE ENGINEER

Note: Except for short single span slabs where top steel is unlikely to receive construction traffic, top bars lighter than #4 at 12 in. are not recommended.

Reproduced with permission of American Concrete Institute from its publication, ACI Detailing Manual — 1988

03200   CONCRETE REINFORCEMENT

Basic Concrete Reinforcing Steel Details

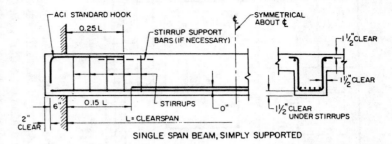

SINGLE SPAN BEAM, SIMPLY SUPPORTED

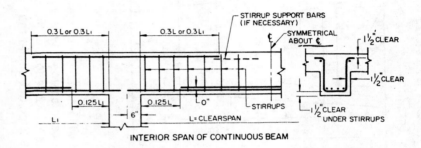

INTERIOR SPAN OF CONTINUOUS BEAM

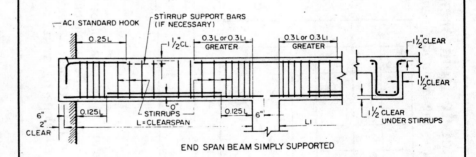

END SPAN BEAM SIMPLY SUPPORTED

TYPICAL REINFORCING STEEL DETAILS FOR BEAMS

Note: Check available depth, top and bottom, for required cover on ACI standard hooks. At each end support, add top bar 0.25L in length to equal the area of bars required.

Reproduced with permission of American Concrete Institute from its publication, ACI Detailing Manual — 1988

03200   CONCRETE REINFORCEMENT

Basic Concrete Reinforcing Steel Details
Reinforcing Around Openings In Slabs

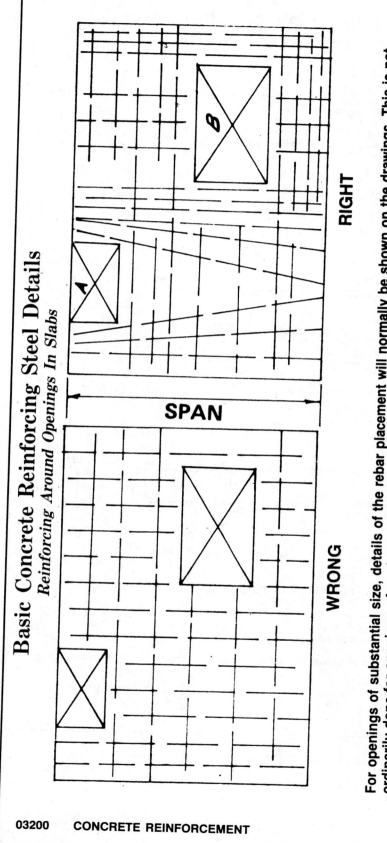

For openings of substantial size, details of the rebar placement will normally be shown on the drawings. This is not ordinarily done for openings of moderate size. Rebar should not be interrupted where it is practical to place the rebar as shown for opening "A." Where it is not practical, equivalent rebar for length of the span is added on either side of the opening as shown at "B." Some additional transverse rebar is used to transfer the load to adjacent spanning rebar.

03200   CONCRETE REINFORCEMENT

Common Types Of Steel Reinforcing Bars

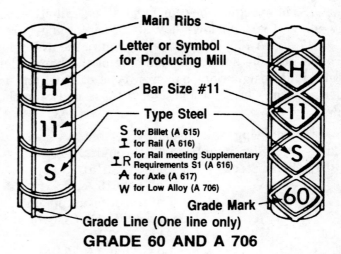

GRADE 60 AND A 706

- Main Ribs
- Letter or Symbol for Producing Mill
- Bar Size #11
- Type Steel
  - S for Billet (A 615)
  - I for Rail (A 616)
  - IR for Rail meeting Supplementary Requirements S1 (A 616)
  - A for Axle (A 617)
  - W for Low Alloy (A 706)
- Grade Mark
- Grade Line (One line only)

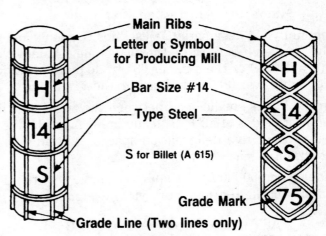

GRADE 75

- Main Ribs
- Letter or Symbol for Producing Mill
- Bar Size #14
- Type Steel
  - S for Billet (A 615)
- Grade Mark
- Grade Line (Two lines only)

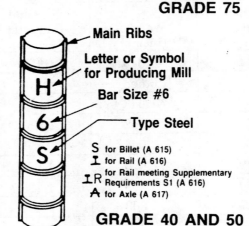

GRADE 40 AND 50

- Main Ribs
- Letter or Symbol for Producing Mill
- Bar Size #6
- Type Steel
  - S for Billet (A 615)
  - I for Rail (A 616)
  - IR for Rail meeting Supplementary Requirements S1 (A 616)
  - A for Axle (A 617)

VARIATIONS:

Bar identification marks may be oriented as illustrated or rotated 90°. Grade mark numbers may be placed within separate consecutive deformation spaces. Grade line may be placed on the side opposite the bar marks.

03200   **CONCRETE REINFORCEMENT**

Identification Marks On Steel Reinforcing Bars

The ASTM specifications for billet-steel, rail-steel, axle-steel and low-alloy steel and low-alloy steel reinforcing bars (A 615, A 616, A 617 and A 706, respectively) require identification marks to be rolled into the surface of one side of the bar to denote the producer's mill designation, bar size, type of steel and minimum yield designation. Grade 60 bars show these marks in the following order:

    1st – Producing Mill (usually a letter)
    2nd – Bar Size Number (#3 through #18)
    3rd – Type Steel: S for Billet (A 615)
                       I for Rail (A 616)
                       IR for Rail meeting Supplement
                          Requirements S1 (A 616)
                       A for Axle (A 617)
                       W for Low-Alloy (A 706)
    4th – Minimum Yield Designation

Minimum yield designation is used for Grade 60 and Grade 75 bars only. Grade 60 bars can either have one (1) single longitudinal line (grade line) or the number 60 (grade mark). Grade 75 bars can either have two (2) grade lines or the grade mark 75.

A grade line is smaller and between the two main ribs which are on opposite sides of all U.S. made bars. A grade line must be continued at least five deformation spaces. A grade mark is the fourth mark on a bar.

Grade 40 and 50 bars are required to have only the first three identification marks (no minimum yield designation).

ASTM Standard Reinforcing Bars

Bar Size	Nominal Area (Sq. Inches)	Weight (Pounds per Ft.)	Nominal Diameter (Inches)
# 3	0.11	0.376	0.375
# 4	0.20	0.668	0.500
# 5	0.31	1.043	0.625
# 6	0.44	1.502	0.750
# 7	0.60	2.044	0.875
# 8	0.79	2.670	1.000
# 9	1.00	3.400	1.128
#10	1.27	4.303	1.270
#11	1.56	5.313	1.410
#14	2.25	7.650	1.693
#18	4.00	13.600	2.257

03200   CONCRETE REINFORCEMENT

Steel Bar Reinforcing — Field Erection

Tolerances in Placement

Unless otherwise specified, reinforcing bars should be placed within the following tolerances:

(a) Tolerance for depth, d and minimum clear concrete cover in flexural members, walls and columns should be as follows:

	Tolerance on d	Tolerance On Minimum Concrete Cover
$d \leq 8$ in.	$\pm \frac{3}{8}$ in.	$-\frac{3}{8}$ in.
$d > 8$ in.	$\pm \frac{1}{2}$ in.	$-\frac{1}{2}$ in.

Except that the tolerance for the clear distance to formed soffits should be $-\frac{1}{4}$ in., and the tolerance for cover should not exceed minus one-third of the minimum cover required on the design drawings or in the specifications.

Note: "d" is the specified effective depth.

(b) Tolerance for longitudinal location of bends and ends of bars should be ± 2 in. except at discontinuous ends of members where the tolerance should be $\pm \frac{1}{2}$ in.

(c) As long as the total number of bars specified is maintained, a reasonable tolerance in spacing individual bars is ± 2 in., except where openings, inserts, embedded items, etc., might require some additional shifting of bars.

(d) Tolerance for length of laps in lap splices should be ± 1 in.

(e) Tolerance for embedded length should be ± 1 in. for #3 through #11 bars, and 2 −2 in. for #14 and #18 bars.

Concrete Protection for Reinforcement

The following minimum concrete cover should be provided for reinforcing bars. For bundled bars, the minimum cover should be equal to the equivalent diameter of the bundle but need not be greater than 2 in.; except for concrete cast against and permanently exposed to earth, the minimum cover should be 3 in.

(a) **Cast-In-Place Concrete (nonprestressed)**

	Minimum Cover, In.
Concrete cast against and permanently exposed to earth	3
Concrete exposed to earth or weather:	
#6 through #18 bars	2
#5 bars, W31 or D31 wire, and smaller	1½
Concrete not exposed to weather or in contact with the ground:	
Slabs, walls, joists:	
#14 and #18 bars	1½
#11 bars and smaller	¾

Steel Bar Reinforcing — (Cont.)

	Minimum Cover, In.
Beams, girders, columns:	
Primary reinforcement, ties, stirrups or spirals	1½
Shells and folded plate members:	
#6 bars and larger	¾
#5 bars, W31 or D31 wire, and smaller	½

(b) Precast Concrete (manufactured under plant control conditions)

	Minimum Cover, In.
Concrete exposed to earth or weather:	
Wall panels:	
#14 and #18 bars	1½
#11 and smaller	¾
Other members:	
#14 and #18 bars	2
#6 through #11 bars	1½
#5 bars, W31 or D31 wire, and smaller	1¼
Not exposed to weather or in contact with the ground:	
Slabs, walls, joists:	
#14 and #18 bars	1¼
#11 bars and smaller	⅝
Beams, girders, columns:	
Primary reinforcement ... d_b but not less than	⅝
and need not exceed	1½
Ties, stirrups or spirals	⅜
Shells and folded plate members:	
#6 bars and larger	⅝
#5 bars, W31 or D31 wire, and smaller	⅜

(c) In corrosive atmospheres or severe exposure conditions, the amount of concrete protection should be suitably increased, and the denseness and nonporosity of the protecting concrete should be considered, or other protection should be provided.

03200   **CONCRETE REINFORCEMENT**

Mechanical Requirements For Standard Deformed Reinforcing Bars

Type of Steel and ASTM Specification No.	Size Nos. Inclusive	Grade[1]	Tensile Strength Min., psi	Yield[2] Min., psi	Percent Elongation in 8" Minimum	Cold Bend Test[3] Pin Diameter (d = nominal diameter of specimen)
Billet-Steel A 615-84a	3-6	40	70,000	40,000	#3 ... 11 #4, #5, #6 ... 12	Under Size #6 ... 4d [4]per S1 #6 ... 3½d #6 ... 5d
	3-11 14, 18	60	90,000	60,000	#3, #4, #5, #6 ... 9 #7, #8 ... 8 #9, #10, #11, #14, #18 ... 7	Under Size #6 ... 4d [4]per S1 #6 ... 3½d #6 ... 5d #7, #8 ... 6d #9, #10, #11 ... 8d [4]per S1 #9, #10, #11 ... 7d #14, #18 ... 9d (90°)
Rail-Steel[5] A 616-84	3-11	50	80,000	50,000	#3, #7 ... 6 #4, #5, #6 ... 7 #8, #9, #10, #11 ... 5	Under Size #9 ... 6d #9, #10 ... 8d #11 ... 8d (90°)
	3-11	60	90,000	60,000	#3, #4, #5, #6 ... 6 #7 ... 5 #8, #9, #10, #11 ... 4.5	Under Size #9 ... 6d #9, #10 ... 8d #11 ... 8d (90°)
Axle-Steel[5] A 617-84	3-11	40	70,000	40,000	#3, #7 ... 11 #4, #5, #6 ... 12 #8 ... 10 #9 ... 9 #10 ... 8 #11 ... 7	Under Size #6 ... 3½d #6 and Larger ... 5d
	3-11	60	90,000	60,000	#3, #4, #5, #6, #7 ... 8 #8, #9, #10, #11 ... 7	Under Size #6 ... 3½d #6, #7, #8 ... 5d #9, #10, #11 ... 7d
Low-Alloy Steel[5] A 706-84a	3-11 14, 18	60	80,000[6]	60,000[7]	#3, #4, #5, #6 ... 14 #7, #8, #9, #10, #11 ... 12 #14, #18 ... 10	Under Size #6 ... 3d #6, #7, #8 ... 4d #9, #10, #11 ... 6d #14, #18 ... 8d

(1) Minimum yield designation.
(2) Yield point or yield strength. See specifications.
(3) Test bends 180° unless noted otherwise.
(4) Under supplementary requirement (S1) of A 615.
(5) Complete specifications for Rail-Steel (A 616), Axle-Steel (A 617) and Low-Alloy Steel (A 706) Reinforcing Bars can be obtained from the American Society for Testing and Materials, 1916 Race Street, Philadelphia, Pennsylvania 19103.
(6) Tensile strength shall not be less than 1.25 times the actual yield strength (A 706 only).
(7) Maximum yield strength 78,000 psi (A 706 only).

03200   **CONCRETE REINFORCEMENT**

Basic Concrete Reinforcing Data

BAR LAP SCHEDULE IN FEET – MINIMUM LAP 1' 0"

Bar Size	12D	15D	17D	20D	24D	27D	30D
#11	1-5	1-9	2-0	2-4	2-10	3-2	3-6
#10	1-3	1-7	1-11	2-1	2-6	2-10	3-2
#9	1-2	1-5	1-7	1-11	2-3	2-6	2-10
#8	1-0	1-3	1-5	1-8	2-0	2-3	2-6
#7		1-1	1-3	1-6	1-9	2-0	2-2
#6			1-1	1-3	1-6	1-8	1-11
#5				1-1	1-3	1-5	1-7
#4					1-0	1-2	1-3
#3							

Bar Size	330D	35D	40D	45D	50D	60D
#11	3-11	4-1	4-8	5-3	5-11	7-1
#10	3-6	3-8	4-3	4-9	5-4	6-4
#9	3-1	3-3	3-9	4-3	4-8	5-8
#8	2-9	2-11	3-4	3-9	4-2	5-0
#7	2-5	2-7	2-11	3-3	3-8	4-5
#6	2-1	2-2	2-6	2-10	3-2	3-9
#5	1-9	1-11	2-2	2-4	2-7	3-2
#4	1-5	1-6	1-8	1-11	2-1	2-6
#3	1-0	1-1	1-3	1-5	1-7	1-11

03200   CONCRETE REINFORCEMENT

Welded Wire Fabric — Common Stock Styles Of Welded Wire Fabric

Style Designation	Steel Area sq. in. per ft.		Weight Approx. lbs. per 100 sq. ft.
	Longit.	Transv.	
Rolls			
6x6—W1.4xW1.4	.03	.03	21
6x6—W2xW2	.04	.04	29
6x6—W2.9xW2.9	.06	.06	42
6x6—W4xW4	.08	.08	58
4x4—W1.4xW1.4	.04	.04	31
4x4—W2xW2	.06	.06	43
4x4—W2.9xW2.9	.09	.09	62
4x4—W4xW4	.12	.12	86
Sheets			
6x6—W2.9xW2.9	.06	.06	42
6x6—W4xW4	.08	.08	58
6x6—W5.5xW5.5	.11	.11	80
4x4—W4xW4	.12	.12	86

Certain styles of welded wire fabrics as shown in the Table have been recommended by the Wire Reinforcement Institute as common stock styles. Use of these styles is normally based on empirical practice and quick availability rather than on specific steel area designs. Styles of fabric produced to meet other specific steel area requirements are ordered for designated projects, or, in some localities, may be available from inventory.

ASTM SPECIFICATIONS

Welded wire fabric used for concrete reinforcement consists of cold-drawn wire in orthogonal patterns, square or rectangular, resistance welded at all intersections. Welded wire fabric (WWF) is commonly but erroneously called "mesh" which is a much broader term not limited to concrete reinforcement. Welded wire fabric must conform to ASTM A 185 if made of smooth wire or A497 if made of deformed wire. These Specifications require shear tests on the welds essential to proper anchorage for bond in concrete. ASTM yield strength is 65,000 psi for smooth fabric (A 185) and is 70,000 psi for deformed fabric (A 497).

Unless otherwise specified, welded wire fabric conforming to ASTM A 185 will be furnished.

An example style designation is: WWF 6x12-W16xW8. This designation identifies a style of fabric in which:

Spacing of longitudinal wires . . . . = 6"
Spacing of transverse wires . . . . = 12"
Longitudinal wire size . . . . = W16
Transverse wire size . . . . = W8

A deformed fabric style would be designated in the same manner with the appropriate D-number wire sizes.

It is very important to note that the terms "longitudinal" and "transverse" are related to the method of fabric manufacture and have no reference to the position of the wires in a completed concrete structure.

03200   **CONCRETE REINFORCEMENT**

Basic Concrete Data

SLUMP

Aggr. Size	2"	2½"	3"	4"	5"	6"
¾"	301/36.1	306/36.7	310/37.2	319/38.3	327/39.2	333/40.0
⅞"	296/35.5	301/36.1	305/36.6	314/37.7	322/38.6	330/39.6
1"	291/34.9	296/35.5	300/36.0	309/37.1	317/38.0	327/39.2
1¼"	281/33.7		290/34.8	299/35.9	307/36.8	316/37.9
1½"	271/32.5		280/33.6	289/34.7	297/35.6	305/36.6

Water requirement in pounds/gals per cubic yard
Rule of thumb: 1 gal. will change slump approximately 1 inch

TIME TABLE OF CEMENT STRENGTHS

Cem. Type	3 days	7 days	14 days	28 days	60 days
I	40	60	80	100	
II	33	55	65	80	
III	60	80	100	120	100
IV	20	40	55	75	
V	20	40	60	80	

Compressive strength rated on percentage of Type I 28 day strength
All types increased in one year 133%

EFFECT OF WATER CONTENT ON COMPRESSIVE STRENGTH OF CONCRETE
(Non Air Entrained Concrete)

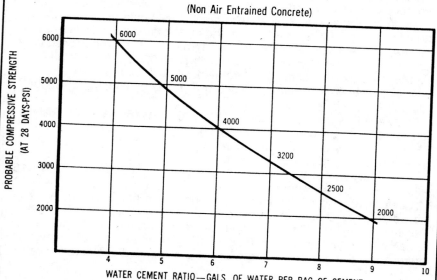

From ACI 613: ACI Recommended Practice for Selecting Proportions for Concrete

03200   CONCRETE REINFORCEMENT

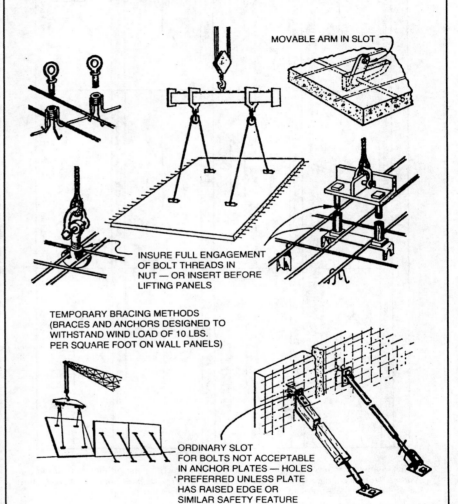

Standard Welding Symbols

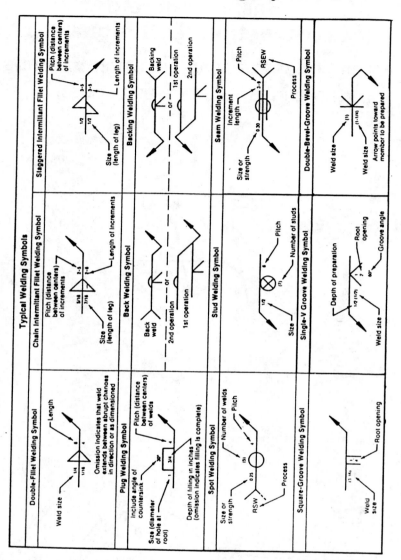

The welding symbols reproduced in this section are from the publication, *Standard Symbols for Welding, Brazing and Nondestructive Examination*, published and copyrighted by the American Welding Society, Miami, Florida 33135.

05050   METAL FASTENING

Standard Welding Symbols

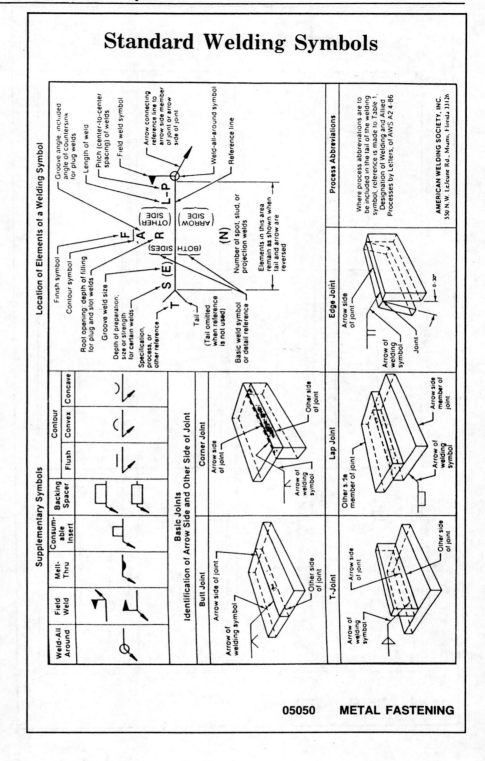

05050 METAL FASTENING

Standard Welding Symbols

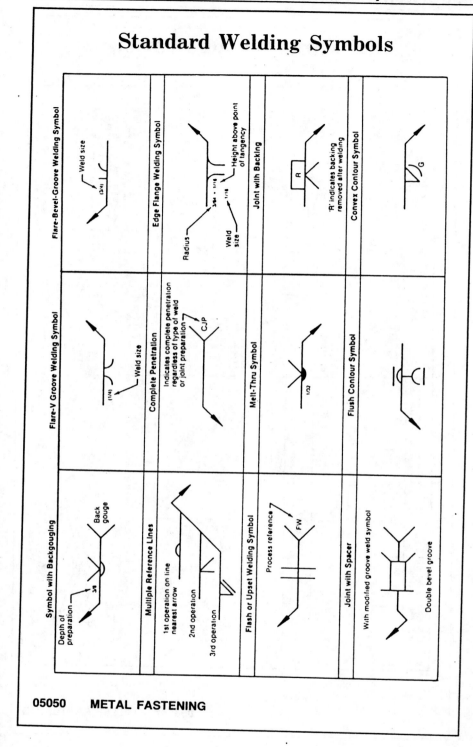

05050 METAL FASTENING

Construction Inspection Manual

Standard Welding Symbols

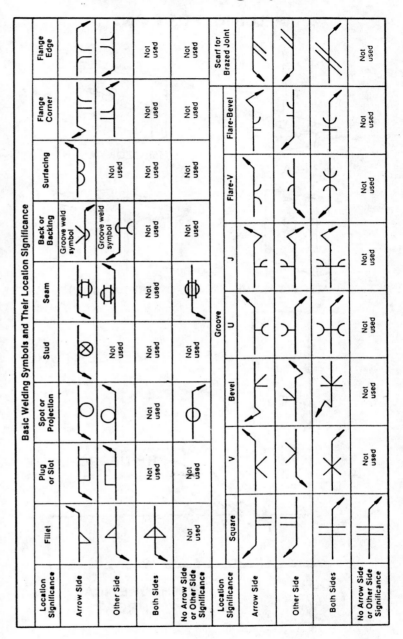

05050   **METAL FASTENING**

General Information On Welding

WELDING POSITIONS

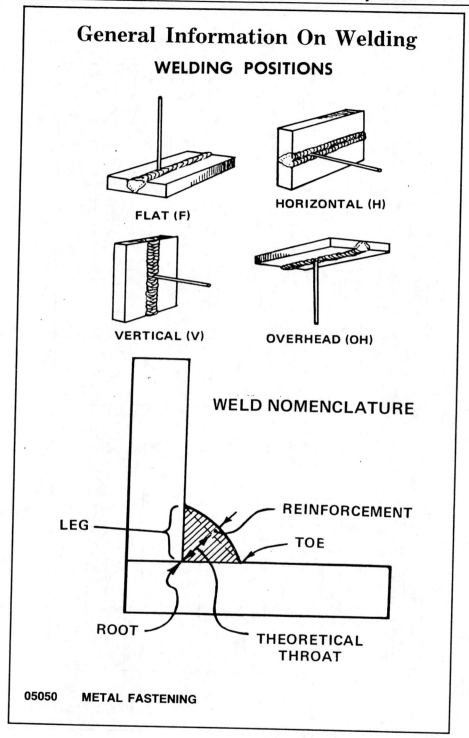

FLAT (F)

HORIZONTAL (H)

VERTICAL (V)

OVERHEAD (OH)

WELD NOMENCLATURE

- LEG
- REINFORCEMENT
- TOE
- ROOT
- THEORETICAL THROAT

05050   METAL FASTENING

General Information On Welding

WELD CHARACTERISTICS

TOO SLOW — Overlap

TOO FAST — Spatter, Porosity

NOT ENOUGH HEAT — overlap

TOO MUCH HEAT — Pits and Spatter

SATISFACTORY — Penetration

INSUFFICIENT THROAT

EXCESS CONVEXITY

UNDERCUT

OVERLAP

INSUFFICIENT LEG

INSUFFICIENT THROAT

EXCESS CONVEXITY

UNDERCUT

OVERLAP

SATISFACTORY — Reinf. 1/8" Max.

05050   METAL FASTENING

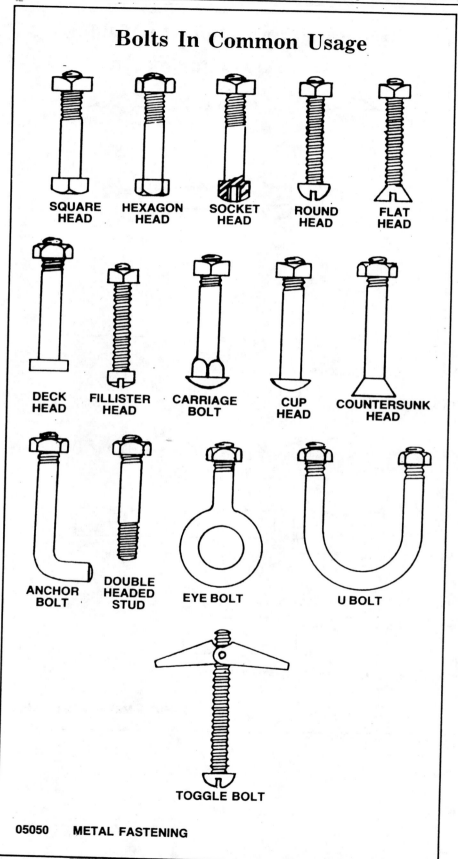

Construction Inspection Manual

High Strength Bolts Tightening Procedures

Equipment and Power

Installation equipment in good working order is vital to proper tightening of high-strength bolts in structural connections. The type of wrench to be used is not designated in the *Specifications for Structural Joints Using ASTM A 325 or A 490 Bolts* as published by the Research Council. However, three methods of tightening are recommended by the Council Specifications.

1. **Calibrated Wrench Tightening** requiring the use of a torque-controlled wrench that cuts off when a pre-set torque is reached.

2. **Turn-of-Nut Tightening** which can be accomplished with either a hand wrench or a standard impact wrench.

3. **Tightening by use of a Direct Tension Indicator** which requires use of any device allowing accurate direct measurement of bolts tension, such as the load indicator washer.

Bolts installed with a load indicator washer are tightened with a standard impact wrench.

The usual source of power is compressed air. There must be an adequate pressure at the tool — an absolute minimum of 100 psi for bolts ⅞" diameter and smaller. For larger bolts, pressure must be higher. Hose lines should be adequate for the number and size of wrenches used.

Calibration

Whether bolts are installed by the calibrated wrench method or by the turn-of-nut method, the use of a calibrating device to check out tools and equipment and to provide a means of reliable inspection is essential.

Fig. 1 – The bolt-tension calibrator is a hydraulic load cell which measures bolt tension created by tightening. As the bolt or nut is turned, the internal bolt tension or clamping force is transmitted through the hydraulic fluid to a pressure gauge which indicates bolt tension directly in pounds. The dial of the gauge may be marked to show the required minimum tension for each bolt diameter. (See Bolt Tension Table as published in the current Research Council Specifications.)

When torque-control wrenches are used, they must be calibrated at least once each working day by tightening not less than three bolts of each diameter from the bolts to be installed. The average torque determined by this calibration procedure may then be used to pre-set the cut-off device built into torque-control wrench. The torque-control device must be set to provide a bolt tension 5 to 10 percent in excess of the minimum bolt tension. The

Construction Inspection Manual

High Strength Bolts — (Cont.)

bolt-tension calibrator is also necessary to calibrate the hand-indicator torque wrenches used by inspectors for checking torque as a measure of tension after tightening by either the calibrated wrench method or the turn-of-nut method.

Proper Tightening Procedure

Regardless of the method used to tighten high-strength structural bolts, the sequence of operations is basically similar. Accordingly, we show below the step-by-step procedure for tightening a simulated beam-to-girder connection by the turn-of-nut method, using 1-in. diameter A 325 bolts. For clarity, the beam is omitted and only the clip angles which had been previously bolted to the girder web are shown.

Fig. 2 – First, holes are "faired up" with enough drift pins to maintain dimensions and plumbness. Next, sufficient high-strength bolts of the proper grade and size are installed to hold the connection in place. Only hand tightening is required at this point. Since these bolts will remain in place as permanent fasteners, washers, if required, should be installed with the bolts during fitting-up.

Fig. 3 – The balance of the holes are now filled with bolts and assembled with nuts and washers. Note the gap between the angles which will be drawn together during the "snugging" operation. In a true connection, this gap would be filled by the beam web. However, it may be considered representative of "difficult" fitting-up conditions.

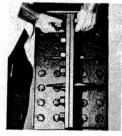

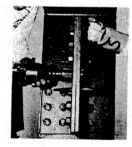

Fig. 4 – The bolting crew starts to "snug" the bolts and nuts. ("Snug" is defined as the point at which the wrench begins to impact solidly). Note that the "snug" condition creates sufficient tension to draw the top half of the angles tightly together while the bottom half still remains open because those bolts are only "hand tight."

Fig. 5 – The crew completes "snugging" the entire connection. As a result, the gap between the angles has entirely disappeared.

Construction Inspection Manual

High Strength Bolts — (Cont.)

Fig. 6 – Now the drift pins are knocked out and the remaining holes filled with bolts and torqued up to "snug." The connection is now ready for final tightening.

Fig. 7 – In this view, the bolts have been numbered to show the suggested tightening sequence. Bolts and nuts should always be tightened progressively away from the fixed or rigid points to the free edges.

Fig. 8 – Here, the wrench operator is just starting final tightening of Bolt No. 2 to the required half-turn beyond "snug" condition. (See Nut Rotation Council Specifications.) In final tightening, a hand wrench is used to hold the end not being torqued to insure that the true required turn measurement is not lost.

Fig. 9 – Here, the operator has completed the half-turn on Bolt No. 12, as shown by the twin double lines on the wrench socket located 90 degrees apart. Notice in Fig. 8 (showing the beginning of a required half-turn) that there are twin single lines 90 degrees apart. These socket markings enable the operator to easily measure nut rotation.

05050 **METAL FASTENING**

High Strength Bolts — (Cont.)

Fig. 10 – This is a close-up of a nut after final tightening. Close examination will disclose slight burrs or peening marks near the edge of each nut "flat." These marks are caused by the "hammering" action of the wrench as it impacts. If A 325 nuts show no such markings, a thorough inspection should be made to insure that the bolts were properly tightened. Nuts furnished with A 490 bolts may not show any distortion because of their greater hardness. However, a slight burnishing of the edges should be evident.

Load Indicator Washer

The Bethlehem Load Indicator Washer (LIW) is a hardened flat circular washer with protrusions on one face. In use, the LIW is placed on the bolt with the protrusions bearing against a hardened surface of the bolt-nut assembly, usually the underside of the bolt head (Figure 1). As the bolt is tightened the protrusions are flattened and the gap reduced (Figure 2).

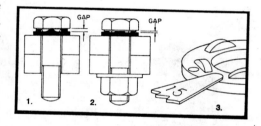

While tightening, be sure the bolt head does not spin on the load indicator protrussions. At a specified average gap, measured by feeler gauge as in Figure 3, the induced bolt tension will not be less than the minimum required by various standards. If connection details require placing the LIW at the nut end, or if the fastener must be tightened from the end where the LIW is located, a supplemental hardened flat washer must be used against the protrusions. The maximum gaps between load indicator washer and bolt head (or hardened flat washer) after tightening are shown in Table 1. Table 2 shows the induced bolt tension which correspond to these gaps. The LIW is available for both A 325 and A 490 bolts, in two distinct configurations.

Inspection and Verification

Inspection is accomplished by checking the average gap of the LIW bolt assembly with a metal feeler gauge (Figure 3). Two important rules for inspection should be emphasized:

  1. Inspection should be based upon the average gap because the bolt will never be perfectly centered in the LIW, therefore, the protrusions will not collapse uniformally.

  2. The feeler gauge is used as a "no go" inspection tool; that is, if the gauge does not enter the gap (but a gap is evident) the installation is considered satisfactory.

High Strength Bolts — (Cont.)

TABLE 1 Load Indicator Gaps To Give Required Minimum Bolt Tension		
Load Indicator Fitting	A 325	A 490
Under Bolt Head Black Finish Bolts	0.025 in.	0.015 in.
Under Nut with Hardened Flat Washer Black Finish	0.015 in.	0.010 in.
With the gaps shown above, required minimum bolt tensions will be induced as given in Table 2.		

TABLE 2 Minimum Bolt Tensions		
In thousands of pounds (Kips) Bolt Dia. (in.)	A 325	A 490
½	12	—
⅝	19	—
¾	28	35
⅞	39	49
1	51	64
1⅛	56	80
1¼	71	102

Turn-Of-Nut Tightening

When the turn-of-nut method is used to provide the bolt tension specified, there shall first be enough bolts brought to a "snug tight" condition to insure that the parts of the joint are brought into good contact with each other. Snug tight is defined as the tightness attained by a few impacts of an impact wrench or the full effort of a man using an ordinary spud wrench. Following this initial operation, bolts shall be placed in any remaining holes in the connection and brought to snug tightness. All bolts in the connection shall then be tightened additionally by the applicable amount of nut rotation specified in table below, with tightening progressing systematically from the most rigid part of the joint to its free edges. During its operation there shall be no rotation of the part not turned by the wrench.

Nut Rotation[a] From Snug Tight Condition

Bolt Length (As measured from underside of head to extreme end of point)	Disposition of Outer Faces of Bolted Parts		
	Both faces normal to bolt axis	One face normal to bolt axis and other face sloped not more than 1:20 (bevel washer not used)	Bolt faces sloped not more than 1:20 from normal to bolt axis (bevel washers not used)
Up to and including 4 diameters	1/3 turn	1/2 turn	2/3 turn
Over 4 diameters but not exceeding 8 diameters	1/2 turn	2/3 turn	5/6 turn
Over 8 diameters but not exceeding 12 diameters[b]	2/3 turn	5/6 turn	1 turn

[a] Nut rotation is relative to bolt, regardless of the element (nut or bolt) being turned. For bolts installed by 1/2 turn and less, the tolerance should be plus or minus 30°; for bolts installed by 2/3 turn and more, the tolerances should be plus or minus 45°.

[b] No research work has been performed by the Council to establish the turn-of-nut procedure when bolt lengths exceed 12 diameters. Therefore, the required rotation must be determined by actual tests in a suitable tension device simulating the actual conditions.

High Strength Bolts — (Cont.)

Commentary On The 1985 Changes To The Specification For Structural Joints Using A325 Or A490 Bolts

The terms *friction connection* and *bearing connection* are now discontinued.

Slip critical connection is the new term for joints where slip between connected parts is undesirable, such as in fatigue loading or where there is significant stress reversal.

Paint is now permitted on the faying surfaces of slip critical connections, but must be qualified by test; and the slip coefficient of the paint must be determined.

Bolts in connections identified as not being slip-critical nor subject to direct tension need not be inspected for bolt tension other than to assure that the plies of the connected elements have been brought into snug contact. *Snug tight* connections must be clearly identified on the drawings.

Fasteners must be protected at the jobsite from dirt and moisture. Unused fasteners must be returned to protected storage at the end of the day. Slip-critical fasteners must be cleaned and relubricated if dirty.

Calibrators are required with all tightening methods.

In *turn of the nut* installations check three-bolt assemblies in a calibrator at the start of the work, and when changes occur.

If using a calibrator wrench, calibrate daily with calibrator and whenever changes in bolt condition occur, using three-bolt assemblies. Hardened washers are required under the turned element. Verify that rotation of the turned element does not exceed rotation shown in Table 3. Use of "standard" torque values is prohibited.

With *alternate design* bolts and *direct tension indicators*, check three-bolt assemblies in calibrator for tension five percent over Table 2 requirements per the manufacturer's instructions, at the start of work and when changes in the bolt condition occur.

A new bolt is available in lengths not more than four times the bolt's diameter that is threaded full length. These bolts are identified as A325T.

Based on tests which demonstrated that the slip resistance of joints was unchanged or slightly improved by the presence of burrs, burrs which do not prevent solid seating of the connected parts in the *snug tight* condition need not be removed. In the case of *arbitration inspection*, the inspection torque is developed by five-bolt assemblies in the calibrator; formerly only three were required.

In the case of *delayed verification inspection*, exposure conditions may affect the validity of torque testing to confirm proper bolt tension after installation. Specific situations need to be individually evaluated and appropriate procedures must be developed.

Types Of Nails

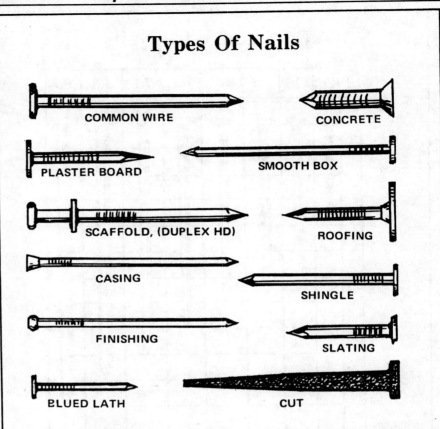

Cut Nails. Cut nails are angular-sided, wedge-shaped with a blunt point.

Wire Nails. Wire nails are round shafted, straight, pointed nails, and are used more generally than cut nails. They are stronger than cut nails and do not buckle as easily when driven into hard wood, but usually split wood more easily than cut nails. Wire nails are available in a variety of sizes varying from two penny to 60 penny.

Nail Finishes. Nails are available with special finishes. Some are galvanized or cadmium plated to resist rust. To increase the resistance to withdrawal, nails are coated with resins or asphalt cement (called cement coated). Nails which are small, sharp-pointed, and often placed in the craftsman's mouth (such as lath or plaster board nails) are generally blued and sterilized.

06050   FASTENERS AND ADHESIVES

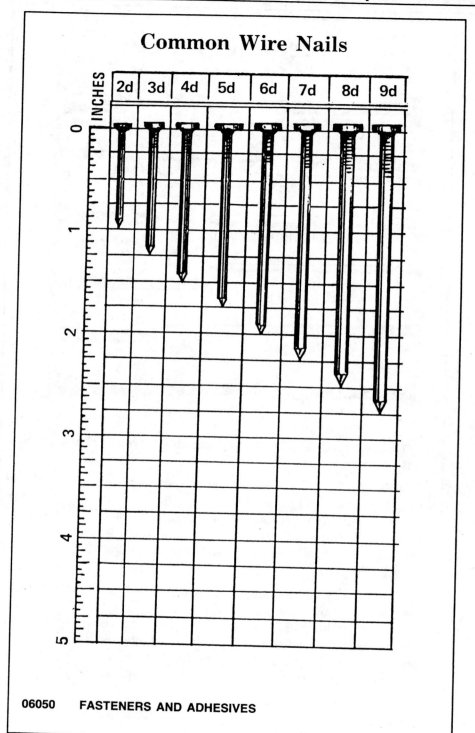

06050   FASTENERS AND ADHESIVES

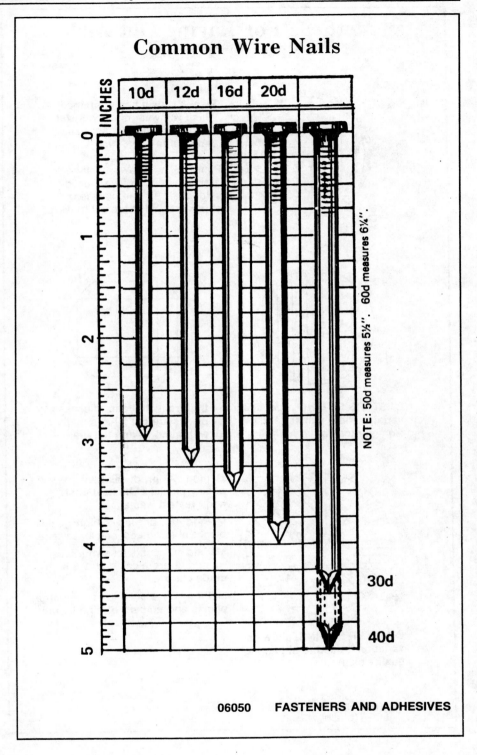

06050 FASTENERS AND ADHESIVES

Methods For Laying Out And Sawing Timber

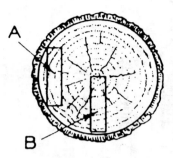

Plank "A" will be flat grained which will wear quickly, sliver easily, have greater shrinkage, and tend to cup or warp.

Plank "B" will be vertically grained which will wear evenly, seldom sliver, and have less shrinkage and tendency to warp.

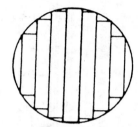

BASTARD SAWING produces variable width planks ranging from vertical to flat grain.

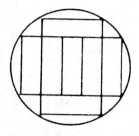

Uniform width and thickness planks are cut by this method, with most having flat grain.

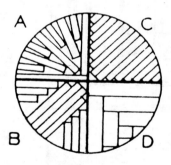

QUARTER SAWING, showing variations. Used to obtain high quality planks.

Method "A" produces best quality planks at highest cost and greatest waste.

Method "B" produces high quality planks at lower cost and less waste.

Method "C" costs less with least waste but produces only 1 5 high grade planks.

Method "D" is used to secure large planks and timbers of high quality.

06100   ROUGH CARPENTRY

Lumber Grading

GRADING-MARK ABBREVIATIONS

GRADES
(Listed alphabetically — not by quality)

COM	Common
CONST	Construction
ECON	Economy
No. 1	Number One
SEL-MER	Select Merchantable
SEL-STR	Select Structural
STAN	Standard
UTIL	Utility

ALSC TRADEMARKS

CLIS	California Lumber Inspection Service
NELMA	Northeastern Lumber Manufacturers Association, Inc.
NH&PMA	Northern Hardwood & Pine Manufacturers Association, Inc.
PLIB	Pacific Lumber Inspection Bureau
RIS	Redwood Inspection Service
SPIB	Southern Pine Inspection Bureau
TP	Timber Products Inspection
WCLB	West Coast Lumber Inspection Bureau
WWP	Western Wood Products Association

SPECIES GROUPINGS

AF	Alpine Fir
DF	Douglas Fir
HF	Hem Fir
SP	Sugar Pine
PP	Ponderosa Pipe
LP	Lodgepole Pine
IWP	Idaho White Pine
ES	Engelmann Spruce
WRC	Western Red Cedar
INC CDR	Incense Cedar
L	Larch
LP	Lodgepole Pine
MH	Mountain Hemlock
WW	White Wood

MOISTURE CONTENT

S-GRN	Surfaced at a moisture content of more than 19%.
S-DRY	Surfaced at a moisture content of 19% or less.
MC-15	Surfaced at a moisture content of 15% or less.

06100   ROUGH CARPENTRY

Redwood Grade Marks
California Redwood Association

CONSTRUCTION GRADES

ALL HEARTWOOD

SELECT HEART — Tight-knotted, heartwood grade with face free of splits or shake. **Sample uses:** decks, posts, garden structures, industrial tanks where decay hazards exist.

```
MILL ONE
SEL HT  R|S®
REDWOOD
```

FOUNDATION GRADE — High quality heartwood grade, specially selected from Construction Heart for durability and resistance to insects. Always grade stamped. **Sample uses:** sill plates, crib walls.

```
MILL ONE
FDTN  R|S®
REDWOOD
```

CONSTRUCTION HEART — All purpose heartwood grade, contains knots of varying sizes and other slight imperfections. **Sample uses:** decks, posts, retaining walls or other uses on or near soil.

```
MILL ONE
CONST HT  R|S®
REDWOOD
```

MERCHANTABLE HEART — Economical grade, allows some holes, splits and slightly larger knots than Construction Heart. **Sample uses:** fences, retaining walls, farm uses, structures where decay hazard exists.

```
MILL ONE
MERCH HT  R|S®
REDWOOD
```

MAY CONTAIN SAPWOOD

SELECT — Same general characteristics as Select Heart, but contains sapwood and some imperfections not allowed in Select Heart. **Sample uses:** decking, fence boards and other above ground uses.

```
MILL ONE
SEL  R|S®
REDWOOD
```

CONSTRUCTION COMMON — Versatile grade with same general characteristics as Construction Heart, but contains sapwood. **Sample uses:** decking, railings, and other above ground uses.

```
MILL ONE
CONST COM  R|S®
REDWOOD
```

MERCHANTABLE — Same general characteristics as Merchantable Heart, but contains sapwood. **Sample uses:** fence boards, railings temporary construction and other above ground uses.

```
MILL ONE
MERCH  R|S®
REDWOOD
```

06100 ROUGH CARPENTRY

Redwood Grade Marks — (Cont)

FINISH GRADES

ALL HEARTWOOD

CLEAR ALL HEART VERTICAL GRAIN — Finest architectural grade, specially selected for grain. Free of defects one face. **Sample uses:** siding, paneling, mill work, processing tanks.

Grade mark: MILL ONE / CLR HT VG / REDWOOD

CLEAR ALL HEART — Same quality as Clear All Heart Vertical Grain, except contains flat grain pieces. **Sample uses:** siding, trim, fine garden structures, industrial storage.

Grade mark: MILL ONE / CLR HT / REDWOOD

MAY CONTAIN SAPWOOD

CLEAR VERTICAL GRAIN — Same general quality as Clear All Heart Vertical Grain, except contains sapwood. **Sample uses:** siding, cabinetry, garden shelters and other above ground uses.

Grade mark: MILL ONE / CLEAR VG / REDWOOD

CLEAR — Same general quality as Clear All Heart, except contains sapwood. Accepts some imperfections not permitted in Clear All Heart. **Sample uses:** paneling, soffits and other above ground uses.

Grade mark: MILL ONE / CLR / REDWOOD

B-GRADE — Quality grade, contains sapwood, limited knots and other characteristics not permitted in Clear. **Sample uses:** siding, molding, fascia and other above ground uses.

Grade mark: MILL ONE / B / REDWOOD

Additional marks: CRA CERTIFIED KILN DRIED CLR RWD; S-GRN RWD CONST HT

Redwood Grademarks

Redwood grades are established by the Redwood Inspection Service in the Standard Specifications for Grades of California Redwood Lumber. Properly grade-marked lumber will bear the RIS mark or that of another accredited inspection bureau. Grademarks may be on seasoned or unseasoned lumber on face, edge or end of piece. "Certified Kiln Dried" marks lumber kiln dried to accepted standards. CRA trademark is on products of member mills of the California Redwood Association only and is an additional assurance of quality.

06100   ROUGH CARPENTRY

Lumber Grading

INTERPRETING GRADE MARKS

Most grade stamps, except those for rough lumber or heavy timbers, contain five basic elements:

(a) Certification mark of certifying association of lumber manufacturers.

(b) Mill identification. Firm name, brand or assigned mill number.

(c) Grade designation. Grade name, number or abbreviation.

(d) Species identification. Indicates species by individual species or species combination.

(e) Condition of seasoning. Indicates condition of seasoning at time of surfacing:

S-DRY — 19% maximum moisture content
MC-15 — 15% maximum moisture content
S-GRN — Over 19% moisture content (unseasoned)

INSPECTION CERTIFICATE

When an inspection certificate issued by a certifying association is required on a shipment of lumber and specific grade marks are not used, the stock is identified by an imprint of the association mark and the number of the shipping mill or inspector.

CERTIFIED AGENCIES AND TYPICAL GRADE MARKS

California Lumber
Inspection Service
1790 Lincoln Avenue
San Jose, California 95125
(408) 297-8071

```
        MILL 467
 CL   STAND
 IS   S-GRN  WCLB
  ®   DOUG. FIR
```

S-GRN
CONST
EASTERN
HEM-TAM

NELMA

Northeastern Lumber
Manufacturers
Association, Inc.
4 Fundy Road
Falmouth, Maine 04105
(207) 781-2252

06100   ROUGH CARPENTRY

Lumber Grading — (Cont)

CERTIFIED AGENCIES AND TYPICAL GRADE MARKS
(Continued)

Agency	Grade Mark
Northern Hardwood and Pine Manufacturers Association, Inc. Suite 501, Northern Building Green Bay, Wisconsin 54301 (414) 432-9161	110 NH&PMA STUD S-DRY BALSAM FIR
Pacific Lumber Inspection Bureau, Inc. 1411 Fourth Avenue Building (Suite 1130) Seattle, Washington 98101 (206) 622-7327	PLIB® HEM-FIR W-10 CONST S-GRN WCLB RULES
Redwood Inspection Service 591 Redwood Highway, Suite 3100 Mill Valley, California 94941 (415) 381-1304	(50) FDTN S-GRN REDWOOD RIS 
Southern Pine Inspection Bureau 4709 Scenic Highway Pensacola, Florida 32504 (904) 434-2611	SPIB® No. 1 KD 15 (7)
Timber Products Inspection P.O. Box 919 Conyers, Georgia 30207 (404) 922-8000	TP® NO. 1 KD-15 000 SYP
West Coast Lumber Inspection Bureau Box 23145 Portland, Oregon 97223 (503) 639-0651	WCLB® MILL 10 NO. 2 DOUG FIR S DRY
Western Wood Products Association 1500 Yeon Building Portland, Oregon 97204 (503) 224-3930	12 WWP® 2 S-DRY D.FIR

06100   ROUGH CARPENTRY

Lumber Grading — (Cont.)

CANADIAN GRADE MARKS

A.F.P.A.® 00
S—P—F
S-DRY STAND

Alberta Forest Products Association
11710 Kingsway Avenue, #204
Edmonton, Alberta T5G OX5, (403) 452-2841

 CONST
S-GRN
D FIR

Mac Donald Inspection
211 SchoolHouse Street
Coquitlan, B.C. V3K 4X9, (604) 520-3321

CLA
S-P-F
100
No. 1
S-GRN.

Canadian Lumberman's Association
27 Goulburn Avenue,
Ottawa, Ontario K1N 8C7, (613) 233-6205

Maritime Lumber Bureau
P.O. Box 459, Amherst.
Nova Scotia B4H 4A1, (902) 667-3880

LMA 1 S-GRN 1
1 D FIR (N)

Cariboo Lumber Manufacturers Association
301 Centennial Building
197 Second Avenue North, Williams Lake,
B.C. V2G 1 Z5. (604) 392-7778

O.L.M.A.® 01-1
CONST. S-DRY
SPRUCE - PINE - FIR

Ontario Lumber Manufacture Association
159 Bay Street, Suite 414
Toronto, Ontario M5J 1J7. (416) 367-9717

CFPA® 38
S-P-F S-GRN
CONST

Central Forest Products Association
P.O. Box 1169, Hudson Bay, Saskatcewan
S0E 0Y0. (306) 865-2595

PLIB. NLGA RULE
No 1
00 S-GRN
HEM-FIR-N

Pacific Lumber Inspection Bureau
Suite 1130, 1411 Fourth Avenue Building
Seattle, Washington 98101
B.C. Division: 1460-1055 West Hastings St.
Vancouver, B.C. V6E 2E9. (604) 689-1561

COFI. S-P-F
S-GRN
100   No 1

**Council of Forest Industries
of British Columbia**
1500-1055 West Hastings Street,
Vancouver, B.C. V6E 2H1, (604) 684-0211

Quebec Lumber Manufactures Association
3555 Boul, Hamel-ouest, Suite 200
Quebec, Canada G2E 2G6, (418) 872-5610

ILMA S-DRY 1
00 S-P-F

Interior Lumber Manufacturers Association
203-2350 Hunter Road
Kelowna, B.C. V1X 6C1. (604) 860-9663

06100   ROUGH CARPENTRY

Construction Inspection Manual

Lumber Grading — (Cont)

GRADE MARKS FOR PRESSURE TREATED LUMBER

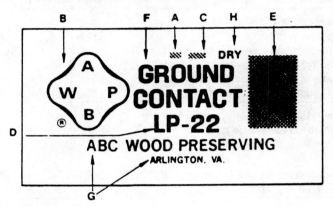

- A Year of treatment
- B American Wood Preservers Bureau trademark or trademark of the AWPB certified agency
- C The preservative used for treatment
- D The applicable American Wood Perservers Bureau quality standard
- E Trademark of the AWPB certified agency
- F Proper exposure conditions
- G Treating company and plant location
- H Dry or KDAT if applicable

Jason
Jason Associates, Inc.
Fort Collins, CO 80522

Timber Products Inspection and Testing Services
Conyers, GA 30207
Timber Products Inspection and Testing Services
Portland, OR 97220

Southwestern Laboratories
Houston, TX 77249

Southern Wood Products Inspection Company
Pensacola, FL 32504

McCutchan Inspections, Inc.
Portland, OR 97203

California Lumber Inspection Services
San Jose, CA 95150

C. M. Rou Service, Inc.
Mobile, AL 36606

Bode Inspection, Inc.
Pake Oswego, OR 97034

PFS Corporation
Madison, WI 53704

Florida Lumber Inspection Service
Perry, FL 32347

Thurlow Inspection
Sandpoint, ID 83864

06100 ROUGH CARPENTRY

Lumber Grading — (Cont.)

AWPB Standards For Software Lumber, Timber and Plywood

Preservative	Standard For Use*	
General Purpose Standards		
Waterborne preservatives	LP-2	LP-22
Light-hydrocarbon-solvent/penta	LP-3	LP-33
Volatile-hydrocarbon-solvent (LPG)/penta	LP-4	LP-44
Creosote or creosote/coal-tar solutions	LP-5	LP-55
Heavy-hydrocarbon-solvent/penta	LP-7	LP-77
Special-Purpose		
Waterborne preservatives for use in residential and light commercial foundations	FDN*	FDN*
All preservatives for use in marine (saltwater) exposure	LMP*	LMP*

*Ground-contact grades are treated to a higher degree than above-ground grades which, in some instances, utilize preservatives not permitted for ground or fresh-water contact. In all cases, materials permitted for ground contact are suitable for fresh-water installation. And, because of their higher retention requirements, ground-contact materials may always be safely used where above ground grade is required. Material designated for above-ground use should not be substituted for ground or water contact. None of these materials should be specified for salt-water installations. (See special-purpose standards, above and below.)

AWPB Standards For Marine Piling And Construction Poles

Preservative	Standard Designation
Marine Piling	
Waterborne and cresote (dual treatment)	MP-1
Creosote and creosote/coal-tar solution	MP-2
Waterborne	MP-4
Construction Poles	
All preservatives	CP

06100   ROUGH CARPENTRY

Lumber Grading — (Cont)

PLYWOOD — BASIC GRADE MARKS SPECIALTY PANELS

| HDO · A-A · G-1 · EXT-APA · 000 · PS1-83 |

Plywood panel manufactured with a hard, semi-opaque resin-fiber overlay on both sides. Extremely abrasion resistant and ideally suited to scores of punishing construction and industrial applications, such as concrete forms, industrial tanks, work surfaces, signs, agricultural bins, exhaust ducts, etc. Also available with skid-resistant screen-grid surface and in Structural I. **Exposure Durability Classification:** Exterior. **Common Thicknesses:** ⅜, ½, ⅝, ¾.

| MARINE · A-A · EXT-APA · 000 · PS1-83 |

Specialty designed plywood panel made only with Douglas fir or western larch, solid jointed cores, and highly restrictive limitations on core gaps and face repairs. Ideal for both hulls and other marine applications. Also available with HDO or MDO faces. **Exposure Durability Classification:** Exterior. **Common Thicknesses:** ¼, ⅜, ½, ⅝, ¾.

```
    APA                          APA
  PLYFORM                    M. D. OVERLAY
  B-B CLASS I                   GROUP 1
  EXTERIOR                      EXTERIOR
    000                           000
   PS 1-83                       PS 1-83
```

APA proprietary concrete form panels designed for high reuse. Sanded both sides and mill-oiled unless otherwise specified. Class I, the strongest, stiffest and more commonly available, is limited to Group 1 faces, Group 1 or 2 crossbands, and Group 1, 2, 3 or 4 inner plies. Class II is limited to Group 1 or 2 faces (Group 3 under certain conditions) and Group 1, 2, 3 or 4 inner plies. Also available in HDO for very smooth concrete finish, in Structural I, and with special overlays. **Exposure Durability Classification:** Exterior. **Common Thicknesses:** 19/32, ⅝, 23/32, ¾.

Plywood panel manufactured with smooth, opaque, resin-treated fiber overlay providing ideal base for paint on one or both sides. Excellent material choice for shelving, factory work surfaces, paneling, built-ins, signs and numerous other construction and industrial applications. Also available as a 303 Siding with texture-embossed or smooth surface on one side only and Structural I. **Exposure Durability Classification:** Exterior. **Common Thicknesses:** 11/32, ⅜, 15/32, ½, 19/32, ⅝, 23/32, ¾.

06100   ROUGH CARPENTRY

Lumber Grading — (Cont.)

APA B-C GROUP 1 EXTERIOR 000 PS 1-83

APA B-C
Utility panel for farm service and work buildings, boxcar and truck linings, containers, tanks, agricultural equipment, as a base for exterior coatings and other exterior uses. *Exposure Durability Classification:* Exterior. *Common Thicknesses:* 1/4, 11/32, 3/8, 15/32, 1/2, 19/32, 5/8, 23/32, 3/4.

APA B-D GROUP 2 INTERIOR 000 PS 1-83

APA B-D
Utility panel for backing, sides or builtins, industry shelving, slip sheets, separator boards, bins and other interior or protected applications. *Exposure Durability Classifications:* Interior, Exposure 1. *Common Thicknesses:* 1/4, 11/32, 3/8, 15/32, 1/2, 19/32, 5/8, 23/32, 3/4.

APA A-C GROUP 1 EXTERIOR 000 PS 1-83

APA A-C
For use where appearance of one side is important in exterior applications such as soffits, fences, structural uses, boxcar and truck linings, farm buildings, tanks, trays, commercial refrigerators, etc. *Exposure Durability Classification:* Exterior. *Common Thicknesses:* 1/4, 11/32, 3/8, 15/32, 1/2, 19/32, 5/8, 23/32, 3/4.

APA A-D GROUP 1 EXPOSURE 1 000 PS 1-83

APA A-D
For use where appearance of only one side is important in interior applications, such as paneling, builtins, shelving, partitions, flow racks, etc. *Exposure Durability Classifications:* Interior, Exposure 1. *Common Thicknesses:* 1/4, 11/32, 3/8, 15/32, 1/2, 19/32, 5/8, 23/32, 3/4.

06100   ROUGH CARPENTRY

Construction Inspection Manual

Lumber Grading — (Cont)

PLYWOOD — BASIC GRADE MARKS
AMERICAN PLYWOOD ASSOCIATION (APA)

The American Plywood Association's trademarks appear only on products manufactured by APA member mills. The marks signify that the product is manufactured in conformance with APA performance standards and/or U.S. Product Standard PS 1-83 for Construction and Industrial Plywood.

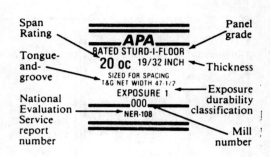

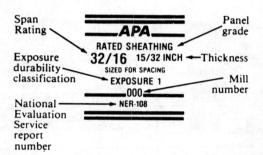

Unsanded and touch-sanded panels, and panels with "B" or better veneer on one side only, usually carry the APA trademark on the panel back. Panels with both sides of "B" or better veneer, or with special overlaid surfaces (such as Medium Density Overlay), carry the APA trademark on the panel edge, like this:

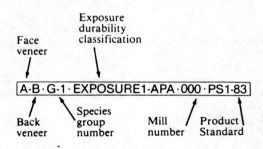

06100   ROUGH CARPENTRY

Identification Index[a]
Table For Sheathing Panels

Species of face and back	Grade		
Group 1	C-C Str. I C-C, C-D Str. II C-C, C-D[c] C-D [b]		
Group 2 [d]	C-C Str. II C-C, C-D C-D	C-C Str. II C-C, C-D C-D [b]	
Group 3		C-C Str. II C-C, C-D C-D [b]	
Group 4		C-C C-D [d]	C-C C-D [b]

Nominal Thickness			
5/16	20/0	16/0	12/0
3/8	24/0	20/0	16/0
1/2	32/16	24/0	24/0
5/8	42/20	32/16	30/12
3/4	48/24	42/20	36/16
7/8		48/24	42/20
[e]			

(a) Identification Index refers to the numbers in the lower portion of the table which are used in the marking of sheathing grades of plywood. The numbers are related to the species of panel face and back veneers and panel thickness in a manner to describe the bending properties of a panel. They are particularly applicable where panels are used for subflooring and roof sheathing to describe recommended maximum spans in inches under normal use conditions and to correspond with commonly accepted criteria. The left hand number refers to spacing of roof framing with the right hand number relating to spacing of floor framing. Actual maximum spans are established by local building codes.

(b) Panels of standard nominal thickness and construction.

(c) Panels manufactured with Group 1 faces but classified as Structural II by reason of Group 2 or Group 3 inner plys.

(d) Panels conforming to the special thickness and panel construction provisions of PS 1-74 (3.8.6.).

(e) Panels thicker than 7/8 inch shall be identified by group number.

06100   ROUGH CARPENTRY

Wire And Sheet Metal Gages

(In Decimals of an Inch)

Name of Gage	American Wire Gage (A.W.G.) (Corresponds to Brown & Sharpe Gage)	Birmingham Iron Wire Gage (B.W.G.)	United States Standard Gage (U.S.S.G.)	
Principal Use	Electrical Wire & Non-Ferrous Sheet Metal	Iron or Steel Wire	Ferrous Sheet Metal	
Gage No.				Gage No.
00 00000				00 00000
0 00000	.5800			0 00000
00000	.5165	.500		00000
0000	.4600	.454		0000
000	.4096	.425		000
00	.3648	.380		00
0	.3249	.340		0
1	.2893	.300		1
2	.2576	.284		2
3	.2294	.259	23.91	3
4	.2043	.238	.2242	4
5	.1819	.220	.2092	5
6	.1620	.203	.1943	6
7	.1443	.180	.1793	7
8	.1285	.165	.1644	8
9	.1144	.148	.1495	9
10	.1019	.134	.1345	10
11	.0907	.120	.1196	11
12	.0808	.109	.1046	12
13	.0720	.095	.0897	13
14	.0641	.083	.0747	14
15	.0571	.072	.0673	15
16	.0508	.065	.0598	16
17	.0453	.058	.0538	17
18	.0403	.049	.0478	18
19	.0359	.042	.0418	19
20	.0320	.035	.0359	20
21	.0285	.032	.0329	21
22	.0253	.028	.0299	22
23	.0226	.025	.0269	23
24	.0201	.022	.0239	24
25	.0179	.020	.0209	25
26	.0159	.018	.0179	26
27	.0142	.016	.0164	27
28	.0126	.014	.0149	28
29	.0113	.013	.0135	29
30	.0100	.012	.0120	30
31	.0089	.010	.0105	31
32	.0080	.009	.0097	32
33	.0071	.008	.0090	33
34	.0063	.007	.0082	34
35	.0056	.005	.0075	35
36	.0050	.004	.0067	36
37	.0045		.0064	37
38	.0040		.0060	38
39	.0035			39
40	.0031			40

07600   FLASHING AND SHEET METAL

Locks And Locksets

Levels Of Control In Master Keying

The establishment of the proper level of control for a master key system is of paramount importance. Only when this has been determined are you ready to lay out the system. One of the main reasons many master key systems are allowed to disintegrate is the fact that they were not established at the proper level when originally planned. The chart shown below will assist you in making this determination.

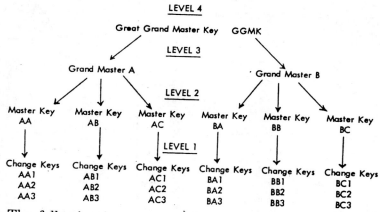

The following is a suggested guide in determining or selecting the proper level of control:

One Level - Change Key: All locks operated by change keys only, and keyed different or alike as required. Example: Home, stores.

Two Levels - Master Key: All locks operated by change keys and master key. Examples: Small school; apartments.

Three Levels - Grand Master Key: All locks operated by change keys, master keys and grand master key. Examples: Office buildings, hospitals.

Four Levels - Great Grand Master Key: All locks operated by change key, master keys, grand master keys and great grand master key.

Five Levels - Great Great Grand Master Key: All locks operated by change key, master key, grand master key, great grand master key, and great great grand master key. Example: Large universy complexes, large industrial complexes.

Cross Keying

The term "Cross keying" can be divided into two classifications: "Controlled cross keying" and "Uncontrolled cross keying."

"Controlled cross keying" is where two or more change keys under the same master key must operate one cylinder.

"Uncontrolled cross keying" is where two or more change keys under different master keys are set up to operate one cylinder. This greatly reduces the amount of available changes in a keying system by putting these unrelated keys together. It also reduces the security of the cylinder itself by requiring additional master split pins to accommodate these added keys.

Cross keying of either type should be held to a minimum, but uncontrolled cross keying definitely should be discouraged.

Locks And Locksets — (Cont.)

It has been common practice to use the terminology "To Pass" or "To Be Passed By" in indicating certain interkeying situations. This terminology encourages misinterpretation and we recommend the use of the following:

To Operate — Identifying a key or keys to operate other cylinders having different key changes.

To Be Operated By — Identifying a cylinder to be operated by one or more individual keys other than its own key.

Explanation Of Coding System

Where grand master keys are used, double letter symbols should be used to identify the grand and master key sets. The grand master symbol should be the first letter, followed by the master symbol.

```
GMK A - Master A = AA
    A - Master B = AB
    A - Master C = AC
    B - Master D = BD
    B - Master E = BE
```

Key symbols using this new key code system automatically indicate the function of each key in the keying system, without having to write any further explanation. Each key has a different key symbol. Key symbol AA1 indicates a lock operated by AA1 change key, AA Master, A Grand Master, and GGM - Great Grand Master. AA1 in this case is a keyed diffenre change. In the case of an alike change AA2, etc., this change or symbol is merely repeated next to each set using this change key.

This key code system allows for exceptions. Examples of a few are listed herein.

1. Single Master Key Systems:
   a. Always use symbol "AA" for the master key and Prefix the change key number (i.e.: 1AA, 2AA, etc.)
2. Grand Master Key Systems:
   a. Always use symbol "AA" for the master key and Suffix the change key number (i.e.: AA1, AA2, etc.)
3. Symbol "A" only is subject to the "A" grand only.
4. Symbol "AA" only is subject to the "AA" master and "A" grand only.
5. Symbol "A1," "A2," etc. These changes are under the "A" grand only. (**Note:** Always start these changes with the number "1")
6. Symbol "A1," "A2," etc. These changes are under the GGM only. (**Note:** Always start these changes with the number "1")
7. Symbol "1AA," "2AA," etc. used in a great grand master key system. The change numbers are prefixed on all locks operated by master keys under the great grand master key only — no grand master.
8. Symbol "SKD1," "SKD2," etc. used for locks in a master, grand or great grand master key system, but not master keyed. Example: Narcotics cabinets, food storage.

Pipe Weights

CAST IRON PIPE

Service Weight

Size, Inches	Weight of Pipe	Weight of Water	Total Weights-Lbs.
2"	3.8	1.45	5.3
3"	5.6	3.2	8.8
4"	7.5	5.5	13.0
5"	9.8	8.7	18.5
6"	12.4	12.5	24.9
8"	18.5	21.7	40.2

Extra Heavy

2"	4.3	1.45	5.8
3"	8.3	3.2	11.5
4"	10.8	5.5	16.3
5"	13.3	8.7	22.0
6"	16.0	12.5	28.5
8"	26.5	21.7	48.2

Steel Pipe

Pipe Size	W/40	H_2O/Lbs.	Total Lbs./LF
2"	3.65	1.45	5.1
2½"	5.79	2.07	7.86
3"	7.57	3.2	10.77
3½"	9.11	4.28	13.39
4"	10.8	5.51	16.31
5"	14.6	8.66	23.26
6"	18.8	12.5	30.5
8"	28.6	21.66	50.26
10"	40.5	34.15	74.65
12"			

15400   PLUMBING

Construction Inspection Manual

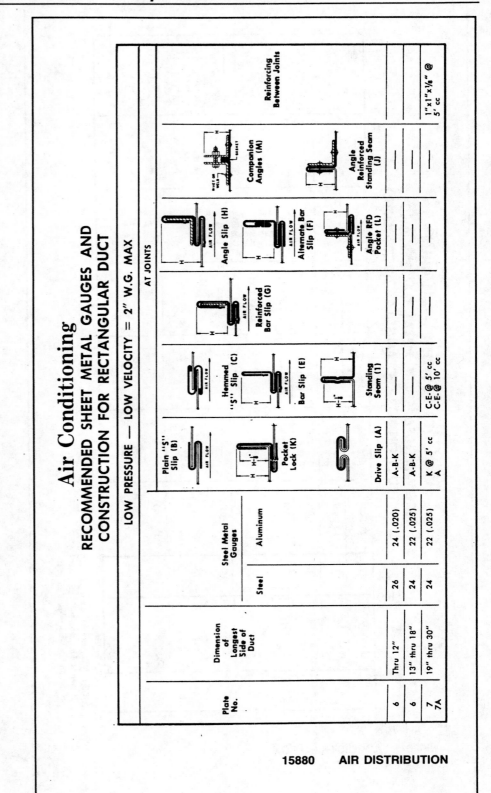

Air Conditioning
RECOMMENDED SHEET METAL GAUGES AND CONSTRUCTION FOR RECTANGULAR DUCT

LOW PRESSURE — LOW VELOCITY = 2" W.G. MAX

Plate No.	Dimension of Longest Side of Duct	Steel Metal Gauges		At Joints				
		Steel	Aluminum					Reinforcing Between Joints
6	Thru 12"	26	24 (.020)	A-B-K				
6	13" thru 18"	24	22 (.025)	A-B-K	C-E @ 5' cc / C-E @ 10' cc			
7 / 7A	19" thru 30"	24	22 (.025)	K @ 5' cc / A				1"x1"x⅛" @ 5' cc

15880 AIR DISTRIBUTION

Air Conditioning — (Cont.)
RECOMMENDED SHEET METAL GAUGES AND CONSTRUCTION FOR RECTANGULAR DUCT (CONT.)

8	31" thru 42"	22	20 (.032)	K @ 5' cc	E-G-K @ 5' cc E-G-K @ 10' cc	—	—	1"x1"x⅛" @ 5' cc
9	43" thru 54"	22	20 (.032)	K @ 4' cc K @ 8' cc	E- @ 4' cc E- @ 8' cc	G - @ 4' cc G - @ 8' cc	—	1½"x1½"x⅛" @ 4' cc
9	55" thru 60"	20	18 (.040)	K @ 4' cc K @ 8' cc	E- @ 4' cc E- @ 8' cc	G - @ 4' cc G - @ 8' cc	—	1½"x1½"x⅛" @ 4' cc
10	61" thru 84"	20	18 (.040)	—	—	G - @ 4' cc	—	1½"x1½"x⅛" @ 2' cc
11	85" thru 96"	18	16 (.051)	—	—	G - @ 5' cc	H- @ 4' cc F- @ 4' cc L- @ 4' cc	1½"x1½"x⅛" @ 2'-6" cc
							H- @ 5' cc F- @ 5' cc L- @ 5' cc	
							H- @ 4' cc L- @ 4' cc	J- @ 2' cc	1½"x1½"x3/16" @ 2' cc
							H- @ 5' cc L- @ 5' cc	M- @ 4' cc	1½"x1½"x3/16" @ 2'-6" cc
								M- @ 5' cc	1½"x1½"x3/16" @ 2' cc
12	Over 96"	18	16 (.051)	—	—	—	H- @ 4' cc L- @ 4' cc	J- @ 2' cc	2"x2"x¼" @ 2' cc
							H- @ 5' cc L- @ 5' cc	M- @ 4' cc	2"x2"x¼" @ 2'-6" cc
								M- @ 5' cc	2"x2"x¼" @ 2'-6" cc
								J- @ 2' cc	2"x2"x¼" @ 2' cc

H (height dimension)—up to 42" = 1"
H (height dimension)—43" to 96" = 1½"
H (height dimension)—over 96" = 2"

15880 AIR DISTRIBUTION

Air Conditioning — (Cont.)

LONGITUDINAL SEAMS FOR SHEET METAL DUCTWORK

Fig. "N"
PITTSBURGH LOCK

Fig. "Z"
BUTTON PUNCH SNAP LOCK

Fig. "O"
ACME LOCK-GROOVED SEAM

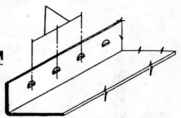

Approximately 2" Spacing Between "Buttons"

DETAIL NO. 1
MALE PIECE-SNAP LOCK

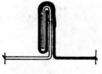

Fig. "T"
DOUBLE SEAM

15880   AIR DISTRIBUTION

Air Conditioning — (Cont.)

TYPICAL DUCT CONNECTIONS
CROSS JOINTS FOR SHEET METAL DUCTWORK
(NOT TO SCALE)

H = HEIGHT REFERRED TO IN DIMENSIONS

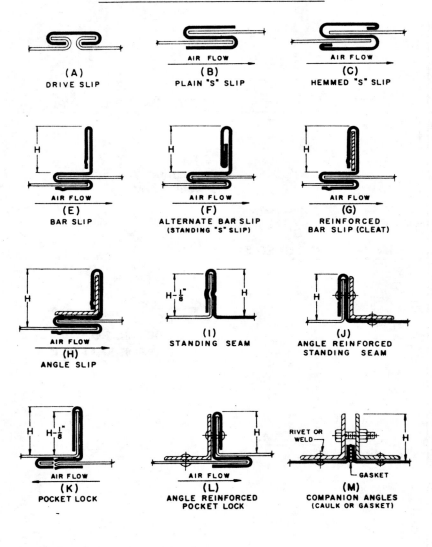

15880   AIR DISTRIBUTION

Electrical Systems
TYPES OF A.C. DISTRIBUTION
600 VOLTS OR LESS

A. SINGLE PHASE, 2-WIRE, 115 VOLTS

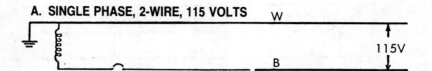

B. SINGLE PHASE, 3-WIRE 115/230 VOLTS

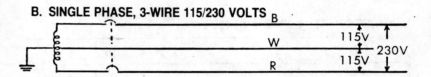

C. THREE PHASE, 4-WIRE WYE 120/208 VOLTS

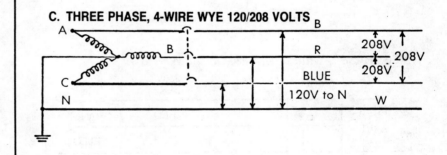

16300   MEDIUM VOLTAGE DISTRIBUTION

Electrical Systems — (Cont.)
TYPES OF A.C. DISTRIBUTION — 600 VOLTS OR LESS

D. THREE PHASE, 4-WIRE DELTA 120/240 VOLTS

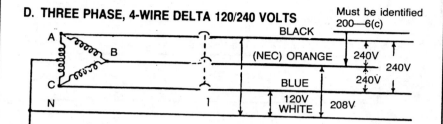

Must be identified 200—6(c)

NEC - This must be different than 120/208. Art. 210-5(b) & (c)

E. THREE PHASE; 4-WIRE WYE 480/277

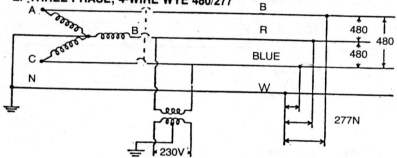

F. THREE PHASE, 3-WIRE DELTA 230 OR 460 VOLTS

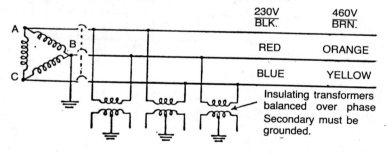

Insulating transformers balanced over phase Secondary must be grounded.

16300   MEDIUM VOLTAGE DISTRIBUTION

APPENDIX I — MATHEMATICS

Mensuration

Area of a square = length x breadth or height.
Area of a rectangule = length x breadth or height.
Area of a triangle = base x ½ altitude.
Area of parallelogram = base x altitude.
Area of trapezoid = altitude x ½ the sum of parallel sides.
Area of trapezium = divide into two triangles, total their areas.
Circumference of circle = diameter x 3.1416.
Circumference of circle = radius x 6.283185.
Diameter of circle = circumference x .3183.
Diameter of circle = square root of area x 1.12838.
Radius of a circle = circumference x .0159155.
Area of a circle = half diameter x half circumference.
Area of a circle = square of diameter x .7854.
Area of a circle = square of circumference x .07958.
Area of a sector of circle = length of arc x ½ radius.
Area of a segment of circle = area of sector of equal radius — area of triangle, when the segment is less, and plus area of triangle, when segment is greater than the semi-circle.
Area of circular ring = sum of the diameter of the two circles x difference of the diameter of the two circles and that product x .7854.
Side of square that shall equal area of circle = diameter x .8862.
Side of square that shall equal area of circle = circumference x .2821.
Diameter of circle that shall contain area of a given square = side of square x 1.1284.
Side of inscribed equilateral triangle = diameter x .86.
Side of inscribed square = diameter x .7071.
Side of inscribed square = circumference x .225.
Area of ellipse = product of the two diameters x .7854.
Area of a prabola = base x ⅔ of altitude.
Area of a regular polygon = sum of its sides x perpendicular from its center to one of its sides divided by 2.
Surface of cylinder or prism = area of both ends plus length and x circumference.
Surface of sphere = diameter x circumference.
Solidity of sphere = surface x ⅙ diameter.
Solidity of sphere = cube of diameter x .5236.
Solidity of sphere = cube of radius x 4.1888.
Solidity of sphere = cube of circumference x .016887.
Diameter of sphere = cube root of solidity x 1.2407.
Diameter of sphere = square root of surface x .56419.
Circumference of sphere = square root of surface x 1.772454.
Circumference of sphere = cube root of solidity x 3.8978.
Contents of segment of sphere = (height squared plus three times the square of radius of base) x (height x .5236).
Contents of a sphere = diameter x .5236.
Side of inscribed cube of sphere = radius x 1.1547.
Side of inscribed cube of sphere = square root of diameter.
Surface of pyramid or cone = circumference of base x ½ of the slant height plus area of base.
Contents of pyramid or cone = ara of base ⅓ altitude.
Contents of frustum of pyramid or cone = sum of circumference at both ends x ½ slant height plus area of both ends.
Contents of frustum of pyramid or cone = multiply areas of two ends together and extract square root. Add to this root the two areas and x ⅓ altitude.
Contents of a wedge = area of base x ½ altitude.

Metric Conversions

Metric System SI (Systeme International d'Unites) with exception noted.*

Table 1 - Abbreviation Of Units

ENGLISH		S.I. (Metric)	
Name	Symbol	Name	Symbol
Inch	In	Meter	m
Feet	Ft	Liter	l
Yard	Yd	Hertz	Hz
Mile	Mi	gram	g
Ounce	Oz	newton	N
Pound	lb	radian	rad
Kip	k	pascal	Pa
Second	s	Joule	J
Minute	m	°Celsius	°C
Hour	h		
Second	"		
Minute	'		
Degree	°		
°Fahrenheit	°F		
Pint	pt		
Quart	qt		
Gallon	gal		
British Thermal Unit	BTU		

Table 2 - S.I. Prefixes For Magnitude

Amount	Expotential Factor	Prefix	Symbol
1 000 000 000 000	10^{12}	tera	T
1 000 000 000	10^{9}	giga	G
1 000 000	10^{6}	mega	M
1 000	10^{3}	kilo	k
100	10^{2}	hecto	h
10	10^{1}	deca	da
0.1	10^{-1}	deci	d
0.01	10^{-2}	centi	c
0.001	10^{-3}	milli	m
0.000 001	10^{-6}	micro	μ
0.000 000 001	10^{-9}	mano	n

Table 3 - Metric Conversion

millimeters	mm			
centimeters	cm =	10 mm		
meter	m =	1,000 mm =	100 cm	
Kilometer	Km =	1,000,000 mm =	100,000 cm =	1000 m
Square mm	mm² =			
Square cm	cm² =	100 mm²		
Square m	m² =	1,000,000 mm² =	10,000 cm² =	
Square km	km² =	1,000,000,000,000 mm² =	10,000,000,000 cm² =	1,000,000 m²
Cubic mm	mm³ =			
Cubic cm	cm³ =	1,000 mm³ =	liter =	1000 g (water)
Cubic m	m³ =	1,000,000 cm³(cc)		
Cubic km	km³ =	1,000,000 mm³		
milligram	mg =			
centigram	cg =	10 g		
gram	g =	1,000 mg	100 cg	
kilogram	kg =	1,000,000 mg	100,000 cg =	
megagram	Mg =	1,000,000,000 ng =	100,000,000 cg = 1,000,000 g = 1000 kg	

Metric Conversions — (Cont.)

Metric System S.I. — Table 4 - Conversion Tables

English to Metric	Metric to English

Length
- 1 inch .... 2.54 cm
- 1 foot .... 0.3048 m
- 1 yard .... 0.9144 m
- 1 mile - 4280 ft. ...... 1.6093 km

- 1 cm .... 0.3937 inch
- 1 m .... 3.2808 feet
- 1 m .... 1.0936 yards
- 1 km .... 0.6214 mile

Area
- 1 square inch .... 6.452 cm^2
- 1 square foot .... 0.0929 m^2
- 1 square yard .... 0.8361 m^2
- 1 square mile .... 2.5899 km^2
- 1 acre - 43560 ft^2 .... 4046.9 m^2

- 1 cm^2 .... 0.1549 in^2
- 1 m^2 .... 10.7643 ft^2
- 1 m^2 .... 1.196 yd^2
- 1 m^2 .... 0.3861 $mile^2$
- 1 km^2 .... 0.000247 acre

Volume
- 1 $inch^3$ .... 16.387 cc
- 1 $foot^3$ .... 0.2832 m^3
- 1 $yard^3$ .... 0.7646 m^3
- 1 gallon .... 3.7854 liter
- 1 quart .... 0.94635 liter
- 1 pint .... 0.47318 liter
- 1 ounce .... 29.574 cc

- 1 1 cc .... 0.061 in^3
- 1 m^3 .... 35.315 ft^3
- 1 m^3 .... 1.308 yd^3
- 1 liter .... 0.2642 gal
- 1 liter .... 1.0567 qt
- 1 liter .... 2.1134 pt
- 1 cc .... 0.0338 oz

Mass
- 1 ounce .... 28.35 g
- 1 pound .... 453.59 g
- 1 kip .... 453.59 kg
- 1 ton(short) 2000 lb .. 907.18 kg
- 1 ton(long) 2240 lb .. 1016.05 kg

- 1 g .... 0.03527 oz
- 1 g .... 0.002205 lb
- 1 kg .... 0.002205 k
- 1 kg .... 0.0011 ton(s)
- 1 kg .... 0.0010 ton(l)

Density
- 1 lb/ft^3 .... 16018 g/m^3
  16.018 kg/m^3
- 1 lb/in^3 .... 27.68 g/cm^3

- 1 g/m^3 .... 0.00006216 lb/ft^3
- 1 kg/m^3 .... 0.03612 lb/in^3
  or 0.062428 lb/ft^3

Force
- 1 lb force .... 4.448 N
- 1 kip force .... 4.448 kN

- 1 newton .... 0.2248 lb force
- 1 kilo newton .... 224.8 k force

Pressure
- 1 lb/m^2 .... 6894.8 Pa
- 1 kip/m^2 .... 6.895 MPa
- 1 lb/ft^2 .... 47.88 pa

- 1 pascal .... 0.000145 lb/in^2
- 1 megapascal .... 145 lb/in^2
- 1 pascal .... 0.0209 lb/ft^2

Energy
- 1 BTU .... 1054.35 J
  1.054 kJ

- 1 joule .... 0.0009485 BTU
- 100 joule .... 0.9485 BTU

Temp
- °F .... (°F-32) 5/9 °C
- 0°F .... −17.8 °C
- 50°F .... 10 °C
- 70°F .... 21.1 °C
- 100°F .... 37.8 °C

- ° Celsius ..... (1.8°C)+32 °F
- 0°C .... 32 °F
- 100°C .... 212 °F

Metric Conversions — (Cont.)

Table 5 - Metric Information

(a) *Carpet Weight*

20 oz/yd²	678.25 g/m²
30 oz/yd²	1017.38 g/m²
40 oz/yd²	1356.51 g/m²
50 oz/yd²	1695.63 g/m²

(b) *Earth Pressure*

200 lbs/ft²	976.51 kg/m²
250 lbs/ft²	1220.64 kg/m²
300 lbs/ft²	1464.77 kg/m²

(c) *Stress*

2000 lbs/in²	140.60 kg/cm²
2500 lbs/in²	175.76 kg/cm²
3000 lbs/in²	210.91 kg/cm²
3500 lbs/in²	246.06 kg/cm²
4000 lbs/in²	281.21 kg/cm²

(d) *Concrete Strength*

lb/in²	MPa	kgf/cm³ * f = force
2000	14	140
2500	17	175
3000	21	210
3500	24	245
4000	28	280
4500	31	315
5000	34	350

(e) *Illumination*

1 ft.-candle = 10.76 lux   1 lux = 1 lumen/m³
1 lumen per sq. ft. = 10.76 lux

(e) *Refrigeration*

1 ton = 3519 watts (W)

(g) *Power*

1 horsepower = 745.7 W
             = 0.7457 kW

(h) *Heating*

"k" value; 1 BTU in/ft.² h °F = 0.1442 Wm °C; thermal conductivity BTU per inch per square foot hour degree F; watt per degree Celsius; W/m °C "U" value; 1 BTU/ft.²h°F = 5.678 W² °C; Coeff. of heat transfer BTU per square foot hour °F; watt per square meter degree Celsius; W/m² °C.

Heat	BTU	joule	J	1 BTU = 1055 J
		kilojoule	kJ	1 BTU = 1.055 kJ
		megajoule	MJ	1 Therm = 105.5 MJ
Heat flow rate				
	BTU/hr.	watt	W	1 BTU/h = 0.2931 W
		kilowatt	kW	1 BTU/h = 0.0002931 kW

APPENDIX J — HAZARDOUS MATERIALS

In is virtually impossible to cover the vast subject of hazardous materials in just a few short paragraphs, so just some highlights of a few recommended procedures are given.

It should be incumbent upon the owner to make a determination as to whether or not the building site has a potential for containing hazardous materials, i.e., previous occupant was a gasoline service station, paint factory, chemical plant, etc.

If the building site does contain a potential for toxic materials, the site should be cleaned up prior to advertising for bids on the construction project, or the specifications for the project should include site cleanup as a part of the bid, together with recommended procedures for the cleanup.

This will then also make it incumbent upon the owner to conduct a study of what toxic materials could be on or under the building site, and then to contact the proper authority as to what will be required for the cleanup.

After award of the contract, it shall be the duty of the general contractor to monitor use of any hazardous material on the jobsite by proper labeling, proper handling and proper storage.

One of the prime duties of the Construction Inspector is to be fully informed on the dangers of hazardous materials, and to alert the contractor as to any potential violation of OSHA regulations.

There are many hazardous materials associated with construction activities and they exist in various forms: liquid, solid and gaseous. The main categories of toxic materials are corrosives, solvents, and, last but not least, asbestos products.

A great amount of literature is now available in various publications on hazardous materials:

Hazardous Materials Handbook, May 1987 edition, published by the California Chamber of Commerce, P.O. Box 1736, Sacramento, CA 95808.

Other California Chamber of Commerce publications are *Hazardous Waste Management Handbook*, and *Hazard Communication Handbook*.

A Manual for the Construction Industry — Health Hazards Associated with the Use of Asbestos and Other Insulating Materials, published in March 1987, by Associated General Contractors of California, Inc. Its address is 3095 Beacon Boulevard, West Sacramento, CA 95691, and

Hazardous Communication Guide for California Construction, also published by Associated General Contractors of California, Inc.

APPENDIX K—HANDY CONSTRUCTION DATA

Handy Things To Know

1. *To Find—*
   (a) The circumference of a circle, multiply the diameter by 3.1416 (approx. 3$\frac{1}{7}$).

   (b) The diameter of a circle, multiply the circumference by .31831.

   (c) The area of a circle, multiply the square of the diameter by .7854.

   (d) The area of a triangle, multiply the base by ½ the perpendicular height.

   (e) The volume of a sphere, multiply cube of the diameter by .5236.

A gallon of water weights 8½ pounds.

A gallon of water contains 231 cubic ins.

A cubic foot of water contains 7½ gals., 1728 cubic inches and weights 62½ lbs.

In board measure, all boards are assumed to be 1-inch thick. Area of a lineal foot multiplied by length in feet will give the surface contents in square feet.

WIND FORCE
FORCE OF WIND IN POUNDS PER SQUARE FOOT

Miles Per Hour	Force Per Square Foot — In Lbs.	Miles Per Hour	Force Per Square Foot — In Lbs.
1	0.005	20	1.969
2	0.020	25	3.075
3	0.044	30	4.429
4	0.079	35	6.027
5	0.123	40	7.873
6	0.177	45	9.963
7	0.241	50	12.30
8	0.315	55	14.9
9	0.400	60	17.71
10	0.492	65	20.85
12	0.708	70	24.1
14	0.964	75	27.7
15	1.107	80	31.49
16	1.25	100	49.2
18	1.55		

Measures

1 Square Mile	= 640 acres
	= 6400 square chains
1 Acre	= 10 square chains
	= 4840 square yards
	= 43,560 square feet
	= A square, each side of which is 208.7 feet
1 Square Chain	= 16 square rods
	= 484 square yards
	= 4356 square feet
1 Square Rod	= 30.25 square yards
	= 272.25 square feet
	= 625 square links
1 Square Yard	= 9 square feet
1 Square Foot	= 144 square inches
1 U.S. Gallon	= 0.1337 cubic feet
	= 231 cubic inches
	= 4 quarts
	= 8 pints
1 Cubic Foot	= 7.48 U.S. Gallons

Weights Of Materials

Material	Approximate Weight Per Cubic Foot-Lbs.	
Aluminum	166	
Ashes	43	
Asphalt	81	
Brass	524	
Brick (common)	120	(about 3 tons per 1000)
Brick (fire)	145	
Bronze	534	
Concrete	150	(4050 lbs. per cubic yard)
Copper	537	
Crushed Rock	95	(2565 lbs. per cubic yard)
Dry earth, loose	76	(2052 lbs. per cubic yard)
Granite	179	
Iron, casting	450	
Lead	708	
Lumber, Fir	32	(2666 lbs. 1000 ft.)
Lumber, Oak	62	(5166 lbs. 1000 ft.)
Marble	168	
Mortar	100	
Portland Cement	94	(376 lbs. per barrel)
River Sand	120	(3240 lbs. cubic yard)
Steel	490	
Tar	63	
Tile	115	
Water	62.5	
Zinc	437	

INDEX

A

AA	43
AAMA	45
AAN	43
Accessible Usable Buildings	41
Accessories	90
ACI	43
Acoustical Insulating	43
Insulation barriers	170
Society America	231
Tile board	170
Treatment	169
ADAAG	40
Adhesive Sealant Council	231
Advisory Council Historic	231
AFPA	49
AGA	43
Agreement	8
Employment	7
AHA	43
AHDGA	44
AHMA	44
AI	46
AIA	44
AIMA	43
Air Conditioning Contractors	231
Refrigeration	231, 241
Air distribution	189
Hanling units	189
Pollution Control	231
AISC	44
AISI	44
AITC	44
ALA	44
Allied Stone Industries	231
Aluminum Association	43
American Arbitration Association	231
Association Hospital	231
Association Museums	231
Association Nursery	43
Concrete Institute	43, 232
Concrete Pipe	232
Construction Inspectors	232
Forest Council	232
Forestry Association	232
Gas Association	43, 232
Hardware Association	43
Hardware Manufacturer	44, 232
Hospital Association	232
Institute Architects	232
Institute Kitchen	232
Institute Landscape	232
Institute Planners	232
Institute Steel	44
Institute Timber	44
American Insurance Association	44, 232
Iron & Steel	44, 232
Library Association	44, 232
National Standards	44, 232
Petroleum Institute	44
Plywood Association	44, 232
Public Power	233
Public Works	45, 233, 241
Road Transportation	233
Segmental Bridge	233
Society Architecture	45
Society Heating	45
Society Mechanical	45
Society Testing	45
Subcontractors Association	241
Water Works	45
Welding Society	45, 233
Wood Council	49
Wood Preservers	45, 233
Americans with Disabilities	40
ANS	44
ANSI	41, 44
APA	44
API	44
Application payment	212
Apportenances	81
APWA	45
Architect	213
Engineer	15
Architectural Aluminum Manufacturers	45
Precast Association	233
Woodwork	117
Woodwork Institute	46
ASAHC	45
ASHRAE	45
ASME	45
Asphalt Institute	233
Institute Research	46
Paving	82
Shingles	123
Associated Builders Contractors	241
Equipment Distributors	233
General Contractors	233, 241
Plumbing Mechanical	241
Roofing Contractors	241
Specialty Contractor	234
Tile Contractors	241
American Universities	234
Builders Contractors	233
University Architects	234
Wall Ceiling	234
Women Architecture	234
ASTM	45
AWI	46
AWPA	45
AWS	45
AWWA	45

B

Balancing testing ... 192
Ballasts ... 81
Bases ... 81
Bearing elements ... 76
Better Heating Cooling ... 234
BHMA ... 46
BIA ... 46
BOCA ... 39
Boilers Equipment distribution ... 183
Bolting ... 106
Bonding ... 201
Brick Institute America ... 46
Builders Exchange Alameda ... 241
    Exchange Modesto ... 241
    Exchange Monterey ... 241
    Exchange Napa ... 241
    Exchange Peninsula ... 241
    Exchange San Francisco ... 242
    Exchange Stockton ... 242
    Hardware Manufacturer ... 46, 234
Building Congress Exchange ... 234
    Materials Research ... 234
    Officials Code ... 39
    Research Institute ... 234
    Stone Institute ... 234
    Systems Research ... 234
    Thermal Envelope ... 234
Busways ... 194
Butts hinges ... 142

C

Cable systems ... 196
Caissons ... 78
California Association Realtors ... 234, 242
    Building Industry ... 242
    Code Regulations ... 41
    Conference Masonry ... 242
    Contractors Association ... 242
    OSHPD ... 242
    Redwood Association ... 46
    Standard Specifications ... 46
    Wall Ceiling ... 242
Carpentry ... 114, 117
Carpet ... 167
    & Rug Institute ... 46
CCR ... 41
CDA ... 47
Cedar Shake Shingle ... 51
Ceilings Interior Systems ... 234
Cellular Concrete Association ... 234
Cementitious fireproofing ... 132
Ceramic Tile Distributors ... 234
    Tile Institute ... 46, 234, 242
Certificate payment ... 212
Chain Link Fence ... 46
Change order ... 212

Checklist ... 53
    Introduction ... 54
Cleaning testing balancing ... 185
CLFMI ... 46
Climate concerns ... 222
Closeout procedures ... 69, 71
Codes regulations ... 38
Commercial Standard ... 46
Committee Steel Pipe ... 234
Communication ... 208
Compensation ... 7
Concrete ... 90, 93, 96, 99
    Clay tile ... 123
    Forms ... 90
    Masonry Association ... 242
    Paving ... 83
    Pumpers Association ... 242
    Reinforcement ... 93
    Reinforcing Steel ... 46, 235
Conductors ... 195
Construction change directive ... 213
    Contract ... 19
    Financial Management ... 235
    Industry Research ... 242
    Industry Standards ... 40
    Inspector ... 3, 10
    Labor Research ... 235
    Specifications Institute ... 235
    Writers Association ... 235
Contract ... 37
    Definition ... 2
    Documents ... 28, 37
    Responsibilities ... 2
    Specifications ... 37
Contracting Plasterers Research ... 235
Contractor ... 2, 21, 213
Contractors Bonding Association ... 242
Coordination ... 30, 207, 211
Copper Development Association ... 47, 235
Correspondence ... 211
Council Educational Facility ... 235
Council Mechanical Specialty ... 235
CRA ... 46
CRI ... 46
CRSI ... 46
CS ... 46
CSS ... 46
CTI ... 46

D

Description ... 15, 16
Design professional ... 2, 13, 15
DFPA ... 47
Disabilities ... 42
Disability access ... 42
Door closers ... 143
Doors ... 134, 136, 138, 140, 142, 146, 148
Douglas Fir Plywood ... 47

INDEX

Drapery 174
Duct lining insulation 192
Ductwork 191
Duties 10, 13

E

Earthwork 72
El Dorado Builders 243
Electric Contractors California 243
    Power Research 235, 243
Electrical 202
    Association 235
    Communications 204
Employment 7
Engineering Contractors Association 243
Engineers Joint Council 235
Entrances 138
Equipment 177, 187
Exit bolts 144
Experience 6

F

Facing Tile Institute 47
Factory Mutual Systems 47
Fair Housing Accessibility 40
Federal 40
    Housing Administration 40, 235
    Specifications 47
FGMA 47
FHA 40
FHAG 40
Fibrous reflective insulation 120
Field administrator 19
    Inspection 53, 56
    Office concerns 217
Filters screens 190
Fine Hardwoods Association 235
Finish carpentry 115
    Grading 88
Finishes. 150, 154, 158, 162, 163, 165, 167, 169, 171
Finishing curing 97
Fire alarm 204
Fixtures 181
Flashing 128
Flat Glass Marketing 47
Flexicore Manufacturers Association 235
Floor Covering Institute 243
    Tile 160
Flush bolts 144
FM 47
Food Facilities Consultants 235
Forest Products Lab 47
    Products Research 235
Form removal 92
Forms 212
Foundation load 76
FPL 47

Frames 134, 135
Framing furring 150
    System 154
Fresno Builders Exchange 243
FS 47
FTI 47
Fuel Oil Piping System 182
Furnaces 189
Furnishing 173, 174

G

Gardens For All 235
Gas piping system 181
General requirements 60, 61, 63, 65- 67, 69, 71
Geographic concerns 221
Glass Association 47
Glazed curtain wall 148
Glazing 146
Grounding 201
GTA 47
Guild Religious Architecture 235
Gypsum Association 47
    Board 154, 155

H

Hardware 142, 144
Hardwood Plywood Manufacturer 47
Hazardous materials 42
Heat generation equipment 183
High strength bolting 107
Historic American Buildings 236
HPMA 47
Hydraulic Institute 47

I

ICBO 39
ILIA 48
Illuminating Engineers Society 236
Independent Roofing Contractors 243
Indiana Limestone Institute 48
Information Bureau Lath 236
Inspectors daily report 213
Institute Electrical Electronic 236
    Noise Control 236
    Real Estate 236
Insulation 120
Inter Society Color 236
International Association Plumbing 39, 236
    Conference 39
    Conference Building 236
    Council Shopping 236
    Institute Ammonia 236
    Masonry Institute 48, 236
Irrigation 86
    system 86

J

Jobsite safety 218

K

Kern County Builders 243

L

Landscape Architecture Foundation 236
    Construction 88
Lath plaster 150
Lawyers 26
Life safety 42
Lighting 202
Liquid heat transfer 183
Locksets latchsets 143
Loose fill insulation 121
Los Angeles County 243

M

Manufactured casework 173
Manufacturer's rpresentative 27
Maple Flooring Manufacturers 48
    Institute America 48
Marin Builders Exchange 243
Mason Contractors Association 236
Masonry 101
    Contractors Association 243
Matrix 54
Mechanical 177, 183, 186, 189
    Contractors Legislative 243
Membrane roofing 125
Merced-Mariposa Builders 243
Metal Buildings Manufacturers 236
    Deck 109
    Doors 134
    Fabrications 111
    Joists 108
    Lath Steel 48
    Windows 140
Metals 105, 108, 109, 111
MFMA 48
MI 48
MIA 48
Minority Contractors Association 243
MLA 48
Model Codes Standardization 236
Moisture protection 118, 120, 122, 125, 128, 132
Mortar Manufacturers Standards 236
Mortgage Bankers Association 236
Motor control 198
Motors 198

N

N S F International 50
NAAMM 48
National Asphalt Pavement 237
    Association Corrosion 237
    Association of Architectural 48
    Association Realtors 237
    Builders Hardware 48
    Building Granite 48
    Bureau Standards 48
    Concrete Masonry 49, 237
    Construction Association 237
    Crushed Stone 237
    Decorating Products 237
    Electric Code 49
    Electrical Contractors 237, 243
    Electrical Manufacturer 49
    Elevators Manufacturing 49
    Fan Manufacturers 49
    Fire Protection 40, 49, 237
    Housing Conference 238
    Oak Flooring 49
    Paint Coatings 49
    Particleboard Association 49
    Petroleum Council 238
    Ready Mixed 50
    Roofing Contractors 238
    Science Foundation 238
    Slag Association 238
    Terrazzo Mosaic 50
    Wood Flooring 238
    Woodwork Manufacturer 50
NBGQA 48
NBHA 48
NBS 48
NCMA 49
NEC 49
NEMA 49
NEMI 49
NFMA 49
NFPA 40, 49
NOFMA 49
North American Wholesale 238
    Coast Builders 243
Northern California Drywall 244
    California Engineering 244
NPA 49
NPVLA 49
NRMCA 50
NSF 50
NTMA 50
NWMA 50

O

Observations inspections 20
Occupational Safety 40
Office practice concerns 216

On site concerns 218
OSHA 40
Other requirements 6
    Roof flashing 130
Outlets 196
    Diffusers registers 192
Owner 2, 23, 213
Owner's representative 2

P

Pacific Coast Builders 244
Painting Decorating Contractors 244, 238
Paints coatings 171
Parquet flooring 164
Pavements 81
PCA 50
PCI 50
PEI 50
Peninsula Builders Exchange 244
Perlite Institute 50
Personal 5
Pipe pipe fittings 177
Piping 186
Planting 87, 89
Plaster 151
Plastering Information Bureau 238
Plastic Construction Council 238
    Doors 136
    Pipe Institute 50
Plumbing Drainage Institute 238
    Fixtures 177
Porcelain Enamel Institute 50
Portland Cement Association 50, 238
PPI 50
Precast concrete 99
Prefabricated wood 114
Prestressed Concrete Institute 50, 238
Priming tack coats 81
Products Standards Section 50
Project designer 16
    Job captain 17
    Location 215
    Specifier 18
PS 50
Public safety 42

Q

Qualifications 5
    Requirements 4

R

Raceways 193
RCSHSB 51
Recommended duties 1, 21, 23
Red Cedar Shingle 238
Redwood Inspection Service 51

Refrigeration equipment 186
Regional concerns 220
Resilient Floor Covering 51
    Flooring 165
Responsibilities 1, 10, 13, 21, 23
RFCI 51
Rigid insulation 120
RIS 51
RMA 51
Roles 2
Roofing Contractors Association 244
    Industry Council 244
    Tiles 122
Rough carpentry 112
Rubber Manufacturers' Association 51

S

Sacramento Builders Exchange 244
San Francisco Builders 244
Sanitary sewerage 79
Santa Barbara Contractors 244
SBCCI 39
Scaffold Industry Association 238
Scaffolding Shoring Institute 239
Scheduling 210
Screen Manufacturers Association 239
SDI 51
Service 200
Shasta Building Exchange 244
Sheet metal 128
    Air Conditioning 51
Shingles 122
Simplified Practice Recommendations 51
Sitework 72, 78, 79, 81, 86, 87
    Existing Vegetation 87
SJI 51
Slate shingles 123
SMACNA 51
Society Plastic Industry 51, 239
Soil, waste vent 179
Solar Energy Industries 239
Sources 4
Southern Building Code 39
    Cypress Manufacturers 239
    Forest Products 51
SPA 51
Special construction 175
    Consultants 31
SPI 51
SPR 51
Sprinkler system 175
SSPC 52
Stained Glass Association 239
Standards 43
    Codes 35
Starters 198
Steel Deck Institute 51
    Door Institute 51, 239

Steel Joist Institute 51
    Structures Painting 52
    Window Institute 52
Stone City Bank 48
Stops holders plates 144
Storefronts 138
Structural steel 105
Stucco Manufacturers Association 239
Subgrades 81
Summary 6
Suppression 175
Suspension systems 169
SWI 52
Switchboards panelboards 200
Systems equipment 176

T

TCA 52
Technical 5
Terminal units 185
Terrazzo 162
The Energy Bureau 235
Thermal protection 118, 120, 122, 125, 128, 132
Through wall flashing 131
Tile Council America 52
TPI 52
Transformers 200
Transmittal 213
Truss Plate Institute 52, 239

U

UL 52
Underfloor raceways 195
Underwriters Laboratories 52
Unit masonry 101
United Brotherhood Carpenters 239
Urban Institute 239

V

Vermiculite Association 239

W

Wall flashing 131
    Tile mortar 159
    Tile thinset 160
Wallcovering Information Bureau 240
    Wholesalers Association 240
Warranties Guarantees 29
Water supply system 180
Waterboard 239
Waterproofing 118
WCLIB 52
Welding 107
West Coast Lumber 52
Western Lath Plaster 48
    Red Cedar 52, 240
    States Ceramic 245
    Wood Products 52, 240
WFI 52
WIC 52
Window treatments 174
Windows 134, 136, 138, 140, 142, 146, 148
WLPDIA 48
Wood and plastics 114
    Doors 136
    Flooring 163
    Flooring Institute 52
    Plastics 112, 115
    Shingles Shakes 122
    Strip flooring 163
    Synthetic Flooring 240
    Truss Council 240
Woodwork Institute California 52
WRCLA 52
WWPA 52

Notes

ND# Notes

```
TH          Construction inspection
439         manual.
.C66
1998
```

TECHNICAL COLLEGE OF THE LOWCOUNTRY

3 2992 00024546 4